The changing US auto industry

The changing US auto industry

A geographical analysis

James M. Rubenstein

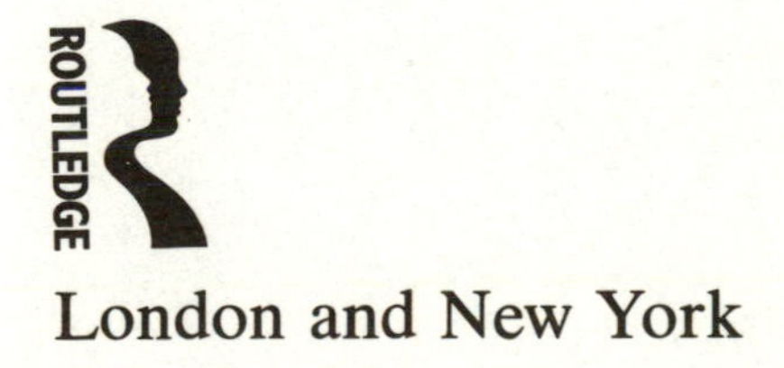

London and New York

First published 1992
by Routledge
11 New Fetter Lane, London EC4P 4EE

Simultaneously published in the USA and Canada
by Routledge
a division of Routledge, Chapman and Hall, Inc.
29 West 35th Street, New York, NY 10001

Typeset in Linotron Times by
Falcon Typographic Art Ltd, Edinburgh
Printed and bound in Great Britain by
Mackays of Chatham PLC, Chatham, Kent

British Library Cataloguing in Publication Data
Rubenstein, James
The changing US auto industry: a geographical analysis.
1. United States. Cars. Production
I. Title
338.4762920973

ISBN 0-415-05544-X

Library of Congress Cataloging in Publication Data
Has been applied for

ISBN 0-415-05544-X

Contents

Figures

Tables

Acknowledgements

I wish to thank the staff at Routledge for their commitment to this project. I am especially grateful to Alan Jarvis, Editor, for his support, and to Jane Mayger, Desk Editor, for her sensitive editing. I also wish to thank at Routledge Eddie Hepper, Sally Mussett and Richard Earney.

Thanks to Mark L. Listermann, who prepared the maps in this book. The work was done at Miami University's Geographic Information Systems Laboratory, directed by Connie McOmber. The cover photograph was taken by Barbara Traub.

I thank my wife, Bernadette Unger, for listening to incessant talk about cars. Finally, I would like to dedicate the book to my parents, who taught me to drive and especially to park a car.

1 The changing geography of automobile production

Henry Ford once said that history is bunk. What he meant was that history could not teach him how to revolutionize the production of automobiles. However, Ford never said that geography is bunk. To the contrary, the Ford Motor Company owed much of its early dominance of North American automotive production to geography. At a time when the location of the North American automotive industry was largely accidental, the Ford Motor Company adopted a strategy for locating plants which proved successful for more than six decades.

Ford actually told a *Chicago Tribune* reporter in 1916 that 'history is more or less bunk.' During a 1919 libel suit he brought against the *Chicago Tribune* after the newspaper called him an anarchist and ignorant idealist, Ford explained 'I did not say it was bunk. It was bunk to me . . . but I did not need it very bad.' Although Ford failed to identify the correct year of the American Revolution, the jury still found the newspaper guilty of libel but levied a fine of only six cents (Nevins and Hill 1957: 137–139).

When the Ford Motor Company was founded in 1903, southeastern Michigan, especially near Detroit, had recently become the national center of automobile production. Under the influence of Ford and later General Motors Corporation, most of the components that go into motor vehicles were built in southeastern Michigan, as well, but most final assembly plants, where components are attached to vehicles, were located elsewhere in the country, especially near population concentrations in the northeast, south, and west coast.

Then in the 1980s, the long-term locational pattern suddenly shifted. As long as General Motors held nearly one-half, Ford one-fourth, and the Chrysler Corporation one-eighth of the North American market by selling almost exclusively 200-inch-long gas guzzlers, there was no reason to tinker with the geography of motor vehicle production. (The US industry continues to measure the length of vehicles in inches.) But the United States and Canada became a battleground for global competitors. Japanese firms, having captured one-fourth of the North American market in the 1970s by importing small, fuel-efficient vehicles, further expanded sales in the 1980s by building plants inside the United States and Canada in communities not traditionally associated with motor vehicle production. Roughly one-third of the vehicles sold by these so-called Japanese transplants represented substitutions for imports but most of the remainder came at the expense of the Big Three American-owned companies, GM, Ford, and Chrysler.

Faced with declining profits and market share, the Big Three reduced their capacity to produce automobiles by more than one-third during the 1980s and early 1990s with more plant closures than even during the Great Depression of the 1930s. Some plants were replaced with new ones nearby, while others were retained for production of trucks, but most were shut permanently. Yet in the 1990s, North American producers were still burdened by overcapacity and the need to close even more plants (Table 1.1).

In the intensely competitive North American automotive market, geography mattered again, as it once had for Henry Ford: finding the optimal plant locations within an immense market area of nearly 20 million km^2 became critical corporate objectives. Japanese firms selected sites for assembly plants or components suppliers after elaborate studies, while American firms shifted products from plant to plant in search of the most efficient distribution. Japanese and American companies reached the same conclusion: the geography of motor vehicle production which had dominated the industry since before World War I had to be scrapped. Embattled American producers closed nearly all of their automobile assembly plants in coastal locations and retrenched in the interior of the country. Similarly, Japanese firms, rejecting Henry Ford's locational pattern, clustered assembly plants and components suppliers in the interior.

Maps clearly display the geographic changes in automobile assembly within the United States during the 1980s and early 1990s. In the 1980 model year, automobiles were assembled at forty-six plants in the United States. These plants were divided almost evenly among three locations: fifteen were located within 100 km of Interstate routes 65 and 75, fifteen were located in interior states more than 100 km from the two interstate highways, and sixteen were located in states abutting the Atlantic and Pacific oceans (Figure 1.1).

Interstate 75 runs nearly 3,000 km between Sault Ste Marie in Michigan's Upper Peninsula and Fort Lauderdale, Florida, passing through the cities of Detroit, Michigan; Toledo, Dayton and Cincinnati, Ohio; Lexington, Kentucky; Knoxville and Chattanooga, Tennessee; Atlanta, Georgia; and Tampa, Florida. Roughly 200 km to the west, Interstate 65 runs 1,500 km between Gary, Indiana, near Lake Michigan, and Mobile, Alabama, near the Gulf of Mexico, passing through Indianapolis, Indiana; Louisville, Kentucky; Nashville, Tennessee; and Birmingham, Alabama.

In the early 1990s, the number of automobile assembly plants located within 100 km of Interstates 65 and 75 increased from fifteen to twenty. Twelve of these twenty plants were built or were under construction during the 1980s and early 1990s, including ten at new locations and two

Table 1.1 Changes in US and Canadian automobile and light truck assembly plants, 1979–1991

	US-owned				*Foreign–US-*	*Total US*
	Chrysler	*Ford*	*GM*	*Total*	*operated*	*and foreign*
All plants						
Open as of 1979	14[a]	21	34	69	1	70
Open in new location	2	1	5	8	13[c]	20[c]
Replacement of old plant	1	0	4	5	0	5
Closed or idled indefinitely	6	3	11[c]	20[c]	1	20[c]
Open as of 1991	10	19	28	57	13	70
Automobile						
Open as of 1979	9	15	27	51	1	52
Additions						
Open in new location	2	0	2	4	13[c]	16[c]
Replacement of old plant	0	0	4	4	0	4
Losses						
Replaced by new plant	0	0	4	4	0	4
Closed or idled indefinitely	6	3	7[c]	16[c]	1	16[c]
Converted to truck production	1[b]	3	4	8	0	8
Open as of 1991	4	9	18	31	13	44
Truck						
Open as of 1979	5	6	7	18	0	18
Additions						
Open in new locations	0	1	3	4	0	4
Replacement of old plant	1[b]	0	0	1	0	1
Converted from automobile production	1[b]	3	4	8	0	8
Losses						
Replaced by new plant	1[b]	0	0	1	0	1
Closed or idled indefinitely	0	0	4	4	0	4
Open as of 1991	6	10	10	26	0	26

Notes: [a] Includes American Motors plants in 1979.

[b] Chrysler's Jefferson Ave assembly plant in Detroit was replaced by a new plant next door, and production was converted from automobile to truck.

[c] Fremont plant was closed by GM and reopened by Nummi; it is counted as a closure for GM and an additional foreign-operated plant but is not included in the total US and foreign changes.

Figures for 1979 include two automobile lines and one truck line at Oshawa (GM), two automobile lines each at Lansing (GM) and Kenosha (American Motors), two truck lines at Pontiac (GM), and one automobile and one truck line each at Lordstown (GM) and Oakville (Ford); figures for 1991 include two automobile lines and one truck line at Oshawa, two automobile lines at Lansing, two truck lines at Oakville, and closures and conversions announced but not yet implemented.

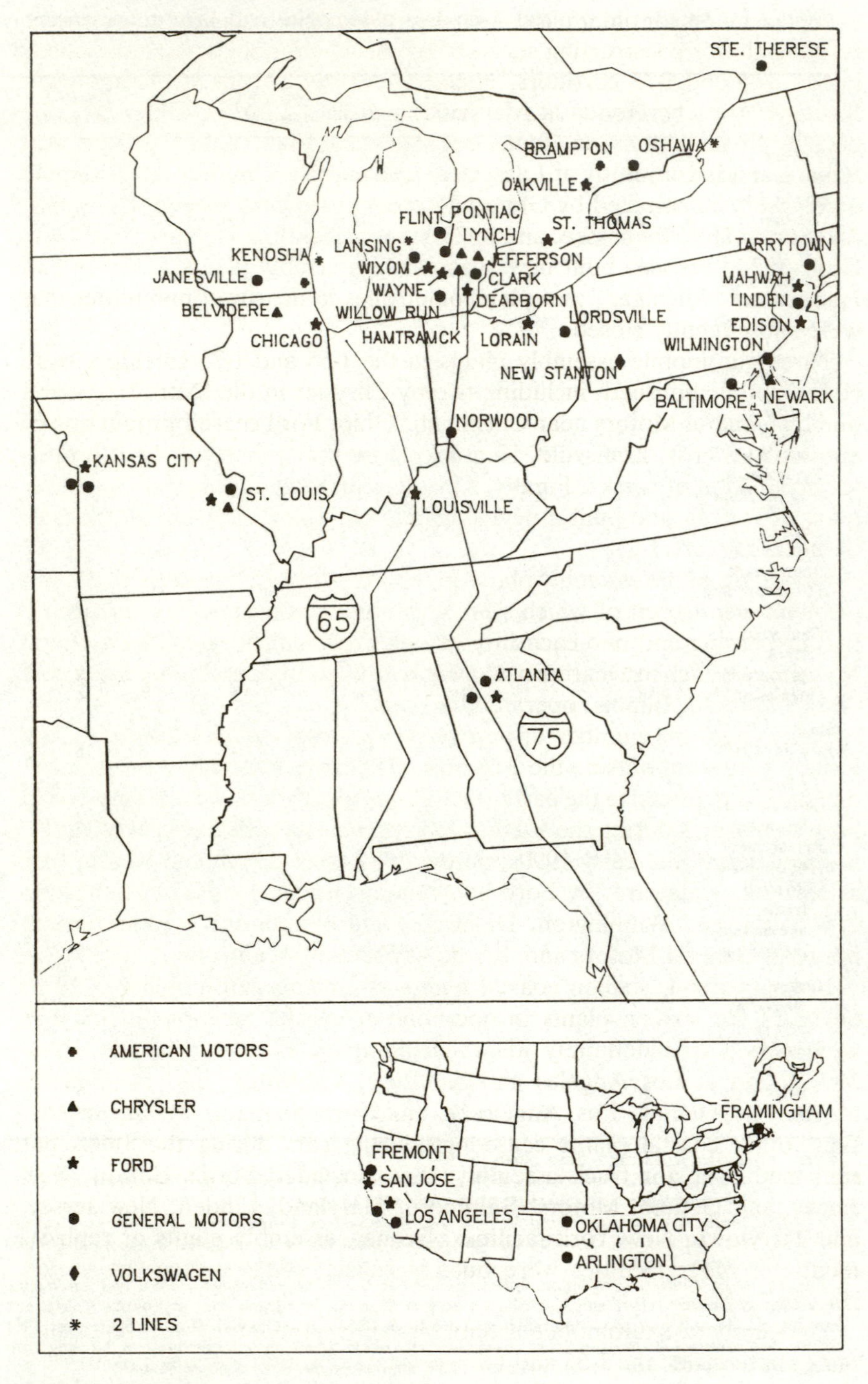

Figure 1.1 Automobile assembly plants, 1979

as replacements for older plants that were then closed. Japanese firms were responsible for constructing seven of the new automobile assembly plants in the I-65 and I-75 corridors, including two by Toyota at Georgetown, Kentucky, two by Honda at Marysville and East Liberty, Ohio, and one each by Nissan at Smyrna, Tennessee, Mazda at Flat Rock, Michigan, and Subaru and Isuzu jointly at Lafayette, Indiana. New automobile assembly plants were constructed by General Motors at Bowling Green, Kentucky, and Spring Hill, Tennessee, and by Chrysler at Sterling Heights, Michigan. General Motors also built new automobile assembly plants at Orion and Hamtramck, Michigan, to replace older ones in nearby communities that were subsequently closed.

Three automobile assembly plants in the I-65 and I-75 corridors were closed and demolished, including two by Chrysler in the Detroit area and one by General Motors near Cincinnati, Ohio. Ford ceased production of automobiles at its Louisville, Kentucky, assembly plant, but kept it open for production of trucks. Finally, Chrysler closed an automobile assembly plant in Detroit and built a new assembly plant next door for production of trucks (Figure 1.2).

Only eight of the assembly plants in the I-65 and I-75 corridors were built prior to 1980, seven of which were clustered in southeastern Michigan – two in Lansing and one each in Flint and Ypsilanti operated by General Motors, one each in Dearborn, Wayne, and Wixom operated by Ford, and one in Chicago, Illinois, operated by Ford.

Meanwhile, the number of automobile assembly plants located in the United States but outside the I-65 and I-75 corridors declined from thirty-one in 1979 to fifteen in the early 1990s. Coastal locations were hit especially hard by closures during the 1980s. Only five coastal plants assembled automobiles during the early 1990s; automobiles were assembled by Chrysler at Newark, Delaware, by Ford at Atlanta, Georgia, by General Motors at Atlanta, and Wilmington, Delaware, and by Nummi, a joint venture between General Motors and Toyota at Fremont, California.

In comparison, sixteen coastal plants assembled automobiles in 1979. Seven of the sixteen plants in operation at coastal locations as of 1979 were closed or indefinitely idled, including by Ford at Mahwah, New Jersey, and at Los Angeles and San Jose, California, and by General Motors at Atlanta, Los Angeles(2), and Framingham, Massachusetts. Four other coastal plants ceased producing cars during the 1980s but remained open for truck assembly; these included Ford's Edison, New Jersey, and General Motors' Baltimore, Maryland, Linden, New Jersey, and Tarrytown, New York facilities. No new assembly plants or replacements for older facilities were built in coastal states during the period (Figure 1.3).

The number of automobile assembly plants in the interior of the United States but outside the I-65 and I-75 corridors declined from fifteen in 1979 to ten in 1991. Five automobile assembly plants closed in the interior region,

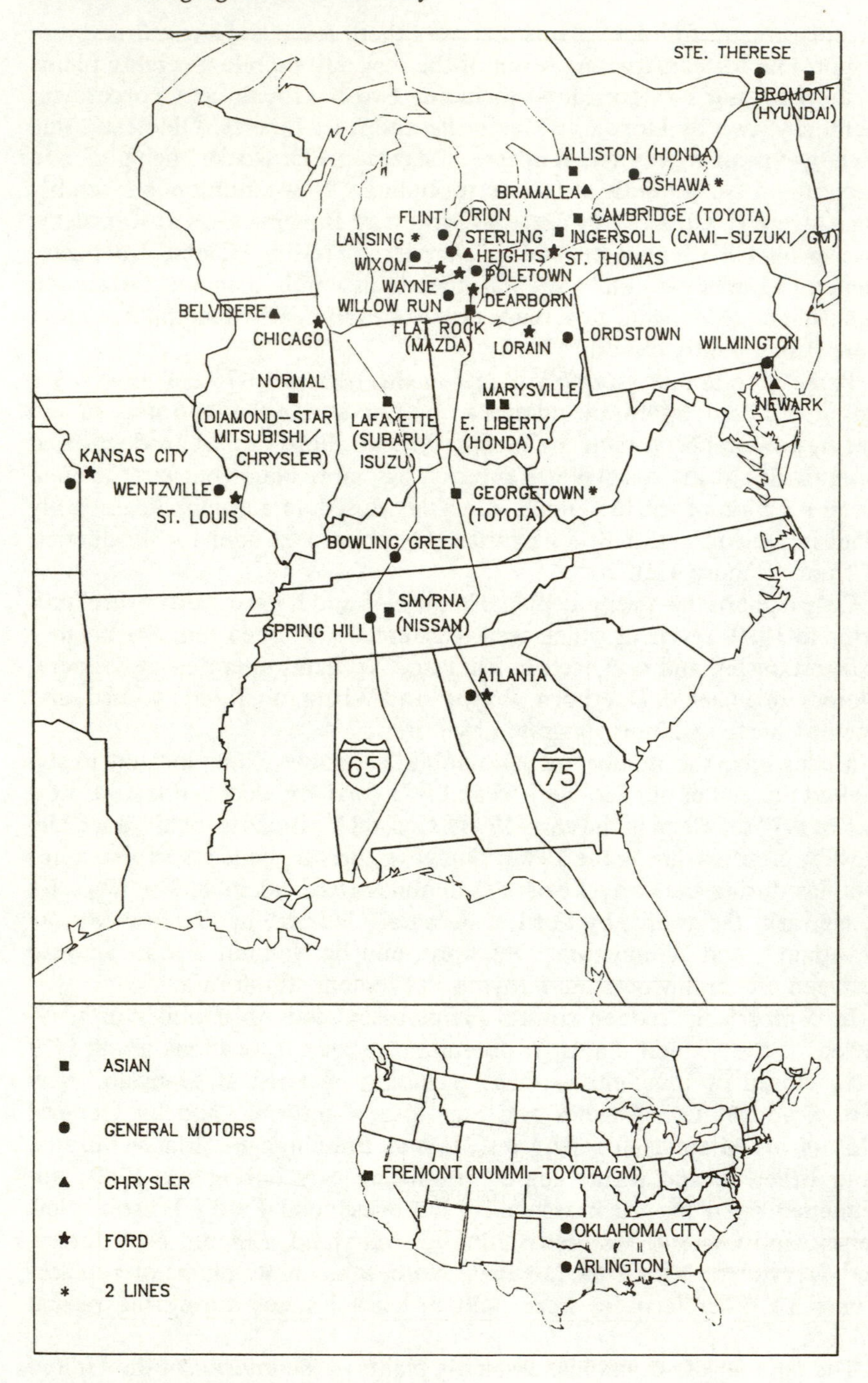

Figure 1.2 Automobile assembly plants, 1991. Included are plants under construction or scheduled to be closed, as of 1991

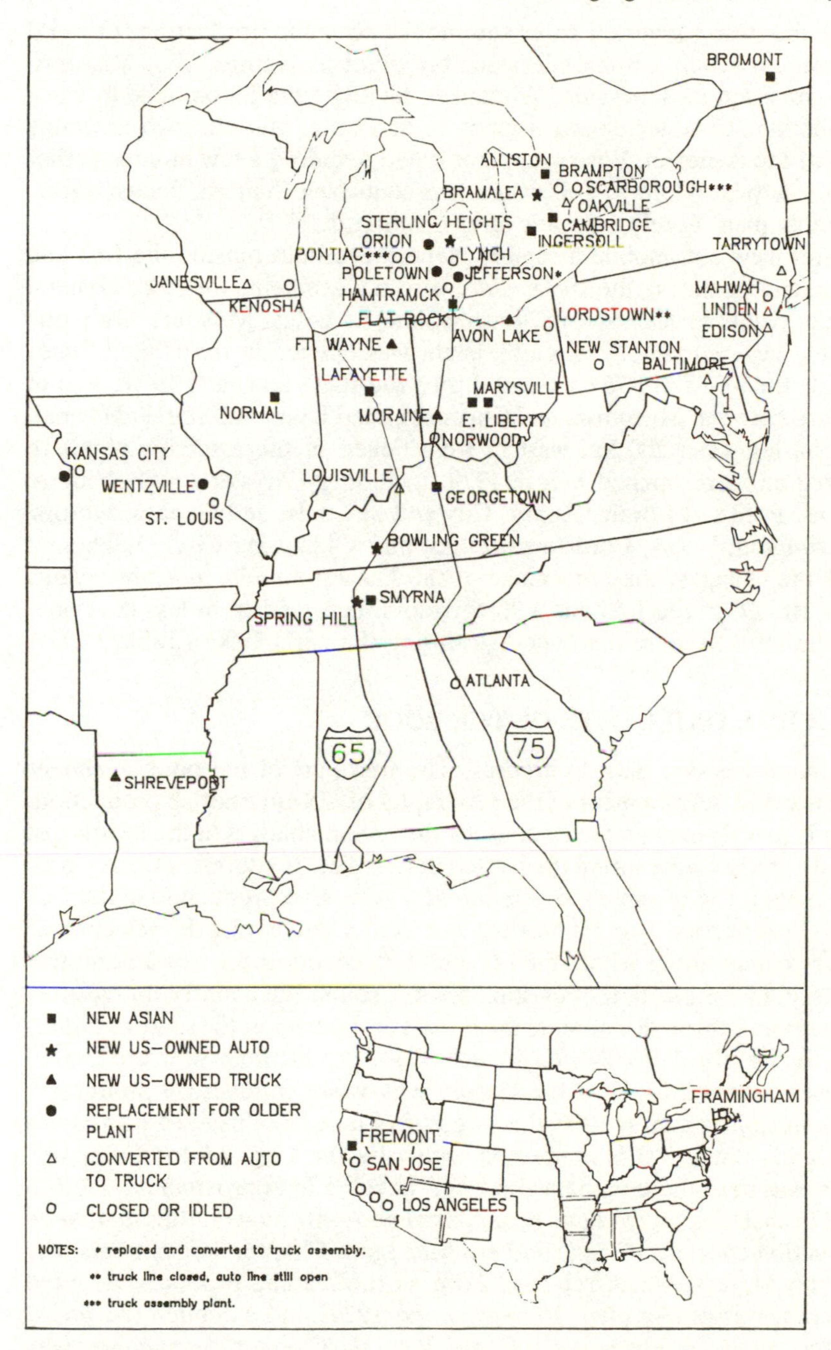

Figure 1.3 Changes in assembly plants, 1979–1991. Included are plants under construction and announced changes, as of 1991. New plants are clustered along the north–south corridor formed by interstate highways 65 and 75, while most closures are located along the east and west coasts

while one was converted from automobile to truck production. General Motors closed an automobile assembly plant at Kansas City, Missouri, and converted its Janesville, Wisconsin, facility from automobile to truck production. Chrysler closed a plant in St Louis, plus the two assembly lines at the Kenosha, Wiscosin, plant it had acquired a few months earlier from American Motors. Volkswagen closed its New Stanton, Pennsylvania, assembly plant barely a decade after opening it.

Three new automobile assembly plants were built outside the I-65 and I-75 corridors during the 1980s. Two were replacements for older General Motors plants in Kansas City, Kansas, and St Louis, Missouri. Only one entirely new automobile assembly plant was opened in the United States outside the I-65 and I-75 corridors during the 1980s and early 1990s, a joint venture between Mitsubishi and Chrysler called Diamond-Star, at Normal, Illinois, less than 200 km west of I-65. Seven of the assembly plants in the region were opened before 1979, including Chrysler's at Belvidere, Illinois, Ford's at Lorain, Kansas City and St Louis, and General Motors' at Arlington, Texas, Lordstown, Ohio, and Oklahoma City, Oklahoma. With the changes, the percentage of the US automobile assembly plants concentrated in the I-65 and I-75 corridors increased from less than one-third in 1979 to more than one-half during the early 1990s (Table 1.2).

PRINCIPAL OBJECTIVES OF THE BOOK

This book has two major purposes. The first part of the book examines the historical development of the geography of US automotive production. Parts II and III explain the reasons for the recent changes in the location of motor vehicle plants within the United States. Part II addresses the regional scale, that is the recent concentration of automotive production in the I-65 and I-75 corridors. Part III looks at the factors underlying the selection of specific communities within the I-65 and I-75 corridors for new automotive plants and identification of existing plants around the country for closure.

Chapters 2 through 5 address the book's first purpose, the reasons underlying the historical distribution of automotive production within the United States. Chapter 2 looks at the reasons why when commercial production began in approximately 1900, the automotive industry clustered in southern Michigan. Chapter 3 is concerned with why the Ford Motor Company, which was then the dominant producer, decided in approximately 1910 to build branch assembly plants near coastal concentrations rather than keep production concentrated in southern Michigan. Chapter 4 documents why General Motors, which replaced Ford as the leading producer, adopted the branch assembly plant strategy in the 1920s and extended the policy to most of its products after World War II. Chapter 5 examines why producers of automotive parts remained clustered in the southern Great Lakes region, even after most final assembly operations were transferred to coastal facilities.

Table 1.2 Location of US and Canadian automobile and light truck assembly plants, 1979–1991

	I–65/75 corridor	*Other interior*	*Coastal*	*Canada*	*Total*
All plants					
Open as of 1979	22	20	17	11	70
Open in new location	12	3	0[c]	5	20[c]
Replacement of old plant	3	2	0	0	5
Closed or idled indefinitely	5	6	7	2	20
Open as of 1991	29	17	10	14	70
Automobile					
Open as of 1979	15	15	16	6	52
Additions					
Open in new location	10	1	0	5	16
Replacement of old plant	2	2	0	0	4
Losses					
Replaced by new plant	2	2	0	0	4
Closed or idled indefinitely	3	5	7	1	16
Converted to truck production	2[b]	1	4	1	8
Open as of 1991	20	10	5	9	44
Truck					
Open as of 1979	7	5	1	5	18
Additions					
Open in new location	2	2	0	0	4
Replacement of old plant	1[b]	0	0	0	1
Converted from automobile production	2[b]	1	4	1	8
Losses					
Replaced by new plant	1[b]	0	0	0	1
Closed or idled indefinitely	2	1	0	1	4
Open as of 1991	9	7	5	5	26

Notes: See notes for Table 1.1

Chapters 6 through 10 explain recent changes in the geography of US motor vehicle production. Part II, which comprises Chapters 6 and 7, examines the reasons why automotive production has clustered in the I-65 and I-75 corridors. Chapter 6 documents the impact on locational patterns of changes in products, especially the proliferation in the number of distinctive models produced by each company. Chapter 7 evaluates the extent to which changes in freight costs – both shipping inputs to plants and hauling finished vehicles to consumers – have influenced clustering of production in the I-65 and I-75 corridors.

Chapters 8 through 10 are concerned with the book's third purpose, the reasons why automotive producers have selected specific sites within the interior for new plants and specific plants around the country for closure. Chapter 8 looks at the impact of governmental policies on specific plant locations. Chapter 9 documents the critical importance for many firms of locating plants in communities along the I-65 and I-75 corridors that have relatively low percentages of unionized workers. Chapter 10 evaluates the extent to which firms decide to retain or close older plants based on employees' willingness to adopt new work rules.

INDUSTRIAL LOCATION THEORY

The remainder of this chapter reviews major geographic theories developed to explain industrial location. Geographers ask two questions: where and why. Where are people and activities located across the earth's surface? What factors explain why people and activities are arranged as they are? Geographers rely on maps to depict where people and activities are located; in the case of the automotive industry, Figures 1.1, 1.2, and 1.3 clearly reveal changes in where assembly plants are located within the United States. Having shown through maps that the most recently built assembly plants are located in the interior of the country, while most coastal plants have been closed, the bulk of this book turns to the reasons why the distribution has changed. What insights can geographic concepts offer to explain the shift?

Geographers do not agree on how to explain why firms select particular locations to open – or in recent years close – factories. Industrial location theory developed early in the twentieth century – termed 'neoclassical' by contemporary geographers – found that firms selected plant sites largely on the basis of minimizing transport costs. In the 1960s, behavioral geographers claimed that industrial location decisions could be explained through understanding the motives of the decision-makers and the goals of the firms. Recently, structuralists have argued that the decision to open, retain, or close a plant must be understood as one of a number of social impacts stemming from global changes in the organization of industrial production.

Neoclassical location theory

Alfred Weber is widely regarded as the first influential advocate of neoclassical industrial location theory. His *Theory of the Location of Industries* appeared in German in 1909, the same year that the Ford Motor Company introduced the Model T, the most popular car in American history, and opened an assembly plant in Kansas City, Missouri, the first automobile plant in the United States situated in accordance with Weber's locational principles. However, the Ford Motor Company's decision-makers were certainly unaware of Weber's work, which was not translated into English until 1929.

Weber argued that the optimal location for a factory was the point that minimized the aggregate costs of bringing in raw materials and shipping out finished products. For a firm that has only one source of inputs and one customer – perhaps a producer of an automotive component made from rolled steel obtained from only one mill and sold only to Honda – the optimal location for the components plant can be easily computed by comparing the two sets of transport costs. If the cost of transporting the finished part to Honda's assembly plant is greater than the cost of bringing in the necessary amount of rolled steel, the optimal plant location is as close as possible to the customer. On the other hand, a firm supplying a relatively compact product made from bulky inputs, such as lubricants or bearings, may find that aggregate transport costs are minimized by locating near the principal source of raw materials.

Building on Wilhelm Launhardt's earlier work (1882), Weber proposed a geometric model to compute the least-cost location for a firm. Weber's model assumed that a firm obtained inputs from two locations, and that all of the firm's customers were located in a third place. Weber computed the optimal location within the triangle formed by the locations of the two inputs and the market as shown in Figure 1.4. The two inputs are obtained from I-1 and I-2, the market is located at M, and the factory's least-cost location is P. Lines a, b, and c represent the distances from P to I-1, I-2, and M, respectively.

The two inputs and market must be weighted to account for variances in shipping costs per mile, as well as in the bulk of the shipments. Consequently, x is a weight attached to I-1 to include the cost per mile of shipping and the amount of the input needed; similarly, y and z are the weights for I-2 and M. The optimal location P is the point that minimizes $ax+by+cz$. If x, y, and z are viewed as weights on pulleys, each pulling in a different direction, the optimal location can also be determined through the application of mechanics (Smith 1981a).

Hoover (1948) refined Weber's least-cost industrial location model by showing that costs per mile vary depending on the mode of transport and distance. In general, trucks have the highest cost per mile and are most efficient for hauling short distances, while boats have the lowest cost per mile and are most efficient for long-distance shipments; rail is most efficient for intermediate distances. When goods must be transferred between modes, the cost per mile increases substantially at the transfer point, because of additional labor and warehousing expenses. To avoid transfer costs, the transfer point – also known as the break-of-bulk point – may be the most optimal location for a factory.

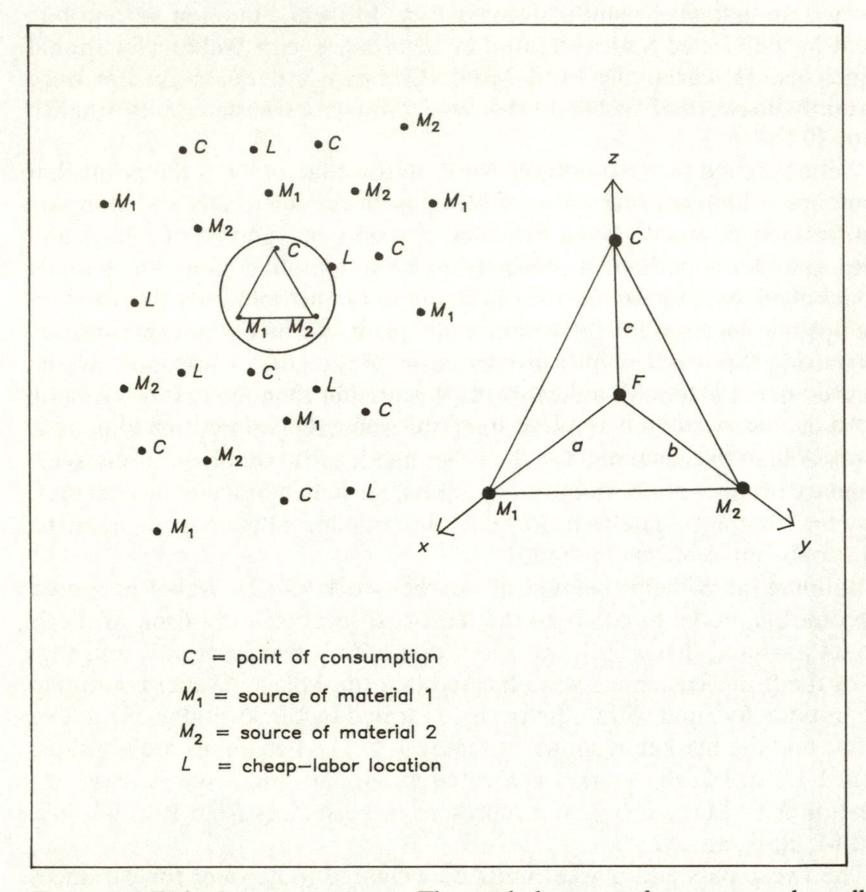

Figure 1.4 Weber's space economy. The symbols x, y, and z represent the pulls exerted by the factors of production at each location
Source: Weber (1929)

Weber's model introduced the effects of low-cost labor on industrial location. Circles drawn at increasing distances around the initially determined least-cost location accounted for added transport costs. Weber then identified other locations that offered low-cost labor. If the money saved at any of the low-cost labor points exceeded the increased transport costs, then that point would become the optimal factory location. Alonso (1967) demonstrated that low-cost land, as well as low-cost labor, can deflect a firm from locating at the point that minimizes aggregate transport costs. Isard (1956) argued that any factor of production – capital, land, or labor – can substitute for higher transport costs.

Weber had recognized the importance of agglomeration economies in plant location, but Hoover (1937) and Tord Palander (1935), a Swedish economist, revised Weber's model to account for the fact that as output rises the optimal location could change. To increase sales, a factory may extend its market area as far as it can until its transport costs exceed those of a factory located elsewhere. The line where the transport costs for the two factories are the same marks the boundary between the two factories' market areas.

August Lösch, whose major work appeared in 1940 in German and in 1954 in English, argued that market demand is the most critical factor in determining industrial location. According to Lösch, a firm looks for the location where profits are maximized rather than the least-cost location based on factors of production, as suggested by Weber. Lösch believed that locational analysis begins with delineating market areas for products based on empirical examination of actual conditions within a particular industry. Manufacturers locate within a market area until enough productive capacity is in place to meet consumer demand. Equilibrium is achieved when each market area contains the same level of production, assuming population, income, and tastes are uniform. Lösch represented market areas as a series of nesting interlocking hexagons.

D.M. Smith tried to integrate market demand with least-cost transport approach (Smith 1966). Smith's model made four assumptions:

1 The cost of producing one unit requires £40 of material, £30 of labor, and £20 of sales costs.
2 The cost of transporting raw materials and labor to the plant and the finished goods from the plant to the market all increase by £1 per mile; the costs of bringing in labor increase perhaps because workers must be enticed from a closer place of employment.
3 Other production costs, such as land, do not vary by location.
4 The locations of raw materials, labor, and market are equidistant from each other, at 30 miles (48 km).

Smith drew a series of concentric circles at 5-mile intervals around the sources of raw materials and labor and the market showing transport costs at each distance. Total transport costs at any point on the plain can be computed by aggregating the transport costs to that point for each of the three factors. Connecting points of equal total cost produces a series of concentric triangular-shaped isopleths around a single point where aggregate transport costs are minimized.

Neoclassical location theory accurately portrayed the distribution of automotive manufacturing in the United States from the World War I era until the late 1960s. For suppliers of components, southeastern Michigan provided proximity to both the most important input – steel – as well as the most important markets – the automobile producers. Nineteenth-century steel producers had been initially attracted to southwestern Pennsylvania near Pittsburgh because of local iron ore and coal fields, but by the time the automobile industry developed, newer steel mills were located further west, along the south shores of lakes Erie and Michigan. Iron ore was brought by barge to the steel mills along the Great Lakes from Minnesota's Mesabi Range, while coal came north by train from Appalachia and southern Illinois. At the same time, parts suppliers located in Michigan because the state contained the largest number of assembly plants, as well as the corporate offices and research facilities of the major auto makers.

While parts suppliers clustered in Michigan, most final assembly plants were located in other parts of the United States. An automobile is much bulkier than the components which go into its construction and consequently is relatively expensive to ship. In order to minimize the cost of shipping the bulky finished products to consumers, most assembly plants were located near concentrations of population, such as New York, Boston, and Los Angeles.

Industrial location theorists became increasingly disenchanted during the 1960s with the neoclassial approach. Unrealistic assumptions concerning market competition, labor mobility, and lack of change in the locations of inputs and markets undermined the validity of the model. Government policies, such as local tax rates and regional development projects, influenced industrial location. Most critically, empirical studies conducted in a variety of industrial sectors reinforced the criticism by showing that plant sites were simply not being selected in order to minimize aggregate transport costs (Massey 1979). Instead, many industries were regarded as 'footloose,' that is able to locate in a wide variety of places without a significant change in aggregate transport costs.

Behavioral location theory

Geographers during the 1960s turned to behavioral theories derived from organizational theory and psychology to explain industrial location decisions. Because in the real world factory sites are selected by individuals, the processes by which these individuals make decisions underlie industrial location patterns. With production increasingly concentrated in the hands of a few large firms in the automotive industry, as well as in other sectors, the structure of decision-making in very large organizations became increasingly important to explaining plant locations (Dicken 1971).

Behavioralists analyzed the process through which plant location decisions were made, including the formulation of corporate goals, the collection and evaluation of information, and the implementation of goals. Locational decisions are made by people who possess varying degrees of entrepreneurial skill and ability to perceive the economic opportunities present in the environment (Smith 1966). Collecting information is an ongoing process, and the quality of information that entrepreneurs receive in order to make locational decisions can range from good to poor (Dicken 1971). Decision-makers may also possess different levels of power and organizational ability to implement the desired locational structure (Hamilton 1978).

Pred placed actual plant sites on a matrix (Pred 1967), with the two axes of the matrix representing the two principal limitations on the ability to select the optimal location: level of knowledge and degree of ability to use the knowledge. Individuals with high levels of knowledge and power could identify and select sites close to the point that maximizes profits, perhaps through systematic comparison of market areas and transport costs at various locations, consistent with neoclassical industrial location theory. Less skilled entrepreneurs would select locations that are economically suboptimal (Smith 1981a).

Even if individuals operate with high levels of knowledge and power, they may still be induced to select suboptimal locations, depending on the specific roles they play within the organization. Instead of profits, a firm's goal may be to maximize levels of production, inventory, sales, market share, growth, or security (Cyert and March 1963; Dicken 1971; Galbraith 1967). More critically, different individuals within one firm may be charged with the responsibility of achieving goals such as growth of the firm, larger control of the market, or diversification of corporate interests that may point to a locational strategy other than the one compatible with maximization of profits (Hamilton 1974). Rather than select a new location, to attain its goals, a corporation can alter the use, size, and number of facilities (Hamilton 1978).

Decision-making is also influenced by the social and administrative environment in which the firm operates (Smith 1966). A firm's decisions and actions are largely shaped by the government's legal and political policies and other organizations' social and economic values (Smith 1966; Hamilton 1978). The constraints of real-world decision inevitably lead to suboptimal locational decisions (Lösch 1954).

The impact of a decision-maker's personal characteristics is especially clear in the location of smaller firms. Socio-economic status, education, friends, hobbies, professional contacts and preference for amenities may influence an owner's locational choice. The location of many manufacturers is dictated by where the owner was born or has lived in recent years (Greenhut 1956). A temperate climate, permitting year-round outdoor activities such as golf, has attracted some owners to the US south and west, a region consequently dubbed 'Sunbelt.' Negative experiences with crime, traffic, or labor unions have driven other manufacturers out of the northeastern 'Frostbelt.'

As a search for the optimal location may be time-consuming and costly, the selected plant site may be the first satisfactory alternative encountered rather than the optimal one (Dicken 1971; Greenhut 1956). Large corporations may operate plants in suboptimal locations inherited through mergers and acquisitions. The importance placed by a corporation in pinpointing the optimal location for a plant depends on the degree of competition within the particular industry, the number of critical factors, and the degree of precision required in the locational decision to be profitable.

Government subsidies and tax breaks can also alter the calculation of the optimal location (Smith 1981b).

The behavioral theory of location fits some of the trends in the US automotive industry in the 1950s and 1960s. As smaller companies like Studebaker, Packard, Hudson, Nash, Crosley, Willys, and Kaiser-Fraser ceased producing automobiles, the surviving firms adopted increasingly oligopolistic – if not monopolistic – practices. General Motors, by capturing more than half of all US car sales, flirted with violations of anti-trust rules and ignored government threats to split it up, as had happened to Standard Oil and US Steel in the past.

US automotive executives dismissed small cars, at that time primarily sold by Volkswagen, as an inconsequential fringe segment of the market. The average American consumer, Big Three executives believed, still overwhelmingly preferred uniformly large cars. More importantly, smaller cars meant smaller corporate profits.

Traditionally, US auto makers had attracted first-time buyers to their lowest-priced models and encouraged subsequent purchases of models shown through advertising to carry higher status. A luxurious model might share the same body and chassis as a basic one but generated a much higher profit because the cost of additional trim was less than the increase in price which could be levied. However, the various models were so similar that they could be produced on the same assembly lines if necessary. Despite their lower profits, though, the Big Three began to produce smaller models in the 1960s as entry-level loss-leaders to capture first-time buyers who would soon trade them in for more profitable larger cars. Smaller cars were not fundamentally redesigned; parts were pirated from larger models to save costs and preserve corporate images which encouraged luxury and power.

As the US automobile industry became more oligopolistic, the locational pattern became stationary. With small cars regarded as little more than a temporary diversion from the main task of building large ones, the Big Three continued to rely on the neoclassical plant location model which had served it so well for half a century. The US automobile industry was a prime contributor to the image of faceless, grey-flanneled, arrogant executives making decisions designed to minimize risk and avoid innovation, especially when these approaches coincided with the principal corporate objective, maximizing profits.

Structuralist location theory

Demand for industrial products made in advanced capitalist countries stagnated during the 1970s and early 1980s, because of both declining domestic markets and slower rates of penetration of new markets in developing countries. Demand was also dampened by problems with petroleum, including shortages as a result of the 1973–74 boycott and 1978–79 revolution in Iran, and a ten-fold price rise between 1973 and 1982. Reflecting the economic slowdown, automobile sales in the United States plunged more than 30 per cent from 1973 to 1982, when fewer than eight million were sold for the first time since 1963.

At the same time, global competition for production increased, especially as firms in newly industrializing Asian countries captured higher percentages of output, breaking long-time oligopolies enjoyed by North American and Western European producers (Bloomfield 1981). Automobile production in the United States declined more than 40 per cent between 1973 and 1982, dropping below six million per year for the first time since 1958, yet sales of imports, mostly from Japan, increased 25 per cent during the same period (Clark 1986).

The global economic crisis of the 1970s sparked dissatisfaction with the ability of both the neoclassical and behavioral theories to explain industrial location. Structuralists argued that locational changes needed to be understood in the context of broader changes in the nature of capitalist production (Massey 1979; Peet 1987; Scott and Storper 1986). Massey and Meegan (1982) identified three types of restructuring of production – intensification, rationalization, and investment in technology – each with distinctive consequences for industrial location.

Intensification referred to changes designed to increase labor productivity without major new investment or changes in production technology (Lever 1985). In Britain, a number of clothing and textile sectors have restructured through intensification, including footwear, leather, finishing, and men's outerwear (Massey and Meegan 1982). Few locational changes result from intensification, because while the overall demand for labor is reduced, entire plants are rarely opened or closed. However, the other two types of restructuring have produced significant spatial changes (Massey and Meegan 1985).

Rationalization, the second type of industrial restructuring, results in a reduction in a firm's total capacity through closure of some existing plants. At a global scale, rationalization has been facilitated by the concentration of economic and social power in the hands of a few large transnational or multinational corporations (Linge and Hamilton 1981). Transnational corporations have invested in overseas plants and acquired controlling or minority intersts in companies in other countries. By spreading production among many areas, transnational corporations can increase their bargaining power with local labor forces and governments. Costs are reduced through global standardization of output, yet product variety can be increased through exchange of components among production units (Storper and Scott 1986).

The growing strength of transnationals has geographic implications, because the corporations have increasingly allocated production in response to characteristics of the local labor force, including level of skills, prevailing wage rates, and attitudes towards unions. Transnationals have removed production from regions of high wage rates and organized labor opposition by relocating facilities or by subcontracting work to independent suppliers. Older plants in the inner cities are especially vulnerable to closure (Massey and Meegan 1982).

The third type of restructuring, investment in technical change, has the most profound impact on regional shifts in employment (Lever 1985; Massey and Meegan 1985). Structuralists have embraced the term 'Fordism' to describe the organization of production that is being replaced with 'post-Fordist' or 'flexible' methods. Foremost among the structural changes with locational impacts is reorganization of labor (Massey and Meegan 1985). Firms are replacing uniform, industry-wide, national labor union contracts with more flexible rules that permit the assignment of workers and managers to various tasks and the hiring of temporary and part-time workers to meet short-term production demands (Storper and Walker 1989). A shift in the location of employment provides the management of a firm with a lever to manipulate the labor force and reduce the strength of unions (Peet 1984). Firms are especially attracted to smaller towns where they can control local labor markets; low levels of unionization and high visibility reduce vulnerability to stoppages even if wages are relatively low and lay-offs become necessary (Lever 1978, 1985).

Changes in production have led to a spatial division of labor, notably a separation of office-based managerial control, business services, and information processing from manufacturing production operations (Scott and Storper 1986). Managerial control functions have clustered in a handful of large metropolitan areas, while branch plants still organized along 'Fordist' principles are shifted to diverse peripheral territories (Sayer 1983). These branch plants may generate relatively few supporting service industries. The restructuring of the division of labor has led to the deskilling of some tasks and the reskilling of others (Braverman 1974; Malecki 1986).

Within the United States, automobile producers responded by distinguishing between operations which required highly skilled crafts people and those which could be performed by unskilled workers. During the 1970s, as many unskilled automotive jobs as possible were relocated to the south, the area of the United States where prevailing wages were lower and unions weaker. When that strategy failed to achieve sufficient cost savings, auto producers built plants in Mexico instead, especially during the early 1980s (Hoffman and Kaplinsky 1988).

In addition to the spatial division of labor, other structural changes have affected industrial location in advanced capitalist countries (Scott 1988). Large capital-intensive factories organized around the continuous flow of the assembly line are being replaced with programmable, electronically controlled robots (Storper and Walker 1989). Automation reduces the number of workers, as well as the degree of physical strength, needed to perform tasks (Peet 1987). Robots give firms the flexibility and versatility to change the mix of outputs in the short run and offer more diverse products. Technical advances in communications, information flow, and data manipulation also allow firms to respond quickly to changing market conditions (Storper and Walker 1989). Innovations in electronics have encouraged the spatial disaggregation of production.

The second structural change has been a shift from standardized output, long production runs, and dedicated nonadaptable capital equipment to a system that permits rapid adjustments in the mix of inputs and outputs and production procedures in response to changing market demands (Storper and Walker 1989). While overall demand stagnated in the 1970s and 1980s, consumers sought a greater variety of products. Markets previously dominated by a handful of essentially similar products, such as motor vehicles, witnessed the proliferation of new products to meet specialized market niches.

US auto makers have struggled to match the ability of Japanese firms to bring new products quickly to the market. The Big Three build cars which compare favorably to Japanese products, but to Japanese products sold several years earlier. Indicative of the internal problems faced by US-owned auto makers, General Motors in the mid-1980s was unable to make a profit selling the two-seated Pontiac Fieros with average annual sales of 75,000. A few years later, Mazda could profitably sell its two-seated Miata at a comparable price, accounting for inflation, with less than half of Fiero's annual sales.

A third structural change has been lower reliance on achieving internal economies of scale through vertical integration of production. Some production has been transferred from once largely self-sufficient corporations to relatively small, specialized subcontractors (Holmes 1986). With increased outsourcing, large mass-production corporations are exercising less hierarchical control over the manufacturing process and instead must rely on cooperation with smaller firms (Storper and Walker 1989).

The geographic result of flexible production is mixed. On the one hand, production has reconcentrated in industrial districts, although perhaps in more peripheral locations within the districts than in the past (Scott and Storper 1986). Concentration enables firms to reduce the costs and difficulties of carrying out transactions with other firms and to maximize access to information available in the production district (Holmes 1986). On the other hand, the international division of labor has led manufacturers to transfer some production from older industrial districts to regions where greater control can be exercised over the local labor market.

RECENT TRENDS

Which of the three perspectives best explains the recent spatial changes in the North American motor vehicle industry? Behavioral theory helps to explain locational decisions in the formative years of the automobile industry, especially the concentration of production in southeastern Michigan after 1900 (Chapter 2), but neoclassical location theory underlies the decisions which soon turned the Ford Motor Company into the country's dominant producer (Chapter 3).

General Motors supplanted Ford as the nation's leading automobile producer by emulating the policy of selecting sites for plants consistent with neoclassical location theory, at a time when locational decisions at Ford were being dictated instead by the behavioral quirks of the company's founder (Chapter 4). Components suppliers were also located according to Weberian principles, although the resulting distribution was more clustered than that of assembly plants (Chapter 5).

Recent upheavals in the automotive industry have resulted in increased concentration of production in the interior. Assembly plants which had previously produced identical models for distribution within a regional market have been converted into specialized plants producing one or two models for national distribution. To minimize the cost of distributing products to a national market – and thereby maximize profits – automotive companies opened new plants in the interior and closed coastal ones (Chapter 6). Parts suppliers then clustered in the interior to maximize access to the new assembly plants, encouraged by the diffusion of just-in-time delivery systems. For geographers, the most distinctive element in the automobile industry's recent return to the interior of the country is the extent to which strategic decisions again are being based on minimizing freight costs, consistent with neoclassical industrial location theory (Chapter 7).

Freight costs may account for the concentration of automotive production in the interior of the country but not the selection of individual communities within the region. Local and national governments have some influence in the precise location of automobile plants (Chapter 8). However, local labor climate is the most critical factor influencing the distribution of automotive production within the I-65 and I-75 corridors. Some producers have selected sites, especially in Mexico, in order to minimize labor costs, but that has not been the principal objective in the location of most new plants. Instead, producers are more likely to select individual communities within the I-65 and I-75 corridors for new plants because of a desire to avoid concentrations of militant and unionized workers (Chapter 9). Similarly, decisions to retain or close existing plants are being influenced by labor climate, although age of plant and marketing success of the plant's products may also prove critical (Chapter 10).

Competition among communities, states, and provinces across the United States and Canada to attract motor vehicle plants lends an urgency to understanding the industry's changing geography. Which communities have realistic prospects of attracting or retaining motor vehicle producers, and which face the likelihood of closures? The stakes are high, because the motor vehicle industry is North America's largest manufacturing sector. More than a million people in the United States and Canada are employed by auto makers such as General Motors, Ford, and Honda to produce components and install them in vehicles. But the automotive industry's economic impact is much more extensive.

Several thousand companies make parts and components which are sold to the auto makers. Thousands of other firms supply the auto makers with materials such as steel, rubber, and plastics, for fashioning parts. Thousands of vendors provide indirect support, such as paper products, food, electricity, and trucking services. These are only the direct suppliers of products and services to auto makers.

As a basic industry – one which exports most of its product outside the community – an automotive plant generates substantial local revenues through a spin-off or multiplier effect. That is, local services, such as shops and restaurants, survive because auto workers and their families patronize the establishments, spending money brought into the community through the sale of vehicles to people living elsewhere.

The opening of a new plant generates rapid economic growth and immigration, producing substantial demand for construction of new housing, roads, utilities, and other infrastructure. Spring Hill, Tennessee, was home to approximately 1,000 people when General Motors selected it in 1985 for the Saturn plant. The plant was expected to attract nearly 6,000 workers to the community from elsewhere in the country. These workers, typically in their thirties, would bring their families to the area, roughly 9,000 additional people. Given the multiplier effect, the total number of immigrants to Spring Hill to work in components manufacturers and local service jobs is likely to be as least as high as the number working at Saturn.

Meanwhile, closure of an older plant may rob a community of its largest taxpayer, causing sharp increases in the level of unemployment, problems stemming from social dislocation, and demand for public services. When GM closed two plants in the Cincinnati, Ohio, area – the Norwood assembly plant and Fairfield stamping facility – the eight-county metropolitan area lost 3.6 per cent of its total manufacturing jobs. The year before the closures GM purchased $111 million worth of supplies from roughly 800 vendors within 80 km of Cincinnati.

The impact of the closures was greater on the two small cities in which the plants were located. Norwood, a city of 26,000, received $2.6 million in payroll taxes from the GM plant in 1987, 24 per cent of the city's $11 million budget. GM accounted for $1.7 million, or 28 per cent, of the $6 million budget in Fairfield, a city of 15,000.

This book does not offer any 'inside' information concerning which plants are likely to be closed and which communities are likely to receive new ones. However, showing the changing spatial patterns and offering geographic perspectives on the reasons for the changing distribution can point to which plants are vulnerable in the future and which communities have prospects of attracting new ones.

Part I

Development of the geography of US automotive production

The location of US automotive plants changed twice in the early years of commercial production. More than half of the producers during the 1890s were located in the northeast, between Philadelphia and Boston. During the first decade of the twentieth century, most production shifted from the northeastern states to southeastern Michigan. Then, a decade later, most final assembly operations relocated to other regions of the United States, while production of most automotive parts, as well as most management functions remained in southeastern Michigan. The locational pattern established during the World War I era remained dominant until the 1980s.

The principal purpose of the first part of the book is to explain the two early shifts in the location of US automotive production. The predominant view among automotive historians is that southeastern Michigan became the center of automotive production primarily through the accident that the most successful innovators and entrepreneurs happened to live near there around 1900.

As Chapter 2 shows, southeastern Michigan in 1900 clearly did not represent the least-cost location according to Weberian analysis. At that time, the nation's market center was near New York City, then by far the largest market in terms of both population and – more critically – purchasing power. The migration of automotive production from the northeastern states to southeastern Michigan after 1900 represented a move away from the least-cost location. While inputs, such as wood and steel, were plentiful near southeastern Michigan, access to these resources did not play a critical role in the relocation either.

Theories that the automotive industry concentrated in southeastern Michigan by accident ignore the locational impacts of factors of production, especially labor and capital. The automotive industry evolved from older industrial sectors, and southeastern Michigan was the production center for two of the three types of manufactures that contributed most heavily to early automotive technology – carriages and gasoline engines. In addition to the talents of carriage-makers and engineers, southeastern Michigan also offered a surplus of unskilled workers, many of whom had recently immigrated from eastern Europe. While the northeast US also contained large numbers of recent immigrants, southeastern Michigan gained a national reputation around 1900 for effectively discouraging unionization.

Venture capital was also available in southeastern Michigan. Although early automotive production did not require heavy capital expenditures, early entrepreneurs lacked even the relatively modest financial resources needed. When bankers in the northeast were reluctant to loan money for such a high-risk operation, wealthy individuals in southeastern Michigan stepped forward.

The Ford Motor Company was largely responsible for the concentration of automotive production in southeastern Michigan. Ford's Model T, introduced in 1909, quickly captured nearly half of all automotive sales in the United States. The company achieved high sales volume by selling the car at a price that was sufficiently low to stimulate consumer demand yet sufficiently high to generate large profits through the application of mass-production technology, such as the moving assembly line. However, as Chapter 3 shows, even while Ford concentrated production of components at Highland Park, Michigan, most final assembly operations were transferred to branch plants elsewhere in the country. Ford's branch assembly plant policy was explicitly based on a Weberian-style analysis of freight costs, and specific plant sites were selected in order to minimize the cost of hauling finished cars to consumers.

General Motors emulated Ford's branch plant strategy for its low-priced Chevrolet, which outsold Ford for the first time in 1927. Chapter 4 presents evidence that minimizing freight costs strongly influenced GM's branch assembly plant strategy, especially after World War II, when the company extended the concept to its medium-priced models, as well. Branch assembly plants were opened near large cities that served as centers of market areas.

The production of components remained highly clustered in the southern Great Lakes region, even after much of the assembly operations were transferred elsewhere, as discussed in Chapter 5. Ford concentrated nearly all components production in southeastern Michigan, while GM's facilities were somewhat more dispersed within the southern Great Lakes states, because many of them had originated as independent companies. Minimizing freight costs was an important factor in the location of components plants, but the resulting pattern varied from that of final assembly plants. For a bulky, fabricated product, such as an assembled car, the critical freight cost is from the assembly plant to the consumers, but access to both inputs and consumers is important for components.

2 Automobile production concentrates in Michigan

Early experimenters with building motor vehicles were scattered across the northeastern United States from Massachusetts to Wisconsin, and when commercial production began in the 1890s, the northeast was the center of activity. In the late 1890s, of the hundreds of firms organized to produce motor vehicles, only one – the Olds Motor Works – located in Detroit. As early as 1903, however, the best-selling cars were made in southeastern Michigan, and by 1913 80 per cent of production was concentrated in that area.

Why did southeastern Michigan become the industry's home early in the twentieth century? The 'great person' view of development is especially popular among historians of automobile production, especially consistent with the approach taken by behavioral geographers. More than most other industries, the history of motor vehicle production in the United States has been explained through the biographies of inventors, written with the benefit of hindsight concerning the handful of firms which survived (Hurley 1959: 416). According to this behavioral perspective, America was carried into the horseless-carriage age by visionary pioneers, some like Ford, Olds, Buick, and Dodge, whose names still grace contemporary cars, and others like Durant, Leland, and Joy less well-known today. Southeastern Michigan's dominance was an accident which resulted from the fact that the area happened to be the home of a remarkable collection of inventive genius.

As John B. Rae, a prominent automotive historian, explains:

> With due allowance for the influence of economic and geographic factors, Detroit became the capital of the automotive kingdom because it happened to possess a unique group of individuals with both business and technical ability who became interested in the possibilities of the motor vehicle.
>
> (Rae 1965: 59)

Behavioral geographers concur in crediting a limited role to neoclassical locational factors in explaining the emergence of southeastern Michigan. Gerald Bloomfield points out 'at its formative stage, motor vehicle manufacturing was not tied to sources of materials or markets, transport costs were low and the industry was not yet capital intensive' (Bloomfield 1978: 122–123). Clearly freight charges were not critical in bringing the automotive industry to southeastern Michigan, because the industry's market center at the turn of the century was in the northeast, near New York City.

The clustering of talented inventors in southeastern Michigan doesn't mean that geographic concepts are irrelevant in explaining the area's rapid dominance in automotive production. Inventive genius doesn't spring in isolation from the local industrial climate. The distinctive features of doing business in southeastern Michigan, especially labor and capital factors, encouraged motor vehicle producers to locate there during the industry's formative years.

EARLY NORTHEASTERN MANUFACTURERS

Controversy surrounds the identity of the builder of the first workable gasoline-powered motor vehicle. Claimants during the early 1890s included Henry Nadig in Allentown, Pennsylvania, and John William Lambert, in Ohio City, Ohio, both in 1891, Gottfried Schloemer and Frank Toepfer in Milwaukee, Wisconsin, in 1892, Charles H. Black, in Indianapolis, Indiana, in 1893, and Elwood P. Haynes in Kokomo, Indiana, in 1894. The multiplicity of claims reveals the dispersed distribution typical of a process of industrial innovation initiated by a number of isolated individuals tinkering with existing technology (see May 1990 for more information about early manufacturers).

The commencement of commercial motor vehicle production reinforced the industry's northeastern orientation. The first company organized for the purpose of producing motor vehicles and then actually making and selling them was the Duryea Motor Wagon Company of Springfield, Massachusetts. In 1895, Duryea sold four cars to lead all US producers.

The Duryea's reputation was enhanced that year by winning a contest organized by the Chicago *Times-Herald* to pick the most practical motor vehicle. The *Times-Herald* contest generated considerable interest among experimenters, as well as the public, and dozens of vehicles were built for the purpose of competing, but few could meet the challenging conditions of the contest. In the end, the Duryea and only five other entrants started on 25 November, the day after a snowstorm paralyzed the city. The other contestants included two from New York and one each from Chicago and Decatur, Illinois, and Philadelphia, Pennsylvania.

The rules of the contest stipulated that the winning vehicle would be selected according to a variety of criteria related to economy and ease of operation, but the public was interested primarily in the speed and durability of competing vehicles. The Duryea was regarded by the public as the winner of the race because it covered the 86-km (54-mile) course through the streets of Chicago and Evanston in a bit over 10 hours, including a 55-minute stop for repairs (Bailey 1971: 86–90).

Leading northeastern manufacturers

Hundreds of firms were organized to build motor vehicles in the late 1890s and the first years of the twentieth century. Automotive historians disagree on the actual number of producers, because a large percentage of them never could put together anything more than an experimental prototype. The best source of information about early producers is the trade journals which first appeared in the 1890s. Researchers have unearthed more than 3,000 firms organized to manufacture automobiles, but few ever entered commercial production or built more than a handful.

More than half of the early automotive firms settled between Boston and Philadelphia, including most of the best-selling models. Springfield, several towns in Connecticut, New York City, northern New Jersey, and Philadelphia were all production centers (Boas 1961: 221–222). The Pope Manufacturing Company's Columbia, built in Hartford, Connecticut, became the first car to sell more than 1,000 in a single year, 1900, accounting for more than one-fourth of all sales that year. Pope was the nation's leading bicycle manufacturer, and its founder Colonel Albert A. Pope, was known as the King of Bicycles. The company expanded during the next few years, buying out manufacturers and their plants in other cities, including the Waverly Company of Indianapolis and H.A. Lozier of Toledo, Ohio. The Columbia Motor Company was sold to Benjamin Briscoe as part of the United States Motor Company in February 1910 (Norbye 1971: 112; Bradley and Langworth 1971: 138). Pope ceased automotive production in 1907, but the Toledo facility remained an active assembly plant, passing through the hands of a number of firms until eighty years later it was assembling Jeeps for the Chrysler Corporation.

The following year, the second company to achieve annual sales in excess of 1,000 was Locomobile, built in two locations, Tarrytown, New York, and Newton, Massachusetts. Amzi L. Barber, an asphalt magnate, and John Brisbane Walker, owner of *Cosmopolitan* magazine, bought the company from the Stanley twins in 1899 but couldn't agree on where to build the cars. Consequently, Barber continued to build Locomobiles in the Stanleys' Newton plant, while Walker organized the Mobile Company of America in Tarrytown to make the identical cars. In 1902 the company switched from steam to gas engines and moved production to Bridgeport, Connecticut. The Locomobile was made until 1929 but became an expensive, low-volume product. Meanwhile, the Stanleys organized a new company in Massachusetts, the last American manufacturer of steam-powered automobiles (Norbye 1971: 113; Bradley and Langworth 1971: 138).

The last large-volume automobile produced in the northeast was the Maxwell. Jonathan D. Maxwell, who had worked for car makers in Kokomo and Detroit, and Benjamin Briscoe, a Detroit metal parts manufacturer, created the Maxwell-Briscoe Motor Company in 1903 and began production in 1904 in the Tarrytown plant once occupied by Brisbane's Locomobile. In 1910, Maxwell-Briscoe bought several struggling companies, including Pope's Columbia, to form the United States Motor Company. Two years later the struggling company was bankrupt, and only its best-selling car, the Maxwell, survived. Control of the renamed Maxwell Motor Company passed to Walter Flanders, who had participated in the production of the Everitt-Metzer-Flanders, or E.M.F., car, which became part of Studebaker. In 1917, Flanders relocated production of the Maxwell from Tarrytown to a Detroit plant owned by the Chalmers Motor Company. The abandoned Tarrytown plant was snapped up by Chevrolet, while Maxwell was reorganized as the Chrysler Corporation during the 1920s. Chalmers' East Jefferson Avenue plant in Detroit became Chrysler's home production center until replaced by a new plant next door in the early 1990s.

The northeast remained the center for production of luxury automobiles through the 1920s. In addition to the Locomobile, other leading luxury cars included the Pierce-Arrow, built in Buffalo, New York, and the Franklin, built in Syracuse, New York. These firms were unable to survive the Depression during the 1930s. As recently as the late 1950s, General Motors, Ford, and Chrysler together operated thirteen branch plants within a 200,000 km^2 triangle formed by Buffalo, Boston, and Norfolk, Virginia. These northeastern plants were responsible for nearly one-fourth of total US passenger car production in the late 1950s. By the 1990s, only two of the plants still assembled automobiles – both in Delaware – although five others remained open to build trucks.

EMERGENCE OF SOUTHEASTERN MICHIGAN

Motor vehicle production increased in Midwestern states other than Michigan early in the twentieth century, especially Indiana, Ohio, Illinois, and Wisconsin. Chicago, Cleveland, and Indianapolis attracted as many producers as Detroit, while other production centers emerged in Anderson, Auburn, Connersville, Elkhart, Kokomo, Muncie, Richmond, and South Bend, Indiana; Decater, Freeport, Moline, Ottawa, and Peoria, Illinois; Dayton and Toledo, Ohio; and Hartford, Kenosha, Milwaukee, and Racine, Wisconsin.

The three highest-volume producers outside of Michigan were located in Kenosha, South Bend, and Toledo. Thomas B. Jeffery bought a plant in Kenosha from the Sterling Bicycle Company to build Ramblers, beginning in 1902. The plant was bought by the Nash Motor Company in 1916 and through mergers passed to American Motors in 1954 and the Chrysler Corporation in 1987 before closing in 1988. Studebaker, a leading wagon builder since 1852, assembled cars at its South Bend plant from 1904 until 1963. Willys-Overland, which bought Pope's Toledo plant, was the second leading producer of automobiles during the World War I era.

Although independent producers outside of Michigan survived until after World War II, motor vehicle production became highly clustered in southeastern Michigan rather than dispersed among other Midwestern states. As early as 1903, the three leading producers were based in Michigan, including Olds, which built 4,000 cars that year, and Ford and Cadillac, which each built approximately 1,700. The 1904 US Census of Manufactures, the first to break out manufacturing of automobiles from parts suppliers, revealed that 42 per cent of all cars were made in Michigan. Five years later, according to the next Census of Manufactures, Michigan's share of national production had climbed to 51 per cent (May 1975: 333–4).

Automotive production concentrated in southeastern Michigan because the industry's most successful pioneers either grew up in the area or migrated there. Surveys of contemporary industrial location decisions yield similar behavioral results. When asked to explain why a business located in a particular place, a spokesperson is most likely to respond that the owner lived nearby. However, founders of new businesses do not operate in a social and economic vacuum; they build on their knowledge of a community's experience in successfully manufacturing other products.

Southeastern Michigan attracted or retained the most successful motor vehicle producers, because industries from which automotive technology derived were already thriving in the region. Mechanics and engineers skilled in other industries based in southeastern Michigan applied their expertise to experimenting with motor vehicles. Then, when early auto makers progressed from the experimental stage to commercial production, they turned to the existing industries for skilled workers.

Topping a long list of problems faced by early motor vehicle producers, practical designs had to be found for three components – an engine to propel the vehicle, a drivetrain to convert engine power to motion, and a platform or chassis strong enough to hold both passengers and the power sources. Two of these three problems were solved largely through technology borrowed from industries which had already clustered in southeastern Michigan. At the turn of the century, southeastern Michigan was a center for machine shops with experience building gasoline-powered engines and for carriage makers with experience building chassis.

Gasoline engines

At first, automobiles powered by gasoline engines were less popular than those driven by steam or electricity. Of the roughly 4,000 automobiles sold in the United States in 1900, 40 per cent were powered by steam, 38 per cent by electricity, and only 22 per cent by gasoline. Within five years the race was over, and the gasoline engine was firmly entrenched as the dominant form of power. Gasoline engines accounted for 83 per cent of the nearly 12,000 sales of vehicles that year, compared to 12 per cent for steam and 5 per cent for electricty.

Had the electric engine proved the optimal choice for motor vehicles, the northeast may have retained its early lead in automotive manufacturing. Electric engines were especially popular in the urban areas of the northeast, because they were quieter and cleaner than the competitors and easier to operate. In an appeal to female consumers, electric cars resembled fashionable horse-drawn carriages in appearance and performance.

Early electric vehicles proved unsuitable for rural driving. Every 30 km or so they had to be recharged, a process which took between two and three hours. At the rate of $15 per recharge, the cost of operating an electric vehicle was roughly two cents per km, compared to less than a penny for steam and gasoline cars. Recharging facilities were scarce in rural areas, although a network of six was established between New York and Philadelphia to permit travel between those two cities. Hills proved difficult to negotiate for the underpowered vehicles. Given the likelihood of limited petroleum supplies and higher emissions standards in the future, automobile producers are facing the prospect of returning to electric power; General Motors, for example, has begun to produce battery-powered cars at an assembly plant in Lansing, Michigan.

Steam-powered cars did not suffer the same handicaps as electric. Powerful enough to climb steep grades, they were especially popular in hilly New England. Steam engines had fewer moving parts, transmitted power to the wheels more smoothly, and were easier to produce than gasoline engines and cost about the same to operate. However, steam engines hit a technological dead-end after 1901 and could no longer generate as favorable a ratio of horsepower to engine weight as the increasingly powerful gasoline engines.

Early gasoline engines were more powerful and mechanically reliable than the competitors and with crude petroleum less than five cents a barrel, inexpensive to operate. Gasoline-powered automobiles were more suitable for rough roads in rural areas, where half of the American population still lived at the turn of the century. Many rural residents were already familiar with gasoline engines, which had been commonly installed in agricultural equipment.

Experience with building gasoline engines gave Michigan automotive manufacturers a critical early technological lead over competitors elsewhere in the country. Michigan had become a leading center for the production of gasoline engines for agricultural and marine uses. Detroit's shipyards were early converts to gasoline engines in part because of dissatisfaction with the heating capacity of locally mined coal.

The two most successful carmakers in the first decade of the twentieth century, Ransom E. Olds and Henry Ford, were both skilled mechanics familiar with gasoline engines. Olds worked in his father's machine shop in Lansing, Michigan, and became a partner in 1885 at the age of twenty-one. He designed one- and two-horsepower engines to meet an increasing demand for small stationary engines which could provide intermittent power for farm implements and industrial machines. To accommodate engine production, Olds built a plant in Lansing and began to tinker with using his engine to propel a carriage. The engine plant was merged with the Motor Works Company in 1899, when Olds began to build automobiles.

Henry Ford had acquired considerable experience in machine shops by the time he launched his motor vehicle companies. He always liked to tinker with machinery and as a child in Dearborn, Michigan, had developed an aptitude for repairing watches. For a time, Ford worked for Westinghouse, travelling around southeastern Michigan repairing machines. Ford moved to Detroit in 1891 and worked for five years for the Edison Illuminating Company, where he was in charge of keeping the generators operating at one of their power plants. His knowledge of machinery is evidenced by the fact that in the evening, he taught a class at the Detroit YMCA to train machinists.

The long-term association of Oldsmobile production with the city of Lansing originated by accident, according to his biographers. In 1899, the Olds Motor Works became the first manufacturer to produce motor vehicles commercially in Detroit, on a two-hectare site on East Jefferson Avenue near the Belle Isle Bridge. When the complex burned down two years later, Olds quickly transferred production to Lansing, where he still maintained an engine plant. According to early automotive industry publications, Olds – anxious to resume production as soon as possible – concentrated exclusively on the only model salvaged from the fire, named the Curved Dash, after the shape of its front. His biographers, challenging this legend, point out that the company could have just as easily built other models by referring to diagrams and photographs which survived (and still exist today). Instead, they claim that Olds chose the Curved Dash because it was relatively inexpensive and easy to manufacture (May 1977: 115–116; Niemeyer 1963: 34–35).

Whether deliberate or accidental, Olds' choice played a critical role in the industry's clustering in southeastern Michigan, because in 1903 the Curved Dash became the first model powered by gasoline and built in Michigan to lead national sales. According to George S. May (1975), when the Curved Dash 'was quickly followed by such other successes as the Cadillac, Ford, and Buick, which were in varying degrees inspired by the Oldsmobile, Detroit's and, to a somewhat lesser degree, Michigan's reputation as an important center of auto production was established' (May 1975: 343).

Machine shops played a critical role in expanding motor vehicle production in southeastern Michigan after the Olds Motor Works fire. Eager to resume production quickly, Olds brought Detroit machine shops into the automotive industry by subcontracting work to them after the fire. The two most important machine shops which supplied Olds beginning in 1901 were Leland and Faulconer Manufacturing Company and Dodge Brothers. Henry M. Leland, arriving in Detroit from Canada in 1890, organized a company to manufacture machine tools, gear cutters, and grinders and gained a reputation as an innovative mechanic skilled at making precision castings. In the late 1890s the company began to manufacture small gasoline marine engines and quickly became the leading producer. When the Olds plant burned in 1901, Leland and Faulconer received orders to make gears and, later, 2,000 engines.

One of Leland and Faulconer's engineers, Alanson P. Brush, was largely responsible for developing an engine with more precision than the one the company was making for Olds. Brush's engine had more horsepower, larger valves, and improved timing, but Olds chose not to use it. Instead, the engine was purchased in 1902 by a company recently taken over by disgruntled backers from its founder Henry Ford. Ford went on to establish another – ultimately successful – company, while the company he left was renamed Cadillac, and eventually hired Henry Leland as general manager.

John and Horace Dodge established a machine shop in Detroit in 1900, initially to make parts for steam engines, bicycles, firearms, and other tools. The following year, when Olds gave them a contract to build transmissions, they became the first machine shop to concentrate on manufacturing automotive parts. In 1903, Henry Ford ordered engines, transmissions, and axles from the Dodge Brothers, but lacking sufficient cash, paid for the first order largely in stock. The Dodge Brothers remained Ford's most important supplier until 1914, when – dissatisfied with Henry Ford's policies – they terminated the relationship – and two years later began to produce their own cars. Dodge regularly ranked among the five best-selling models, appealing primarily to doctors, travelling sales representatives, and other business people who were attracted by its reputation for ruggedness and dependability. After both John and Horace Dodge died in 1920, the company was sold first to Dillon, Read and Company, a New York banking firm, and then in 1928 to the Chrysler Corporation.

Carriage-makers

The popular early name for the automobile – the horseless carriage – reveals the importance of the carriage industry in the development of the motor vehicle industry. Flint, Michigan, was a center for wagon and carriage production in the United States at the turn of the century, including the home of the Durant-Dort Carriage Company, the nation's largest producer in the first decade of the twentieth century. Known until 1895 as the Flint Road Cart Company, the company was organized in 1886 by William Crapo Durant, a local entrepreneur, and J. Dallas Dort, manager of a hardware store. Anticipating the strategy later applied successfully to the automotive industry, Durant built his carriage company into the nation's largest through vertical integration. At a time when carriages sold by competing companies were made under contract with independent manufacturers, Durant-Dort established subsidiaries to manufacture carriage components, such as bodies, wheels, axles, upholstery, springs, varnish, and whip sockets.

Carriage-makers clustered in Flint to take advantage of proximity to Michigan's extensive hardwood forests. Durant's grandfather Henry Howland Crapo moved to Flint from New Bedford, Massachusetts, in 1856 and acquired control over 5,000 hectares of pine forest in Lapeer County, Michigan, east of Flint. The timber was cut and floated down the Flint River to Crapo's Flint saw mills, which soon dominated local sales and expanded to other markets. Crapo was elected mayor of Flint in 1860 and governor of Michigan for two terms beginning in 1864; he died in 1869. Although virgin forests east of Flint were substantially depleted in the 1870s, the city's accessibility to remaining forests in northern Michigan still attracted wagon and carriage manufacturers.

Durant, correctly anticipating the demise of the horse-drawn carriage, took control of a struggling Flint firm, the Buick Motor Company, in 1904 and began to convert his carriage company's subsidiaries to production of automotive parts. David Dunbar Buick, a plumbing parts producer, had started the automobile company bearing his name in Detroit but was soon in debt to his most important supplier, Benjamin and Frank Briscoe, who made sheet metal. The Briscoes sold their interest in Buick to the president of the Flint Wagon Works, James H. Whiting, who moved the company to Flint. Benjamin Briscoe remained in the automotive industry as president of the Maxwell-Briscoe Motor Company. Meanwhile, Whiting, unable to make Buick profitable, turned over its control to Durant. For Durant, Buick was merely his first automotive acquisition: he went on to found General Motors in 1908.

Other Michigan-based carriage-makers played key roles in the development of automotive bodies. A driver of an early automobile sat on an open-air buggy seat attached to the chassis on top of the engine. At first, automotive producers concentrated on reliability and performance rather than passenger comfort, but by around 1904, the seating area was improved. The engine was moved from underneath the driver's bench to the front, and a body was placed over the frame behind the engine, typically with two passenger seats. Carriage-making companies received contracts to produce wooden body shells and leather upholstery.

Michigan's first high-volume automotive producer, the Olds Motor Works, bought its bodies from Byron F. Everitt, a native of Ontario who had moved to Detroit in 1891 to work for a carriage-maker named Hugh Johnson, before establishing his own business in 1899. Nine years later, Everitt, along with William E. Metzger and Walter Flanders, began to produce automobiles, known by the three partners' initials E.M.F., and sold through Studebaker wagon dealers. A wagon maker since the 1840s, the Studebaker Brothers Manufacturing Company made bodies in South Bend for several automotive companies before entering automotive production in 1902 and acquiring E.M.F. in 1911. Studebaker, which merged with Packard in 1954, continued to build cars in the United States until 1963 and in Canada until 1966.

Flooded with orders and interested in producing his own car, Everitt turned over some of his early body-building business to a friend C.R. Wilson, who had begun as a carriage-maker in Detroit in 1873. When the Ford Motor Company was established in 1903, it contracted with the C.R. Wilson Body Company to supply bodies and leather upholstery. Among Wilson's employees were Fred and Charles Fisher, who had learned about carriage-making in their father's Norwalk, Ohio, shop. In 1908, the Fishers set up their own Detroit automotive body company which soon became the nation's largest until acquired by General Motors.

Carriage firms in the northeastern United States were also capable of producing automotive bodies. Amesbury, Massachusetts, competed with southeastern Michigan as a carriage manufacturing center at the turn of the century, and one of the town's firms, Biddle and Smart, became the chief source of bodies for a Detroit automotive producer, the Hudson Company. Biddle and Smart, founded in 1870, absorbed several smaller Amesbury firms during the 1920s, but they ceased production in 1930 when Hudson switched to cheaper sources of bodies produced in Detroit.

Brewster and Company, of Long Island City, New York, was one of the nation's oldest and most prestigious carriage builders, beginning in 1810. The company specialized in luxury cars rather than the mass market, including Rolls-Royces, which were made for a time in the United States. Ford chassis carrying luxury bodies were sold as Brewsters until 1938. However, because bodies were bulky and fragile, automobile producers preferred to obtain them near the final assembly sites. Once southeastern Michigan became the automotive production center, the region's carriage-makers were destined to dominate body making as well.

Bicycle manufacturers

In contrast to the engine and carriage industries, the third major contributor to the early development of the automobile – bicycle manufacturing

– was not concentrated in southeastern Michigan. To the contrary, automotive production began during the 1890s primarily in the northeast largely because the largest bicycle manufacturers were located there. The automobile was regarded as a top-of-the-line bicycle attractive to customers who wanted personal transport without physical exertion. Expansion of the bicycle market was otherwise difficult because of the product's fundamental limiting factor, its reliance on human power.

Bicycling first became popular in the 1870s, especially among the millions of recent immigrants to the nation's rapidly growing cities. For urban residents confined to trains, streetcars, and other collective forms of transport, the bicycle offered freedom to choose a destination and pace of travel. Biking became a major leisure-time activity during the 1890s, spurred by a price reduction to $30 and the invention of the safety bicycle in 1885 by James Kemp Starley, in Coventry, England, where the British automotive industry later clustered. In the twentieth century, the automobile replaced the bicycle as the preeminent symbol of individual control over the means of travel.

Enthusiasts of bicycling were also instrumental in securing better roads, which were in poor condition at the turn of the century. The first census of American roads, conducted by the Department of Agriculture's Office of Road Inquiry in 1904, found that only 7 per cent, or less than 300,000 km, of the country's rural roads were surfaced, mostly with gravel. As a substitute for paying taxes, farmers did most of the rural road work by hauling a drag along the road to smooth out the worst ruts and fill in the chuckholes (Glasscock 1937: 89). Improvements in highway design and construction developed in Britain by John L. McAdam and Thomas Telford early in the nineteenth century were largely unknown in the United States. The early hazards of automotive travel have been graphically preserved in turn-of-the-century photographs depicting automobiles mired in mud.

When commercial production of automobiles began in the late 1890s, the leading manufacturers came from the bicycle industry. The Duryea Brothers, who arguably produced the first automobile for sale, were bicycle mechanics. Colonel Albert A. Pope, who organized the American Bicycle Company trust in 1899, dominated bicycle sales through a network of 15,000 sales agents. Pope led automotive production at the end of the nineteenth century by selling vehicles through his bicycle agents. In the first decade of the twentieth century bicycle manufacturers entering automotive production included Alexander Winton of Cleveland, the leading manufacturer of gasoline-powered vehicles in 1900, and George N. Pierce, a Buffalo producer of the Pierce-Arrow, a highly-regarded luxury car built until the late 1930s.

Midwestern industrialists involved in bicycles who played a longer term role in automotive production included Thomas B. Jeffery, John N. Willys, and the Dodge Brothers. John and Horace Dodge organized the Evans and Dodge Bicycle Company in the 1890s, where they developed a ball-bearing bicycle. After selling the company to a Canadian firm, John Dodge served for a time as general manager, but in 1900 the brothers established a machine shop in Detroit to produce automotive parts. Willys, operator of an Elmira, New York, sporting goods store, wanted to carry automobiles as well. When the Overland Company of Indianapolis could not fill his contracted order, Willys took control of the company and moved its production to the former Pope plant in Toledo. Jeffery was a bicycle manufacturer who began production of the Rambler automobile in Kenosha. The firm, as well as the plant, passed to the Nash Motor Company, before merging with Hudson in 1954 to form the American Motors Corporation, which in turn was taken over by Chrysler in 1987.

Experimental automobiles closely resembled bicycles in appearance and function. Long-term applications of bicycle technology to the automobile chassis included the use of pneumatic tires, the chain and sprocket drive, ball bearings, and steel tubing (Wager 1975: xiv). Although some bicycles were manufactured in the Midwest, the region did not enjoy a competitive advantage over the northeast in their production. Had the automobile continued to be regarded like the bicycle as a faddish consumer good, production might have remained in the northeast near the largest markets. However, Pope and other northeastern bicycle manufacturers who regarded the automobile as a sideline, failed to retain their early production lead in the face of rapid technological innovation. Indicative of the shortsightedness of the northeastern bicycle manufacturers, Pope rejected an offer from the Duryeas to sell him the rights to build their car for $50. Believing the Duryeas' vehicle to be merely a bicycle with an engine attached, Pope refused to pay more than $5 (Glasscock 1937: 45).

AVAILABILITY OF CAPITAL

Early auto makers needed cash to get started. Like other new industries, motor vehicle production was regarded by many financial institutions as too risky to lend money. Availability of capital influenced the location of automotive producers in southeastern Michigan.

Early methods of financing auto production

A confidential 1914 report prepared by C.C. Parlin and H.S. Youker, manager and assistant manager, respectively, of the commercial research division in the Curtis Publishing Company's advertising department, documented the method by which early producers built cars with limited capital.

> A manufacturer who could obtain control of $25,000 hired a designer, got a picture of a car, and advertised to supply the market. Consumers were clamoring for cars, paying $250 bonus for the privilege of standing

> in line to obtain delivery for a favorite car. Dealers knew that they could dispose of all the cars they could get and hastened to place orders accompanied by $100 deposit on each car ordered.
>
> (Parlin and Youker 1914: 7–8)

The $100 deposit amounted to roughly 20 per cent of the cost of a typical automobile.

The Curtis report revealed that a common practice among early manufacturers was to accept $100 deposits for more cars than they knew they could actually produce in a year. A typical example cited by the Curtis report was a manufacturer who expected to build 1,000 automobiles in a year but accepted deposits for 3,000. The practice was justified in part to cover customers who withdrew their orders and lost their deposits, but the principal reason was to raise working capital. In control of $300,000 rather than $100,000, a manufacturer could open a bank account and use the funds, along with the list of orders for 3,000 automobiles, to establish a line of credit.

As most early auto makers confined their manufacturing efforts to final assembly, the next step was to contract with independent suppliers to produce component parts, such as engines, bodies, and transmissions. These parts were not purchased with cash but on the line of credit established at the bank with the deposits and order list. Naturally, suppliers were asked for the longest possible repayment period, normally between thirty and ninety days.

Assembled automobiles were loaded onto flat rail cars for shipment to dealers with a bill of lading attached, as well as a sight draft, meaning that the dealer was required to pay cash on the spot upon receipt of the merchandise. The financially pressed auto maker could use this sight draft to obtain funds from the bank to start repaying suppliers, if necessary. The cash paid by the dealer for the automobile would then be used to repay fully suppliers and other creditors, rent factory space, purchase machinery, hire workers, and secure utilities. If all went well in the process, the receipts were sufficient to provide for new machinery, expanded factory space, and other capital expenditures (Glasscock 1937: 42; Flink 1970: 295–296).

However, innovations came so fast that auto makers had constantly to invest in capital improvements or risk becoming uncompetitive. The low-cost mass-production market was soon ceded to a handful of producers, while others survived for a time selling small quantities of high-priced luxury models. The Curtis commercial research report pointed out the risks involved in this operation.

> If any parts failed to arrive on time the shipment of the car was delayed, for no car can be forwarded until the last pair of mud-guards is in place. But partly finished cars ran rapidly into money; for 100 cars of the value of $2,000 each means $200,000, and the slender resources of the organization were soon overtaxed.

> To avoid bankruptcy it is said that desperate expedients were sometimes used to meet the payroll; such for example as loading freight cars with junk, issuing bills of lading for a new automobile, drawing sight drafts on distant retailers, discounting the said draft and then, when the missing parts arrived reloading the freight car and forwarding.
>
> (Parlin and Youker 1914: 8)

At the turn of the century, the large banks, which were clustered in New York and other northeastern cities, were the main source of financing for industrial development. Given that the northeast was also the center of the market for automobiles at the turn of the century, auto makers naturally looked to that part of the country for capital. However, eastern bankers would not provide adequate financing to the infant automotive industry. They preferred to gamble on the stock market, which was familiar to them, than to support a new industrial venture with an uncertain future. The Curtis report explains their hesitation.

> It was but natural that men with real capital hesitated to enter the competition. As one capitalist, who looked the matter up with interest and then abandoned the project, explains, 'I had so much more capital than all the others in the game that I thought I had better stay out and keep it.' This early lack of capital was not wholly without compensation; for one manufacturer declared that if he had started with $100,000 he would have failed. The fact that he had but $25,000 and had to strain every nerve to make ends meet, he said, enabled him to succeed. But as it happened, while many companies failed, others lived through their early vicissitudes and accumulated enough out of their profits to establish themselves on a sound basis.
>
> (Parlin and Youker 1914: 9)

Support from Michigan financiers

Given the relative difficulty of obtaining financing from eastern banks, fledgling carmakers looked elsewhere. They found generous sources of funds in the Midwest, especially in Detroit. Michigan at the turn of the century was a hotbed of economic and political radicalism. Michiganders blamed the nation's economic problems on a conspiracy of railroad tycoons, eastern bankers, and moneylenders.

The state's powerful agricultural interests especially opposed the tight money policy followed by a succession of Republican national administrations, which were supported by eastern financial leaders. Farmers, who carried heavy mortgages to buy property and equipment, wanted the government to print more money in order to stimulate inflation and thereby reduce the value of their debts. They also wanted the government to regulate railroad freight charges, issue more loans to farmers, and institute a graduate income tax. Michigan voters reinforced these extreme views by electing as governor in 1882 the candidate of the Greenback Party, which supported the farmers' position (Dunbar 1980: 447–450).

In this atmosphere, financial backing was more easily available in Michigan for the production of motor vehicles. However, the leading sources of capital for industrial ventures in Michigan proved to be individual speculators rather than banks. Wealthy investors wanted to be a part of the new industry and had little difficulty diverting funds from other business activities to support early producers.

By the turn of the century, much of Michigan's wealth had been produced in three resource-based industries – copper, iron, and lumber. Copper was discovered in Michigan's Upper Peninsula in 1847, and until the late 1880s the state was the world's leading producer. Detroit became the largest producer of iron ships and a national center for a variety of other iron products. As already noted, Michigan's extensive hardwood forests provided lumber for a variety of industries, including carriage-making. Early automotive producers found that people who had made their fortunes in these and related industries were looking for new investment opportunities. These investors, known as the princes of Griswold Street, after Detroit's miniature version of Wall Street, accurately anticipated that automotive production would prove financially rewarding.

Olds' initial capital came from wealthy citizens of Lansing who had seen his father's business grow from a small machine shop to a leading manufacturer of gasoline engines. However, when Olds failed to obtain the level of capital needed to build automobiles, he was receptive to an offer from East Coast capitalists to build a plant in Newark, New Jersey. Olds considered this the best place for a factory because it was only a few kilometers from New York City, then the nation's largest automotive market. He picked out a suitable factory site, but the New Jersey investors backed out.

Back in Michigan, Olds was able to obtain funds in Detroit from Samuel L. Smith, who had made millions of dollars by investing in the copper and lumber industries through the Michigan Land and Lumber Company. As a former resident of Lansing, Smith had followed Olds' progress in building up a large engine repair and construction shop. After moving to Detroit, Smith invested during the 1890s in the Olds Gasoline Engine Works and the Olds Motor Vehicle Company, which built experimental motor vehicles. When Olds reorganized his firms into the Olds Motor Works Company in 1899 for the purpose of constructing motor vehicles for sale, Smith became the principal investor, holding all but forty of the 20,000 shares originally issued. Smith became president of the new company, and his son Fred secretary and treasurer, while Olds was named vice-president and general manager. Olds' selection of Detroit for the new firm's original factory was strongly influenced by the fact that his leading backer and president lived there (Niemeyer 1963: 27).

After Olds moved the firm to Lansing, he was forced out of the company by the Smith family in 1904. Olds wanted to continue building a low-priced high-volume car like the Curved Dash, but the Smiths believed that the company would earn higher profits by concentrating on a larger, more expensive model (May 1977: 230–232). Later that year, Olds formed the Reo Motor Car Company, also in Lansing, which produced automobiles until 1920. Reo was purchased in 1957 by the White Motor Company, which closed the Lansing plant in 1974; White in turn was taken over in 1984 by Volvo, now Volvo-GM Heavy Truck Corporation.

Henry Ford also secured capital to get started from wealthy individuals in Detroit rather than from banks. Backers of Ford's first venture in 1899, the Detroit Automobile Company, included William Maybury and William H. Murphy, who had both made fortunes by investing in real estate. Maybury was also the mayor of Detroit and a friend of the Ford family. The Detroit Automobile Company was dissolved one year later, but the following year the Henry Ford Company was established, again with backing from Murphy, as well as from James and Hugh McMillan, who controlled a network of shipping lines, railroads, banks, and insurance companies, plus the Detroit Dry Dock Company, where Ford had once worked. Ford's second company also collapsed soon after it was established.

Ford claimed that his backers gave up on him too quickly, and that they were looking for rapid returns. However, the backers attributed both failures to Ford's preference for racing and tinkering with experimental prototypes instead of managing daily corporate operations. He was seen as a man out of his depth, not yet ready to run a successful motor vehicle facility. As a result, Ford had more difficulty securing funds for a third venture. He received most of the support from Alexander Y. Malcomson, a coal merchant, who had sold fuel to Edison Illuminating Company when Ford worked there. The remaining investors were generally either related to Malcomson or tied to him through other business ventures.

Meanwhile, Murphy took control of the second Ford company after Ford was removed in 1902. He brought in Henry M. Leland, who worshipped at the same church, to run the firm, which was renamed the Cadillac Automobile Company. Leland replaced Ford's engine with a more powerful one which he designed in his role as a director and partner in Leland and Faulconer. Three other prominent Detroiters, Lem W. Bowen, Clarence A. Black, and A.F. White, also backed the reorganized company.

By the middle of the first decade of the century, investors outside of Detroit were more eager to invest in the new automotive industry. When he was forced out of the company that bore his name in 1904, Olds had a choice of investors to finance a new company. He chose a Californian, Reuben Shettler, although New York financial interests had also approached him. But by then, southeastern Michigan had grabbed the lead in automotive production which it would not relinquish.

The availability of venture capital from wealthy individuals also helps to explain why Detroit rather than, say, Cincinnati or Cleveland, became the center of motor vehicle production in the Midwest. Detroit investors were lenient and daring, whereas bankers in Cleveland came from conservative New England stock (Wager 1975: xiii). 'Cincinnati bankers were approached [by an unnamed company], but at that time could not see anything but ruin for the country in the progress that the automobile was making . . . [Capitalists in Cincinnati] all were afraid to venture into the automobile business' (Smith 1970: 31).

As one example, a mechanical engineer James W. Packard, who had opened the Packard Electric Company in Warren, Ohio, with his brother William in 1890, looked for financial support to expand into automotive production. Rebuffed by officials in northeastern Ohio, including the president of the Cleveland Chamber of Commerce, Packard accepted support from a group of Detroit businessmen, headed by Henry B. Joy, who was interested in investing in automotive production. Joy, who displaced Packard as the firm's principal decision-maker, moved production to Detroit in 1903. The Packard brothers remained in Ohio to run their prospering electrical equipment company, which was acquired by General Motors in 1932 (May 1975: 285–287).

SOUTHEASTERN MICHIGAN DOMINATES AUTOMOTIVE PRODUCTION

Luxury cars were assembled primarily in the northeastern United States, but the best-selling models in the other classes were produced in southeastern Michigan. The percentage of national automotive production concentration in southeastern Michigan increased from 53 per cent in 1909 to nearly 80 per cent in 1913.

Market segmentation by price

The market for motor vehicles after 1910 fragmented into four segments according to price. Luxury cars, priced over $2,500, accounted for less than 5 per cent of the total market in 1913. Most luxury cars were produced by the industry's northeastern pioneers who chose to continue building a small number of hand-crafted models rather than millions of identical machines. Leading luxury cars included Pierce-Arrow built in Buffalo, White and Winton in Cleveland, Franklin in Syracuse, and Locomobile in Bridgeport, Connecticut. Michigan firms accounted for only one-sixth of the luxury car market, primarily Packard and Oldsmobile.

The Packard factory, at 1580 East Grand Boulevard in Detroit, was one of the most impressive in the country when it opened in 1903, the first designed by Albert Kahn, who would become the country's most important architect of industrial buildings. Its great expanse of glass was

Table 2.1 Market share held by US automobile producers, 1913

Producer	*Assembly plant location*	*Sales in 1913*	*Market share*
Luxury (over $2,500)		18,500	4.79
Packard	Detroit, Michigan	2,300	0.60
Pierce-Arrow	Buffalo, New York	2,000	0.52
White	Cleveland, Ohio	1,500	0.39
Franklin	Syracuse, New York	1,400	0.36
Winton	Cleveland, Ohio	1,300	0.34
Locomobile	Bridgeport, Connecticut	1,100	0.28
Oldsmobile	Lansing, Michigan	1,000	0.26
Others	—	7,900	2.05
Medium ($1,500-$2,500)		62,500	16.19
Cadillac	Detroit, Michigan	15,000	3.89
Chalmers	Detroit, Michigan	8,000	2.07
Hudson	Detroit, Michigan	5,000	1.30
Oakland	Pontiac, Michigan	4,000	1.04
Mitchell	Racine, Wisconsin	3,000	0.78
Cole	Indianapolis, Indiana	3,000	0.78
Rambler	Kenosha, Wisconsin	3,000	0.78
Others	—	21,500	5.57
Moderate ($600-$1,499)		120,000	31.09
Willys-Overland	Toledo, Ohio	35,000	9.07
Buick	Flint, Michigan	26,000	6.74
Studebaker	South Bend, Indiana	25,000	6.48
Hupmobile	Detroit, Michigan	12,000	3.11
Reo	Lansing, Michigan	9,000	2.33
Maxwell	Tarrytown, New York	4,000	1.04
Paige-Detriot	Detroit, Michigan	3,000	0.78
Others	—	6,000	1.55
Inexpensive (under $600)		185,000	47.93
Ford	Detroit, Michigan	178,000	46.11
Others	—	7,000	1.81

Source: Parlin and Youker 1914: 419

visually impressive, but the building was also a landmark for organizing the work places and movement of materials to fit the needs of the new automotive industry.

The three best-selling medium-priced cars – between $1,500 and $2,500 – were produced in Detroit, including Cadillac, Chalmers, and Hudson (Table 2.1). Cadillac assembled vehicles in a three-storey factory at the corner of Cass and Amsterdam avenues, built in 1905, two years after the company emerged from a reorganization of the Henry Ford Company and three years before its acquisition by General Motors. Production of Cadillacs shifted in 1921 to a new plant at 2860 Clark Avenue, which continued assembly operations until 1987.

The Chalmers-Detroit Motor Company was organized in 1908 by Hugh Chalmers, who took over the Thomas-Detroit company, established two years earlier by Edwin Ross Thomas. When Chalmers went bankrupt in 1921 its assets were acquired by the Maxwell Company, which had been leasing space in Chalmers' assembly plant for four years. In 1926, Maxwell was reorganized as the Chrysler Corporation, which retained the assembly plant on East Jefferson Avenue previously occupied by Thomas, Chalmers, and Maxwell.

When Chalmers took over Thomas-Detroit, several employees, including Roy Chapin and Howard Coffin, departed to establish their own firm, named the Hudson Motor Car Company, because much of the financing came from J.L. Hudson, owner of Detroit's largest department store. Hudsons were assembled in a plant acquired from the Aerocar Company, an auto maker between 1905 and 1908, organized by Alexander Malcomson, Henry Ford's principal backer. The Aerocar plant was located at East Jefferson Avenue and Conner Street catercorner from Chalmers and across Jefferson from the Jeep plant opened in the early 1990s. Hudson operations were transferred to the Nash Motors Company plant in Kenosha, Wisconsin, when the two companies merged in 1954 to form American Motors. The Detroit plant was sold to Cadillac and demolished for a parking lot.

Three of the five best-selling moderately-priced cars, which cost between $600 and $1,500, were assembled by Michigan producers, including Buick in Flint, Hupmobile in Detroit, and Reo in Lansing. Buick, General Motors' most profitable product at the time, boasted that it had the country's largest automotive plant. Robert and Louis Hupp started production of Hupmobiles in 1908 at a plant at 3051 E. Milwaukee Avenue in Detroit. The Hupmobile did not survive the Depression years; production was suspended in 1936, revived in 1938, and discontinued permanently in 1941. The Reo was created by Ransom E. Olds after he lost control of the Olds Motor Works. Outside of Michigan, Willys-Overland, second overall in sales in 1913, assembled vehicles in Toledo, Ohio, in a plant obtained from the Pope Manufacturing Company in 1911. Studebaker, a long-time wagon maker, was based in South Bend, Indiana. Though located in other states, both Toledo and South Bend are located within 10 km of the Michigan state line.

Ford dominates production of low-priced models

The Ford Motor Company was the principal contributer to southeastern Michigan's dominance. The Detroit-based firm registered nearly half of all car sales in 1913 and corralled virtually all of the low-priced market.

After two failed attempts to run motor vehicle firms, Henry Ford became the first Michigan-based producer to achieve a dominant position in the industry. The Ford Motor Company, founded in 1903, took over sales leadership in 1906, making 8,700 cars, more than twice as many as second-place Cadillac. When Ford introduced the Model T three years later, sales reached 18,000, although still well under 10 per cent of the total national market. From then on, Model T sales skyrocketed, reaching a peak of 1.8 million in 1923. Ford had the best selling car every year from 1906 until production of the Model T ceased in 1927, soon after the fifteen millionth rolled off the assembly line.

In the first decade of the century, most firms generated profits by selling a small number of automobiles at prices substantially above production costs. This strategy worked as long as demand for the new product outstripped supply. In contrast, Ford chose to maximize sales of low-priced cars, each of which carried a small profit. As production costs declined, Ford lowered prices, in turn stimulating additional sales. This approach proved profitable, because each price reduction expanded the number of consumers who for the first time could afford to buy an automobile, or more specifically, a Ford, the only vehicle within their price range. The price of a Ford Model T declined from $950 in 1910 to $360 in 1917 and $290 in 1924, in an era of high inflation.

An essential element of Ford's strategy was to plow a large percentage of the company's profits into innovative technology rather than pay higher dividends. Consistent with this policy, Ford relocated production to larger plants in the Detroit area four times within its first fifteen years.

The Ford Motor Company's first plant was located on Mack Avenue in Detroit, in a remodelled wagon shop leased from Albert Strelow, one of the city's largest painting and carpentry contractors. Alexander Malcomson, Ford's principal backer, persuaded Strelow to remodel the shop into an automotive assembly plant, in accordance with Ford's design. Strelow agreed to set the rent at $75 a month for three years, although the renovations cost $3,000 to $4,000. On 1 April 1903, Ford brought over his equipment and working force from a shop at 81 Park Place in Detroit, where he and Tom Cooper had been building racing cars since May 1902 (Nevins 1954: 214, 230). In the 1930s, Henry Ford moved the long-abandoned Mack Avenue plant to Greenfield Village, a 100–hectare collection of historic structures he established in Dearborn, now Michigan's most popular tourist attraction.

Precisely one year after Ford moved into the Mack Avenue plant, the company voted to build its own plant on a one-hectare site at the corner of Piquette and Beaubien Avenues, 2.5 km northwest of the original plant. The $100,000 three-storey plant was ready by the summer of 1904. The first floor contained offices, a machine shop, an electrical department, a testing area, and a shipping room. The second floor housed another machine shop, plus designing and drafting areas. Painting and final assembly occupied the top floor (Nevins 1954: 262–266).

Henry Ford set up an independent company, called the Ford Manufacturing Company, in late 1905 to make engines, gears, and other parts for the Ford Motor Company's low-priced cars, while the Dodge Brothers continued to make parts for higher-priced models. A building was leased for the Ford Manufacturing Company from the Wilson and Hayes Manufacturing Company at 773–5 Bellevue Avenue in Detroit. The arrangement was inefficient because parts had to be hauled, frequently by horse-drawn wagon, more than 6 km to the Piquette Avenue plant (Nevins 1954: 279–281).

Ford's motivation was apparently to siphon profits from the Ford Motor Company into the Ford Manufacturing Company in order to force out of the Motor Company his original backer Alexander Malcomson, who disagreed with the policy of concentrating resources on low-priced cars. Malcomson further alienated Ford and the other directors by organizing a competing firm, the Aerocar Company, financed in part with his Ford dividends. Malcomson resigned as Treasurer of the Ford Motor Company in July 1906 and sold most of his shares to Henry Ford, who for the first time controlled more than half of the company bearing his name. Early the following year, Ford merged his two companies and consolidated operations in an expanded Piquette Avenue plant (Nevins 1954: 330–331).

Ford's expansionist policy quickly made the Piquette Avenue plant overcrowded. Ford reportedly started looking for a replacement as early as 1906, settling for a 25-hectare site previously occupied by a sports park and race track in Highland Park, then on the northern edge of the Detroit urban area. The site was located 5 km northwest of the Piquette Avenue plant, north of Manchester Avenue and east of Woodward Avenue, a major artery extending in a straight line from downtown Detroit 13 km away. The first operations were transferred from Piquette Avenue to Highland Park on the first day of January, 1910, although construction was not completed for several years. The Piquette Avenue building, sold to Studebaker in 1911, is currently a warehouse.

The four-storey main building at Highland Park was 264 m long and 23 m wide, made primarily of steel, concrete, and glass. Designed by Albert Kahn, its shape and appearance earned the factory the nickname of Crystal Palace. The first floor contained work stations for manufacturing crankshafts, axles, and transmissions. Cylinders, gears, and other chassis parts were made in an adjacent one-storey machine shop. Fenders, hoods, and other body parts were made on the fourth floor of the main building. The third floor housed stations for producing accessories, such as wheels, tires, and lamps, plus the paint shop. Final assembly took place on the second floor. Later, a separate power plant and foundry were built (Nevins 1954: 452–455).

The Highland Park plant became famous as the birthplace of the moving assembly line, which revolutionized automotive production. Ford and other automotive manufacturers had already begun to standardize parts and tools and arrange workers, machines, and materials in a logical sequence. Ford installed continuously moving lines fed by overhead conveyors, for subassembly operations such as engines and transmissions during 1912 and for final assembly in 1913. By 1914, Ford had reduced the amount of time needed to assemble a vehicle at Highland Park from approximately twelve and a half hours to thirty minutes.

All told, 80 per cent of America's motor vehicles were built in southeastern Michigan in the period immediately prior to the outbreak of World War I, with much of the remaining production in communities only a few kilometers from the Michigan state line. The Ford Motor Company was responsible for much of Michigan's dominant position. However, what Ford gave it would soon take away. Michigan's share of national automotive production fell within a few years from 80 to 40 per cent.

3 Ford revolutionizes the geography of production

The Highland Park plant was not Ford's last or largest investment in the Detroit area. Despite its massive size and moving assembly lines, Highland Park was regarded by Henry Ford as obsolete before it was finished. Ford acquired over 400 hectares of land in an undeveloped area southwest of Detroit along the banks of the Rouge River, now part of the city of Dearborn, and over the next decade constructed the world's largest industrial complex devoted to the production of motor vehicles.

The Rouge complex included a power house, foundry, and steel mill, as well as facilities to build engines, chassis, bodies, and other major components. When Ford switched production from the Model T to the Model A in 1927, final assembly operations were moved from Highland Park to the Rouge. After being used for storage during the Depression, the Highland Park plant was reopened after World War II to produce trucks and buses. Ford still assembles cars at the Rouge plant. However, the principal purpose of the Rouge was to produce automotive parts and components rather than final assembly.

Instead, Ford transferred final assembly operations to several dozen branch assembly plants scattered around the country. Initially, constructing branch assembly plants was only one of a number of ideas to solve distribution problems caused by the rapid growth in sales. Once the network of branch assembly plants proved to be the geographical arrangement which minimized total production costs, the concept moved from an experiment to corporate policy. Through trial and error, the company ended up with a distribution of plants which corresponded closely to the expected model according to neoclassical industrial location theory.

Years later, Henry Ford took credit for inventing the branch assembly plant idea, as he did a great number of things, including the automobile itself. Ford stated through a collaborator, Samuel Crowther:

> Only a small percentage of our automobiles are used in or around Detroit and so we quit being automobile makers excepting for the district and instead began to make automobile parts and ship them out to assembly points through the country. This saved the cost of final

assembly at the factory, of testing, of knocking down the automobile and crating it, and of paying the extra freight on a bulky piece of machinery like an assembled automobile.

(Ford 1926: 40)

In reality, principal credit for creating and implementing the branch assembly plant concept goes to two other Ford officials less well-known today, James Couzens and Norval Hawkins. Hawkins showed that assembling automobiles at branch plants around the country was less expensive than shipping finished vehicles from Michigan, while Couzens selected suitable locations for the assembly plants.

Information concerning early decision-making at the Ford Motor Company comes primarily from thousands of pages of depositions and testimony stemming from two legal cases. First, two minority stockholders John and Horace Dodge brought suit in the Michigan Circuit Court to require Ford to pay higher dividends. The Dodge brothers, who were then in the process of setting up their own automotive production firm after having once been Ford's most important parts supplier, charged that by repeatedly cutting prices and expanding production capability Henry Ford had sacrificed profits and paid low dividends. In 1917, Ford was ordered to distribute over $19 million in dividends, amounting to 50 per cent of the company's accumulated profits, including approximately $1 million to each of the two Dodge brothers (Nevins and Hill 1957: 88–105).

After the decision was sustained by the appeals court, Henry Ford threatened to resign and establish a competing firm. Instead, the minority shareholders – Anderson, Couzens, Horace H. Rackham, the estate of John Gray, and the Dodge brothers – agreed to sell all of their shares to Ford for approximately $106 million, or approximately $2.5 million per share. However, they refused to complete the transaction until they learned their tax liability.

Stock acquired prior to ratification of the Sixteenth Amendment to the US Constitution, which legalized the federal income tax, was subject to a capital gains tax based on the value of shares on 1 March 1913. In view of the magnitude of the proposed sale, the stockholders persuaded the Commissioner of Internal Revenue, Daniel C. Roper, to determine the value of the Ford Motor Company as of 1 March 1913. An expert, P.S. Tablert, was dispatched to review Ford's books and set a figure. Following Tablert's recommendation, Roper wrote to Ford's attorney on 19 May 1919 that a fair valuation as of 1 March 1913 was $9,489.34 per share. Less than two months after the valuation was set, Ford completed all of the transactions and became sole owner of the company (Nevins and Hill 1957: 105–111).

Shortly after the sale was completed, though, the US government initiated a suit against the former minority stockholders, claiming that additional income tax was due. The government claimed that the fair valuation of Ford Motor Company stock as of 1 March 1913 was approximately $5,000 rather than $9,489, because automotive production still carried uncertainties, and Ford in particular was a risky company because it depended on only one product. As the government lawyer argued, Henry Ford's 'reputation in 1913 was that of a genius, but of an eccentric one' (*Dodge et al. v. Commissioner of Internal Revenue* 1927b: 74). The former stockholders claimed that they acted in good faith on the basis of Roper's letter, a position which was upheld in the courts in 1927. The thousands of pages of depositions and testimony from the Dodge and the tax liability cases provide geographers with the best evidence concerning plant location decisions in the industry's formative years.

COUZENS AND HAWKINS INVENT THE BRANCH PLANT

James Couzens, more than any other person, including possibly Henry Ford himself, was responsible for the rapid growth of the Ford Motor Company, including the distribution of plants according to Weberian locational principles. While Ford looked after the design and engineering of the automobiles, Couzens held unchallenged authority in the front office, with responsibility for setting up the company's advertising, sales network, and book-keeping.

Couzens had served as clerk, office manager, and business adviser to Detroit coal magnate Alexander T. Malcomson, who became the largest investor in Henry Ford's third attempt to establish an automobile company, the one that eventually succeeded. Recruited as manager of the new company, Couzens also invested $2,500 to buy twenty-five shares, or 2.5 per cent of the stock, and was named secretary of the board of directors. After Malcomson resigned from the board and sold his shares in 1907, Couzens succeeded him as treasurer and increased his holdings to 11 per cent, second only to Henry Ford.

Ford and Couzens brought different talents to the new company. Holding little patience for day-to-day business management, Ford preferred to spend his time on the factory floor dealing with design problems. Couzens, on the other hand, possessed little understanding of motor vehicle engineering but could impose administrative practices more efficient than those typically found in automotive firms during the industry's formative years. John Anderson, Ford's lawyer and a member of the board of directors, succinctly characterized the difference between Ford and Couzens. 'Mr. Ford knew practically nothing of finance, except in a purely general way. Mr. Couzens was a very keen student of finance' (Lacy and Anderson 1926: 281).

Temperamentally, the two were quite different, as well. Couzens was cold and severe, a stickler for hard work and strict obedience to rules

(Nevins 1954: 268). His bursts of anger were legendary and feared within the company, especially when production delays threatened his sales and distribution plans. In contrast, Henry Ford was quite relaxed and casual with his employees, at least in the early years of the company (Sorenson 1956: 86–87).

Branch agencies to sell and service cars

Couzens' initial marketing step as Ford manager was to line up a network of dealers around the country. At first, bicycle merchants, wagon builders, and blacksmiths sold cars as a sideline to their principal trade, but Couzens recruited dealers to open showrooms devoted exclusively to the sale of Ford products. By the fall of 1905, Fords were sold through 450 dealers, including the prominent department store chain John Wanamaker, which had the agencies in Philadelphia and New York. A dealer was given the exclusive right to sell in a vast territory, typically an entire county and in some cases an entire state (Nevins 1954: 265).

Reliance on dealers proved unsatisfactory for two reasons. First, few dealers knew how to repair the cars. At the directors' meeting of 20 June 1904, Ford and Dodge secured passage of a resolution calling for the company to send mechanics around the country to train dealers in automotive repairs. The second problem associated with selling cars through dealers was that the company could not control the sales price. In the early days of automotive production, when demand exceeded supply, customers paid the full advertised price, rather than negotiate a reduction, as has been common in recent decades. However, for the dealer to make a profit selling cars at the advertised price, they had to obtain them from the company at a lower price.

Couzens' solution to the retailing problem was to establish facilities in a number of cities which were known in the company at the time as branch plants, although branch agencies is a more accurate term. In August 1905, the directors voted to establish branches in New York, Philadelphia, Boston, Buffalo, Chicago, and Kansas City. The following month, a committee comprising Ford, Couzens, and David Gray, son of one of the original investors, visited the latter four cities, as well as Cleveland, in order to set up the branch agencies. In March 1910, Ford's twenty-fifth branch was opened in Fargo, North Dakota.

Branch agencies served three purposes. First, in the city in which it was located, the branch replaced a dealership as the place where customers went to buy a Ford automobile. As Ford employees, the sales force – as many as 75 per branch – received a salary supplemented by bonuses based on the net profits on total sales, in contrast to independent agents who worked on commission. Branch agencies were more costly than contracting with dealers, who paid for their own buildings and employees. However, by owning the retail outlets, the Ford Motor Company pocketed the full retail price and did not have to split the profits on the rapidly increasing sales with independent dealers (Parlin and Youker 1914: 226).

The second function of the branch was to distribute vehicles to dealers who were still responsible for selling automobiles in the surrounding district. The company-owned branch agency sold the vehicles to the dealers at a 20 per cent discount, allowing them to make a profit when individual customers paid the full list price. The arrangement saved Ford money, because prior to setting up the branches the company had to provide vehicles at one-fourth off list price to independent wholesale distributors, who then sold them to dealers at 20 per cent off (Knudsen 1926: 6; Parlin and Youker 1914: 226).

Third, the branch provided Ford automobile owners with replacement parts and performed repairs and routine servicing on the vehicles as necessary. Access to service was a major factor in building up strong customer loyalty to Ford, at a time when most models could be repaired only if they were sent back to the factory. The branch agency maintained an extensive inventory of parts, valued at $300,000 in the case of New York and Chicago. Each day, the branch sent a telegram listing needed parts, which would be sent out from Detroit before four o'clock the next afternoon (Parlin and Youker 1914: 892). The branch also installed accessories selected by the customer, an important function because the Model T left the factory consisting of little more than a chassis, motor, and body, lacking even headlights. A manufactured vehicle was shipped to the branch agency for installation of additional components before it was sent on to the dealer or picked up by the customer.

The early branch agencies were located in rented buildings located in or near the central business district (Rackham 1926: 122). Typical was Cincinnati's branch, a four-storey building at 917 Race Street. The first floor featured a showroom where customers could view new models and place orders to purchase vehicles. The second floor contained a stockroom to store accessories for installation on new vehicles, as well as replacement parts. The third floor provided storage facilities for new vehicles which had not yet been picked up and paid for by customers. The fourth floor housed the repair shop. The building currently houses a restaurant (Ohio Historic Preservation Office 1988).

Innovative sales practices

When he could no longer cope with all of his responsibilities, Couzens in 1907 hired Norval Hawkins, at the time the company's accountant, as general sales manager. Hawkins had run Hawkins-Gies and Company, an accountancy firm which audited Ford's books beginning in 1904 (Hawkins 1925: 189). Described as 'perhaps the greatest salesman that the world ever knew' (Goodenough 1925: 183), Hawkins originated many of the ideas which stimulated the rapid growth in sales of the Model T. Couzens had set up the first handful of branch plants before his arrival, but Hawkins opened several dozen more and expanded their role to include final vehicle assembly.

In those days, the sales manager wielded considerable authority beyond stimulating sales through aggressive advertising and prodding dealers. After gathering orders for future delivery of vehicles placed in the spring by customers, accompanied by 20 per cent down payments, the manager was then responsible for setting the factory schedule so that enough automobiles were produced to fill the year's orders. Further, the sales manager contributed to corporate profits by minimizing the cost of shipping finished vehicles to consumers around the country.

Hawkins' task at Ford was especially challenging given the company's rapid expansion and frequent price reductions. According to J.W. Anderson, an attorney and a minority stockholder:

> When the business of the Ford Motor Company began to expand in the way that it did, so rapidly, there was bound to be a great deal of confusion and a lot of loss due to a failure to have the factory properly organized into departments, for the purpose of taking care of spare parts . . . . Mr. Hawkins stepped in there and organized, I believe in the first instance, the business in the Piquette Avenue plant, and afterwards in the Highland Park plant, all of the departments.
>
> (Lacy and Anderson 1926: 282)

Hawkins by all accounts was a brilliant sales manager at Ford. He was described as 'original in idea, forcible in presenting it, a perfect dynamo for work, and a man who gets the quickest execution of any men I ever knew. He originated a great many ideas which made possible the proper marketing of this car' (Goodenough 1925: 183–184). According to a competitor, C.D. Hastings, president of the Hupp Motor Company, at Ford 'the sales organization and character of the agencies were much better than those of any competitors. Mr. Hawkins was regarded as being the most fertile in sales suggestions of any man in the industry' (Hastings 1926).

Hawkins' first marketing idea at Ford was to expand the number of branch agencies which sold directly to consumers. However, branches were swamped with sales and could not keep up with rapidly increasing demand (Knudsen 1926: 6). Qualified sales people were difficult to retain because they could make more money selling on commission at an independent dealer than working for salary at a branch plant. Company-owned branches persisted into the 1920s among the luxury automobile producers, including Packard, Locomobile, Winton, and other low-volume luxury car producers, who had difficulty finding independent distributors (Parlin and Youker 1914: 228–229). At Ford, Hawkins replaced the branch agencies with a network of thousands of dealers serving local areas. New dealers were recruited to serve prospective customers who had lived too far away from existing dealers (Hawkins 1925: 194–195). Ford's branches remained open to provide service and replacement parts and eventually were converted to regional distribution centers.

Hawkins regarded the construction of plants for the final assembly of vehicles as a logical extension of the sales-oriented branch agency concept rather than an application of industrial location theory. The branch assembly plants were not 'established at rate breaking points to secure the greatest economy in freight rates. Their location was determined more by Mr. Hawkins' plan of selling cars direct to consumers' (Knudsen 1926: 6). However, once Hawkins demonstrated that branch assembly plants also substantially reduced shipping costs, the concept became the basis for the Ford Motor Company's expansion program.

That the initial development of branch assembly plants was motivated by sales rather than minimizing production costs was stipulated as fact at the Additional Tax Case trial during the 1920s.

> The assembly plant system was designed to give every purchaser of a car immediate or very quick supplies in case of broken parts. It gave a miniature factory at a strategic location to serve that location or neighborhood, which stimulated sales because of the investment in that particular plant in that immediate locality, assuring purchasers that they would have their troubles and complaints well looked after. They supplied stocks of parts to dealers and also enabled them to drive away their cars as they wanted them, without handling by freight.
>
> (*Dodge et al. v. Commissioner of Internal Revenue* 1927a: 62)

Ford was already shipping a large supply of parts from Detroit to the branches for repairs or installation on new vehicles. Why not ship all of the parts needed for a motor vehicle and let the branches put the pieces together?

Reducing shipping costs

Hawkins may have been initially attracted to building branch assembly plants as a way of enhancing customer service. However, he soon discovered that they would yield massive financial benefits through reduced shipping costs.

In the early days, Fords, like other automobiles, were distributed by piling as many as could fit into a freight car and shipping them to customers. The normal practice of wedging three or four fully assembled automobiles into a freight car ran up high shipping expenses, for two reasons. First, railroads set rates based on a minimum of 10,000 pounds (4,500 kilograms) per eleven-meter (thirty-six-foot) freight car and 11,200 pounds (5,100 kilograms) per twelve-meter (forty-foot) freight car. Because the early Model T weighed only approximately 1,200 pounds, three or four fully assembled automobiles accounted for less than half the minimum weight. As a result, producers were charged for several thousand pounds they were not actually shipping, and shipping costs in reality amounted to more than double the published rates. Second, railroads published high rates for shipping assembled automobiles, 10 per cent above normal first-class tariffs, because automobiles were fragile and difficult to maneuver in and out of freight cars (see Chapter 7).

Ford was charged somewhat lower rates, because the Model T's small chassis permitted it to be loaded crosswise in freight cars, simplifying the tasks of loading and unloading (*Dodge et al. v. Commissioner of Internal Revenue* 1927a: 62–63; Nevins 1954: 500–501).

Hawkins was not satisfied with these marginal freight cost savings.

> [He] conceived the idea of making a freight car earn its way, by the establishment of assembling plants all over the world, and shipping in knock-down condition from the plant in Detroit the pieces and parts that went to make up this car. Hawkins . . . spent six weeks in loading, unloading and reloading freight cars to find out just how they could pack the stuff in to make the car carry its maximum of weight, so that every dollar they spent for freight meant the transportation of a corresponding amount of weight
>
> (Goodenough 1925: 183)

The assembly plants permitted Ford to fill freight cars completely with up to one hundred engines, or large quantities of gas tanks, tires, wheels, radiators, or other components. The equivalent of twenty-six automobiles could be shipped in knocked-down form, compared to seven or eight fully assembled vehicles, thus substantially reducing the freight bill. Ford realized additional freight cost savings by shipping in knocked-down form, because rail lines charged lower rates to ship parts than assembled automobiles. Components were shipped at between second- and sixth-class rate, whereas assembled automobiles were charged at 10 per cent higher than first class (*Dodge et al. v. Commissioner of Internal Revenue* 1927a: 63; Nevins 1954: 343–345).

Freight charges could be further reduced by shipping parts made by outside suppliers, notably bodies and tires, directly to the branch assembly plants rather than first to Ford's Highland Park plant. A car load of a particular part destined for one assembly plant could be diverted en route to another plant experiencing a shortage of that part (*Dodge et al. v. Commissioner of Internal Revenue* 1927a: 63–4). Branch assembly plants also reduced shipping costs by enlarging the drive-away territory, that is the area within which dealers took possession of automobiles by driving them away from the plant rather than loading them onto freight cars or, in later years, trucks. In the early days of automotive production, the drive-away radius around a plant was about 100 miles (150 km), but with the construction of better roads the radius increased to about 200 miles (300 km). Branch assembly plants placed a high percentage of Ford dealers within drive-away range (Knudsen 1926: 2).

Ford was also able to reduce the economic impact resulting from sharp differences in the selling season in the early years of automotive production. At the time, the standard practice among automotive producers was to shut down production during the winter, when few orders were received. The lack of winter sales is understandable in view of the fact that until the 1920s vehicles provided little protection or traction in poor weather. Each spring, after a sufficient number of orders arrived, producers hired workers and reopened their factories. In contrast, Ford could make parts at its Highland Park plant through the winter and store them at the branch assembly plants, which had more storage space than the overcrowded Highland Park plant. When orders began to arrive in the spring, the branch assembly plants had the parts on hand to restart final assembly operations and deliver vehicles promptly to customers.

All of these freight innovations reduced Ford's shipping costs by one-half. Yet the company still presented buyers with a freight bill FOB Detroit, calculated under the old system of shipping several fully assembled automobiles from Detroit in a freight car rather than the actual cost of shipping from the branch assembly plant, often by drive-away.

The opening of branch assembly plants throughout the country brought Ford a considerable amount of favorable publicity that probably translated into additional sales. With a branch assembly plant in the community the local newspaper would find news about the company's products to be of local interest. According to Knudsen, 'the location of these assembly plants probably forestalled the location of other auto factories in these sections.' Knudsen believed that the Willys-Overland Company tried to establish a branch assembly plant in Minneapolis, as well as other cities, but could not (Knudsen 1926: 2).

Ford's first branch assembly plants

On 28 July 1909 Ford's board of directors voted to build its first branch plant outside of Michigan devoted to final assembly of vehicles, not just sales and servicing, in Kansas City, Missouri, on a 1.5 hectare site at 1025 Winchester Avenue. Ford records show that the land had in fact been acquired one month earlier, on 27 June 1909, so that planning for the new assembly plant could begin within a few days of the board's action (Ford Motor Company 28 July 1909). When the first floor of the new building was completed later in 1909, sales and repair operations were relocated from the existing branch plant at 1608 Grand Avenue. The assembly portion of the Winchester Avenue facility was completed in April 1911 and first used to assemble vehicles in 1912. Two floors were added to the building in 1912, and a second three-storey building immediately to the north was opened in 1916. Ford used the Winchester Avenue facility until 1956, when it was replaced by a new assembly plant in Clay County.

On 4 March 1910 the board approved construction of a building at Long

Island City, New York, originally to serve as another sales and service branch plant for the rapidly growing New York metropolitan area market (Ford Motor Company 4 March 1910). The three-storey building at Jackson Avenue and Honeywell Street was converted to final assembly operations in 1912, and after five storeys were added to the structure in 1914 all New York area sales and service activities were consolidated there. However, just four years later, Ford replaced the building with a new assembly plant in Kearney, New Jersey.

THE SELDEN PATENT SUIT HINDERS EXPANSION

On hindsight, the condition of the Ford Motor Company in 1910 appears healthy. The Model T, introduced the previous year, had already become the best selling automobile in history. The company's new Highland Park plant was the largest and most modern in the automotive industry. As net income in 1909 exceeded $3 million, the company in 1910 declared three dividends totalling $2 million. The other $1 million was put away for future expansion plans, notably implementation of the branch assembly plant policy. 'It is perfectly apparent to anybody that knows anything about the Ford Motor Company what they were saving this money for. They were not paying it all out in dividends. They were saving this money for a great expansion program' (Lacy 1925: 134–135).

However, Ford's expansion plans were on hold in 1910, and the firm's financial prospects were shaky. The reason was that the company faced a legal restraint on the number of automobiles it could sell, if not an outright prohibition, and the prospect of a costly settlement of a damage suit. The problems were caused by an adverse judgment in Federal District Court in the Selden Patent suit.

> We were handicapped and could not proceed with that expansion program in the early years of the adopting of Model 'T', for the reason that the Selden Patent case and the competition of the trust put us in jeopardy . . . The Selden Patent case was always a matter of most serious concern to all of us. We all realized that until it was disposed of it placed the entire fortune of the Ford Motor Company and the rest of us in hazard. I recall that we were advised by our counsel not to expand too much until this matter was disposed of. As a matter of fact until the Selden case was out of the way Mr. Ford and the company could not carry out their expansion program which we always had in our minds . . . . It was in fact a settled policy not to expand because of the Selden Patent case . . . . [The Selden Patent case made expansion] impossible both in our judgment and in the judgment of our patent attorneys.
>
> (Rackham 1926: 3)

Branch plants were leased rather than bought.

Enforcement of Selden's patent

George Baldwin Selden, a Rochester, New York, lawyer and inventor, had filed a patent in 1879 designed to give him the sole right to license and charge royalties on 'a liquid-hydrocarbon engine of the compression type', in other words, the basic elements of a gasoline engine. He used evasive legal tactics to delay implementation of the patent until 1895, when the prospect of motor vehicle production became commercially viable. By delaying the start of the patent, he maintained the claim but deferred the start of the seventeen-year period of exclusive rights granted by a patent.

Selden sold the rights to the patent in 1899 to the Electric Vehicle Company, recently formed by Wall Street investors. The company tried to enforce the patent and began litigation against Winton, then the leading manufacturer of gasoline-powered motor vehicles. Winton agreed to recognize the patent, and other leading carmakers followed suit. Henry Joy of Packard and Roger Smith of Olds led negotiations with The Electric Vehicle Company to establish a trade association in 1903, known as the Association of Licensed Automobile Manufacturers (ALAM).

The ALAM leased to its members the right to manufacture and sell automobiles under patent, awarding licenses to thirty-two established manufacturers the first year. Each licensee paid 1.25 per cent of the retail price of every gasoline-powered automobile they sold. One-fifth of the royalties went to Selden, two-fifths to the Electric Vehicle Company, and two-fifths to the ALAM to finance future enforcement of the patent. The license gave the producer a sense of security and insulation from potential suits. The ALAM ran advertisements (Table 3.1) to discourage people from buying cars from Ford and other companies not covered by the Selden Patent (Lacy 1925: 126–128).

Henry Ford's application was turned down in 1903, because the ALAM claimed that he had not yet constructed an operable gasoline engine. Ford vowed thereafter not to cooperate with the ALAM.

> [The Ford Motor Company] at all times refused to become a member of that Association and refused to lease the right to manufacture under the Selden Patent . . . . The purpose and object of the Association was to limit the output of cars and thus maintain prices which in turn would result in the demand for autos always exceeding the available supply.
>
> (Hawkins 1925: 191)

The ALAM sued several carmakers, most notably Ford, for infringement of the patent, and the District Court upheld the validity of the patent in 1909. The court recognized the validity of Selden's claim that the gasoline-powered engine he had patented uniquely combined several preexisting elements. Several companies, including General Motors, then agreed to

Table 3.1 Advertisement by Association of Licensed Automobile Manufacturers warning against infringement of Selden Patent

NOTICE
To Manufacturers, Dealers, Importers, Agents
and Users of Gasoline Automobiles

United States Letters Patent No. 549,160, granted to George B. Selden, Nov. 5, 1895, controls broadly all gasoline automobiles which are accepted as commercially practical. Licenses under this patent have been secured from the owners by the following named manufacturers and importers:

Electric Vehicle Co.
Winton Motor Carriage Co.
Packard Motor Car Co.
Olds Motor Works
Knox Automobile Co.
The Haynes-Apperson Co.
The Autocar Co.
The George H. Pierce Co.
Apperson Bros. Automobile Co.
Searchmont Automobile Co.
Locomobile Co. of America
The Peerless Motor Car Co.
U.S. Long Distance Automobile Co.
Waltham Manufacturing Co.
Pope Motor Car Co.
J. Stevens Arms & Tool Co.
H.M. Franklin Manufacturing Co.
Charron, Ciaradot & Voight Co.
of America (Smith & Mabley)
The Commercial Motor Co.
Berg Automobile Co.
Cadillac Automobile Co.
Northern Mfg. Co.
Pope-Robinson Co.
The Kirk Mfg. Co.
Elmore Mfg. Co.
E.R. Thomas Motor Co.
Buffalo Gasolene Motor Co.

These manufacturers are pioneers in this industry and have commercialized the gasolene [sic] vehicle by many years of development and at great cost. They are the owners of upwards of four hundred United States Patents, covering many of the most important improvements and details of manufacture. Both the Selden patent and all other patents owned as aforesaid will be enforced against all infringers.

No other manufacturers or importers are authorized to make or sell gasolene automobiles, and any person making, selling or using such machines made or sold by any unlicensed manufacturers or importers will be liable to prosecution for infringement.

ASSOCIATION OF LICENSED AUTOMOBILE MANUFACTURERS
7 East 42nd Street, New York

Source: Lacy 1925: 126–128

pay the ALAM royalties in the amount of 0.8 per cent of all production since 1903, a total of $1 million in GM's case.

Henry Ford refused to pay royalties, pending an appeal of the verdict to the Court of Appeals, but placed $12 million in an escrow account to cover eventual payments to ALAM. An even greater potential liability from an adverse judgment was threatened in the last paragraph of the ALAM advertisement: 'Any person making, *selling* or *using* such machines made or sold by any unlicensed manufacturers or importers will be liable to prosecution for infringement.' To calm dealers, who were afraid to sell Fords, the company arranged with the National Casualty Company to furnish a $6 million bond to cover every purchaser, dealer, and user of Ford automobiles (Lacy 1925: 129–130).

Ford successfully appeals

On 11 January 1911, the US Court of Appeals reversed the lower court decision. The Selden Patent was found valid, but only for a two-cycle engine originally built by George B. Brayton in 1872, not for the so-called Otto-type four-cycle engine, which virtually all automotive manufacturers actually used. The Court of Appeals decision rendered the Selden Patent worthless as it applied to motor vehicles. As the patent's seventeen-year enforcement period was due to expire in 1912 any way, the ALAM did not appeal to the US Supreme Court, and the organization soon disbanded.

> You cannot imagine how freed we all felt after the final decision against the validity of the Selden patent. It was then that we could extend the expansion policy which was Mr. Ford's dream and the sky was then the limit. Prior to that time we could not embark upon a policy of establishing branch houses all over the world, as we did immediately thereafter.
>
> (Rackham 1926: 4)

> In June 1911, [Ford] began to make plans for a great expansion of their business – a tremendous expansion program, one which was probably never equalled in the history of America
>
> (Lacy 1925: 133)

> [Hawkins] immediately upon the rendering of this decision . . . caused to be made a survey of the entire field of the United States for the purpose of expanding the business and increasing the production to take care of the sales that would be possible by more extensive work in the field.
>
> (Hawkins 1925: 192–193)

Ford directors were convinced that as a result of the favorable decision in the Selden Patent case, 'the only thing from then on that held the Ford Motor Company back was their ability to produce machines as fast as they could be sold' (Lacy 1925: 132).

Hawkins told Henry Ford that the opportunity had arrived when sales would be limited only by the ability of the company to finance and manufacture automobiles. He told Ford that he would have no difficulty in selling 75,000 vehicles in 1912, a figure Hawkins believed to be the extreme manufacturing ability of the company that year. Actually, Ford sold 76,150 automobiles in 1912 and held 20,000 orders which could not be filled because of inadequate productive capacity. In the fall of 1912, at the end of that year's selling season, Henry Ford told Hawkins:

> Thereafter [Ford] would not ask the Sales Department for any schedule at all but that he would expand the business just as far as it was possible for him to expand, that he was satisfied that the market could absorb all the cars he could manufacture.
>
> (Hawkins 1925: 195–196)

IMPLEMENTING THE BRANCH ASSEMBLY PLANT POLICY

The first step in Ford's expansion program after the successful ruling on the Selden Patent suit was to purchase the John R. Keim Mills in Buffalo, New York. Active in bicycle manufacturing in the nineteenth century, Keim became a leading producer of pressed steel parts for the automotive industry. Automobile manufacturers originally used frames built with structural steel or wood reinforced with structural steel plates, but by 1909 switched to pressed steel, which was stronger in proportion to weight, smoother, and more attractive than the alternatives. Other pressed steel parts became important, such as rear-axle housings, brake drums, steering columns, and fly wheel covers.

Ford placed large orders with Keim for pressed steel beginning in 1908, and several Ford officials were stationed in Buffalo to help design parts. Ford directors had rejected a proposition to buy Keim in January 1910, because of the adverse Selden Patent ruling, but with the Selden suit settled, Ford bought the plant early in 1911 for $574,000 (Lacy 1925: 134). Ford's original intention was to manufacture bodies at the Keim plant but in 1912, following a wildcat strike, the plant was closed, and the presses and machinery loaded on flat cars and shipped to Highland Park (Ford Motor Company 22 June 1911).

Decisions by Ford's directors

'December 12, 1911, was probably the great red letter day in the history of the Ford Motor Company, outside of the day of its birth and the day that it won the Selden patent litigation' (Lacy 1925: 135). On that date, the board of directors voted to send Couzens to California, 'in the interest of establishing Branch Houses, Warehouses, or to make such other arrangements for the handling of our business as may seem necessary' (Ford Motor Company 12 December 1911). The board gave

Couzens the power 'to sign contracts of purchase or leases as in his opinion seem advisable' (Ford Motor Company 12 December 1911). Couzens 'was on a sleeping car most all of the time for eight or nine months . . . with absolutely carte blanche authority to do anything that was necessary to be done, in order to expand that business to the limit' (Lacy and Anderson 1926: 282).

On 1 May 1912, the board of directors met at 9 a.m. to hear Couzens discuss his trip. Couzens presented his principal conclusion: the company should initiate assembly of automobiles on the west coast instead of shipping vehicles from the Highland Park factory,

> due to the distance between the Factory and Pacific Coast points, and the apparent possibilities for large business on the Pacific Coast . . . and for the further purpose of giving service to our customers, and the general handling of our business at the various points designated.
>
> (Ford Motor Company, 1 May 1912)

Couzens reported that he had actually purchased property in two locations, one near downtown Los Angeles, at Seventh and Santa Fe, for approximately $93,000, and the other in downtown Portland, Oregon, at Seventh and Division Streets, for approximately $30,000. He had also committed the company to buying land in San Francisco, at Harrison and Treat Streets, between Twenty-first and Twenty-second Streets, for $100,000, although not all arrangements had been completed. Couzens also decided to buy land in Seattle, Washington for between $40,000 and $60,000.

> Owing to the rivalry between Seattle and Portland, and the two distinct territories served by these cities, it was deemed advisable to purchase real estate in Seattle as well, although these cities are located but about one hundred and eighty-seven miles apart.
>
> (Ford Motor Company, 1 May 1912)

The board ratified the purchases of the Los Angeles and Portland properties and authorized completion of negotiations for the San Francisco and Seattle properties.

Couzens and Ford were not content with just establishing the four west coast branch assembly plants. They told the board of directors 'that it seemed desirable to complete arrangements for carrying out the plan outlined for the handling of our business on the Pacific Coast in certain other large centers of distribution: particularly points where our leases are now running out' (Ford Motor Company, 1 May 1912). The board voted to not only advise stockholders to accept the plan, but more importantly, to seek authorization from the stockholders 'to spend fifteen per cent of the net earnings of the Company each year, beginning with October 1st, 1911, and until further advised, for the purpose of developing this plan' (Ford Motor Company 1 May 1912). Although the minutes of the board of directors indicate that a 'thorough discussion of the matter' was held at the 1 May 1912 meeting, the discussion could not have been particularly lengthy, nor could the decision have been in doubt, because a special stockholders meeting had been called to start one-half hour after the board meeting, at 9.30 a.m. Furthermore, the stockholders comprised the same group of people as the board of directors.

After another 'thorough discussion of the subject was entered into, and details of the plan, and cost of establishing the various plants discussed,' according to the minutes, the stockholders gave the board what it wanted. Once the stockholders meeting ended, the board of directors reconvened at 10.30 a.m. on 1 May 1912 armed with the authorization to spend 15 per cent of Ford's net earnings. The board then empowered the company's officers to proceed with the development of the plan to establish branch assembly plants.

> In other words, Mr. Couzens had been to the Pacific Coast, and observed the situation, and when he came back they not only approved what he did, but proposed to do that all over the world. That is what they did.
>
> (Lacy 1925: 139)

The decision to siphon millions of dollars from dividends into construction of branch assembly plants did not appear to be controversial at the time, although the Dodge brothers raised strong objections a few years later when they sued to obtain additional dividends allegedly withheld for construction of new plants. John Anderson, who missed the 1 May 1912 meetings, took the 'unusual action' of writing a letter

> in which he said that he was advised of the result of that meeting, and he was very much pleased to learn that they had set aside 15 per cent to the sinking fund of the net earnings of the company for the purpose of expanding their business in that way.
>
> (Lacy 1925: 139–140)

Couzens sent his assistant Frank Klingensmith through the south to set up branch assembly plants (Klingensmith 1926). On 15 April 1913, less than one year after receiving authorization to spend 15 per cent of the company's profits on branch assembly plants, Couzens documented for the board of directors the rapid pace of construction. Plants were under construction at the time in eleven cities, including Cambridge (Massachusetts), Chicago, Denver, Los Angeles, Memphis, Minneapolis, Philadelphia, Portland, St Louis, San Francisco, and Seattle (Ford Motor Company 15 April 1913). By 1914, all eleven were in operation, as were three others which had not yet been started as of April 1913, in Columbus (Ohio), Dallas, and Houston.

In cities which did not yet have branch assembly plants, including Charlotte, Cleveland, Dallas, Oklahoma City, and Omaha, Ford began to deliver partially assembled automobiles to the branch sales and service plants. Bodies, fenders, and tops for Model T's were shipped separately from the chassis, an arrangement known as semi-knocked-down kits. Because sales and service branch plants were located in or near downtown they did not have always direct access to a rail line. Consequently, final assembly sometimes had to be done in the open at the rail yards where the kits were unloaded rather than at the branch plant. Once branch assembly plants were built in these cities, semi-knocked-down kits were no longer delivered, although they were exported until the early 1960s.

During 1915, eight more assembly plants were ready, in Atlanta, Cincinnati, Cleveland, Fargo, Indianapolis, Louisville, and Pittsburgh, plus a new building in Buffalo to replace the assembly operations in the former Keim mill. In 1916, five more opened, in Charlotte, Milwaukee, Oklahoma City, Omaha, and Washington DC, although the Washington building was used only for sales and service until the following year. By 1917, Ford was assembling Model T's in twenty-nine US cities outside of Michigan (Figure 3.1).

Characteristics of early branch assembly plants

Most of Ford's assembly plants were designed by Albert Kahn (1869–1942), a German-born architect who started a practice in Detroit in 1892. After several years of working primarily on residences, Kahn turned his attention to industrial buildings, beginning with the Boyer Machine Company in Detroit in 1901. Kahn's first automobile plant, for Packard in 1903, was the largest factory in Detroit until supplanted in 1910 by his Highland Park plant for Ford.

The Packard factory was noteworthy as one of the first reinforced concrete industrial buildings in the United States. Traditional factory building methods, characterized by heavy timber framing and wooden floors, were considered unsuitable for automotive assembly plants because engine lubricants and oils posed extreme fire hazards. Reinforced concrete construction allowed greater floor loads, reduced the effects of vibration from the machinery, and permitted a reduction in the number of interior posts and columns. By 1938, Kahn's firm employed more than 400 people and was responsible for 20 per cent of all architect-designed industrial buildings in the United States. A few of the plants, including Seattle and Milwaukee, were designed by John Graham of Seattle.

The man most responsible for setting up production in the new branch assembly plants was William S. Knudsen, who had been manager of the Keim Mills in Buffalo when it was acquired by Ford in 1911. He moved to Highland Park in 1913 to oversee installation of the machinery removed from the Keim Mills but made no secret of his reluctance to remain there. A few weeks later, Knudsen was sent around the country

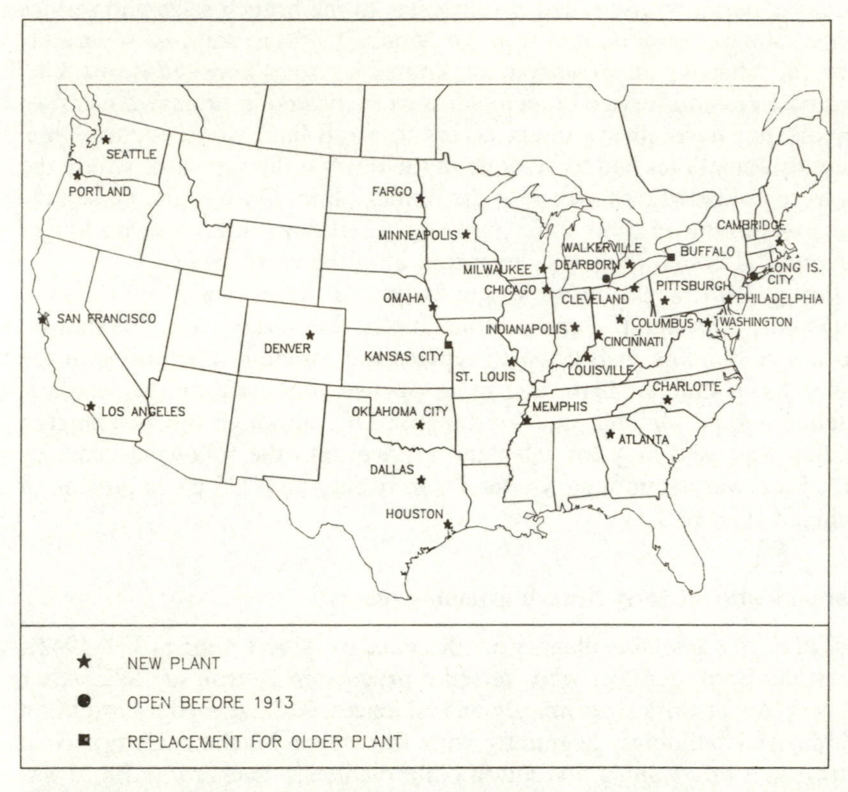

Figure 3.1 Ford Motor Company assembly plants opened between 1913 and 1917. Branch plants were opened to assemble the popular Model T close to major population centers

installing machinery in the new branch assembly plants. Over the next few years, he supervised the opening of fourteen Ford branch assembly plants. During World War I Knudsen was placed in charge of production of Liberty engines at Ford's Highland Park plant and then the Eagle boat project at the new Rouge complex (Beasley 1947: 54–55, 85; Doss 1926; Nevins and Hill 1957: 168).

Automotive production at a number of branch plants stopped temporarily, because of World War I, permanently in the case of the Washington branch. The War Department took over the Washington plant in late 1917, less than one year after assembly operations had started. When it was returned in late 1918, Ford reopened the plant only for sales and service; these functions were transferred to a new building in Alexandria in 1927. The Washington branch was sold to the District of Columbia in 1930 as part of the development of the Municipal Center. New assembly plants were opened in 1918 in Kearny, New Jersey, to replace the one in Long Island City, and in Des Moines, while assembly operations were dropped in Fargo and Portland, as well as Washington, reducing the number to twenty-eight, including Highland Park, in 1920.

A report in 1916 by Frank Klingensmith, who replaced Couzens as Ford treasurer in October 1915, revealed the extent of the financial commitment in the first wave of branch assembly plant construction. Between 1 October 1912 and 1 December 1915, 15 per cent of Ford's profits amounted to $16,321,274.25. Of this, approximately $13 million was spent on branch assembly plants and much of the rest on expansion of the Highland Park plant. Estimated expenditures for the eleven new branch assembly plants, plus additions to the two already in existence in Kansas City and Long Island City totaled $1.1 million for land and $3.3 million for buildings. Land acquisition costs ranged from $16,000 in Denver to $176,000 in Chicago, with a median of approximately $80,000, while building costs varied from $130,000 in Portland to $415,000 in Chicago, with a median of $230,000 (Ford Motor Company 15 April 1913).

The early branch assembly plants were for the most part multi-storey buildings, typically between four and six storeys. The Oklahoma City facility, for example, contained 17,000 m^2 in a four-storey building 122 m long and 38 m wide. The six floors of the 16,240-m^2 Cincinnati plant were known by letters. The main entrance level, on the 'C' floor, contained a showroom, garage, and stock room for servicing vehicles. The rail siding, boiler room, and maintenance shop were on the sub-basement or 'A' level, while paint dipping tanks and vehicle storage space were on the basement or 'B' level. The second or 'D' floor housed general offices, the service department, and paint ovens. Final assembly took place on the third or 'E' level, while the top or 'F' level contained stock bins and body conveyers (Ohio Historic Preservation Office 1988: 7).

At first, automobiles were assembled by hand in the branch assembly plants, but within eighteen months of the first experiments at Highland Park, moving assembly lines were installed in almost all of the plants. Daily production in the branch plants increased from a handful to several hundred. The manager of the Oklahoma City branch plant reminisced:

> There was quite a contest between the different branches . . . . It was actually a competitive situation. If one branch made a better production performance than another there would be an exchange through the grapevine to see how the better method was handled. Detroit manufacturing would let you know anytime they picked up a better, quicker way of doing anything from some other branch.
>
> (Doss 1926)

FORD'S SECOND GENERATION OF BRANCH ASSEMBLY PLANTS

Ford launched a new wave of branch assembly plant construction during the 1920s even more extensive than the effort of a decade earlier. Only three of the twenty-seven plants outside of Michigan open in 1920 continued assembly operations after 1932.

Location of Ford's new assembly plants

Even though they were barely a decade old, eight of the plants were replaced by new ones during the mid-1920s, including Cambridge, Charlotte, Chicago, Dallas, Louisville, Memphis, Minneapolis, and Philadelphia. Five others, in Buffalo, Kearney, Los Angeles, San Francisco, and Seattle, were replaced in the early 1930s, and eleven were permanently closed by 1932, including Cincinnati, Cleveland, Columbus, Denver, Des Moines, Houston, Indianapolis, Milwaukee, Oklahoma City, Omaha, and Pittsburgh. The three World-War-I-era branch plants still operating after 1932 – Atlanta, Kansas City, and St Louis – were all replaced with new facilities after World War II.

Ford's post-World-War-I construction program began in Chicago. Land for the plant at 12600 Torrence Avenue was purchased in 1920, construction started in 1921, and the plant was ready for occupancy in 1923. Two replacement plants opened in 1924, at 1820 Statesville in Charlotte and Riverside Drive in Memphis. Replacement plants were opened in 1925 at 5200 East Grand Avenue in Dallas and Western Parkway in Louisville. In addition, a plant at 966 South Mississippi River Boulevard in St Paul opened in 1925 to replace the downtown Minneapolis facility. In December 1926, an assembly plant was opened at 183 Middlesex Avenue, in Somerville, Massachusetts, to replace one in nearby Cambridge. The following year, the Philadelphia branch was replaced by a new facility in Chester, Pennsylvania. Existing plants were also remodeled during the early 1920s, including those in Cleveland, Atlanta, Denver, Los Angeles, and Portland.

Production was also initiated during the 1920s in four southern and western cities in order to accommodate increasing demand in those regions. New branch assembly plants were built in Arabi, St Bernard Parish, east of New Orleans, Louisiana, in 1923, on Wambolt Street in Jacksonville, Florida, and at 414 West Third Street in Salt Lake City, Utah, both in 1924, and 2424 Springfield Avenue in Norfolk, Virginia, in 1925. The four new facilities gave Ford an all-time high of thirty-two branch assembly plants in 1925, including Highland Park (Figure 3.2).

With the termination of Model T production in 1927, every Ford assembly plant was closed. However, all but two of the thirty-two branch assembly plants – Portland and Salt Lake City – reopened a few months later to produce Ford's new car, the Model A. Even after the Stock Market crash in 1929, not only was automotive production maintained in all of the branch assembly plants, the replacement program initiated in the 1920s continued. In May 1930, a plant in Long Beach, California, replaced the one recently renovated in downtown Los Angeles. Six months later,

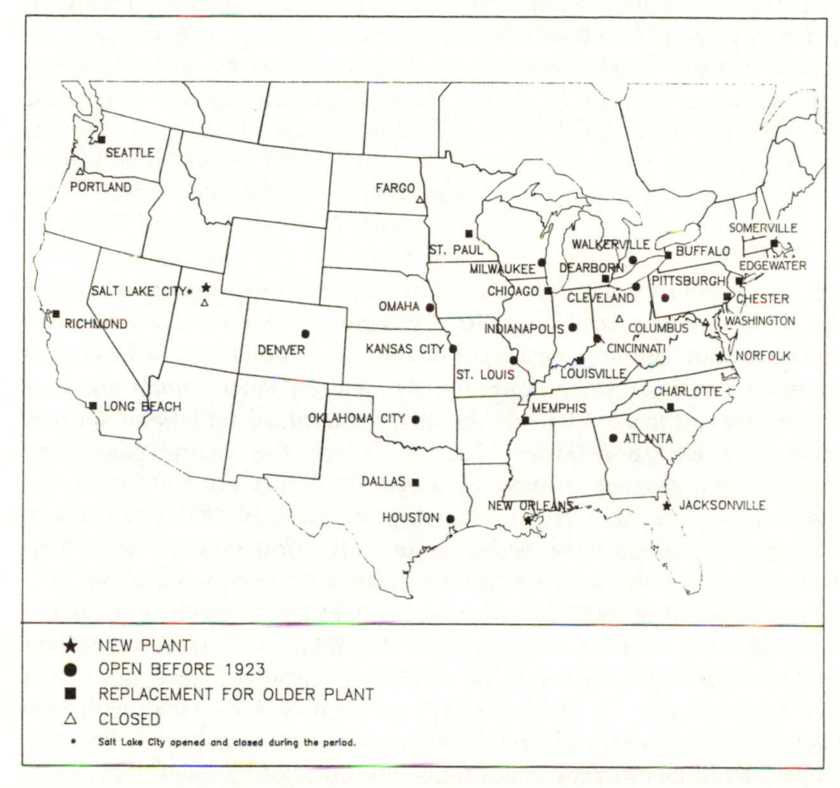

Figure 3.2 Ford Motor Company assembly plants opened and closed between 1923 and 1932

assembly operations in the New York metropolitan area were transferred from Kearney to Edgewater, New Jersey. The following year, replacement plants were opened on Fuhrmann Boulevard in Buffalo and in Richmond, California, to replace the one in downtown San Francisco. Finally, in January 1932, a replacement plant was opened in Seattle at 4735 East Marginal Way.

The short life-span of the first generation of branch assembly plants reflected the rapid changes in motor vehicle production, mostly initiated by Ford. When the original branch assembly plants were built, the moving assembly line had been recently installed at Highland Park, but fewer than one hundred automobiles a day were assembled in the typical branch plant, mostly by hand. Enjoying almost unlimited financial resources for capital improvements, Ford quickly drew up plans to replace the recently built branch assembly plants with new ones capable of accommodating moving assembly lines and other machinery, although the program was delayed a few years by the US entry in World War I. Whereas the plants built between 1913 and 1916 were multi-storey structures with less than 25,000 m^2, the replacement plants after World War I had more than 70,000 m^2 spread out over one or perhaps two floors. These plants employed 1,000 workers and produced 25,000–50,000 vehicles per year. All of the plants, with one or two exceptions, were designed by Albert Kahn (National Register of Historic Places 1982; Nevins and Hill 1957: 687; Ohio Historic Preservation Office 1988: 8).

Site selection process

Locations for the second generation of branch assembly plants were more haphazardly selected than the first, because the innovators of Ford's branch plant concept – Couzens, Hawkins, and Knudsen – had all left the company, having dared to question Henry Ford's increasingly absolute and arbitrary control. Whereas the first generation had been located in accordance with neoclassical location theory, the second generation reflected the increasingly irrational behavior of Henry Ford.

Couzens was the first to go, resigning in October 1915 over Ford's unwillingness to support the Allied cause at the outbreak of World War I. He remained on the board of directors for a few more years, but sold his shares to Ford in 1919 in order to enter Michigan politics. Couzens was elected mayor of Detroit in 1919 as a Republican and three years later US Senator, serving until his death in 1936. Ironically, Couzens was first elected to the Senate to fill the unexpired term of T.H. Newberry, who had been forced to resign as a result of charges of fraud in his narrow 1918 victory over the Democratic candidate – Henry Ford.

Hawkins, never a personal favorite of Henry Ford, was replaced as sales manager on the last day of 1918, although he had already been relegated to a minor role when automotive production stopped after US entry in World War I. Offered the job of special sales representative for Europe and South America – probably in order to move him far away from Henry Ford – Hawkins resigned instead. He moved on to General Motors, where he served for three years as General Consultant to the Executive Committee for advertising, selling, and servicing (Hawkins 1925: 190).

Knudsen resigned on 28 February 1921 following a dispute with Charles Sorenson concerning control of Ford's European operations. Knudsen joined General Motors in 1922, rising within months to be president of Chevrolet and in 1937 president of General Motors. While at Chevrolet, Knudsen directed its branch plant program which helped the division pass Ford in sales in the late 1920s. He left General Motors in 1940 to run the Office of War Production.

In the early 1920s, Ernest C. Kanzler and Henry Ford's son Edsel were largely responsible for choosing which branch assembly plants should be replaced. Kanzler, trained as a lawyer, went to work for Ford in 1917, as assistant to Sorenson, and became General Manager of the Highland Park plant. In 1924 he was elected second vice-president of the Ford

Motor Company, while Edsel served as president. Kanzler and Edsel were good friends and had married sisters. Kanzler, though, admitted that the selection process was unscientific. Past sales in various areas and the opinions of branch managers served as guides (Nevins and Hill 1957: 255–257).

Kanzler incurred the anger of Henry Ford by writing a six-page letter in January 1926 advocating that production of the outdated Model T be terminated. Six months later, he left the company. By the late 1920s, with Kanzler gone, not to mention Couzens, Hawkins, Knudsen, and the other architects of the earlier generation of branch plants, Henry Ford was making unilateral decisions concerning the location of branch assembly plants.

Obsessed with the high cost and unreliability of shipping by rail, Henry Ford was convinced that the optimal location for branch assembly plants was adjacent to, or – in a number of cases – in a major body of water. All of Ford's branch assembly plants built in the 1920s and 1930s, with one exception, were adjacent to deep water docks. The pattern was set during World War I, when the government agreed to dredge the River Rouge from the Detroit River to Ford's new industrial complex. After the war, Ford had the Calumet River dredged from Lake Michigan to his new assembly plant on Chicago's south side, so that large freighters could dock. Ford built plants on landfill in Richmond Inner Harbor of San Francisco Bay, Seattle's Puget Sound, Lake Erie near Buffalo, and the Hudson River at Edgewater, New Jersey.

Ford built two 187-m freighters, the Henry Ford II and the Benson Ford, during the 1920s, and bought several barges to haul coal and iron ore through the Great Lakes. In the early 1930s, he added a pair of 2,000-ton freighters, approximately 90 m long and 13 m wide, named the Edgewater and the Chester, to deliver components from Michigan to east coast assembly plants by way of the Great Lakes, Erie Canal, and Hudson River (National Register of Historic Places 1982).

Henry Ford's obsession with waterfront sites was revealed most graphically by an incident in the Netherlands. In 1930, Ford travelled to Europe to lay cornerstones for several new plants, including an assembly plant in Rotterdam. When he learned that the company's long-time head of European operations Sir Percival Perry had not selected a waterfront site in Rotterdam, because local government officials refused to make any available, Ford cancelled his visit and terminated construction plans in that city. The Dutch plant was ultimately built instead in Amsterdam, where officials made available a waterfront site (Nevins and Hill 1957: 554).

Ford's decline

The Ford Motor Company maintained its branch assembly plant network into the 1930s, but as sales plummeted 85 per cent in three years – from

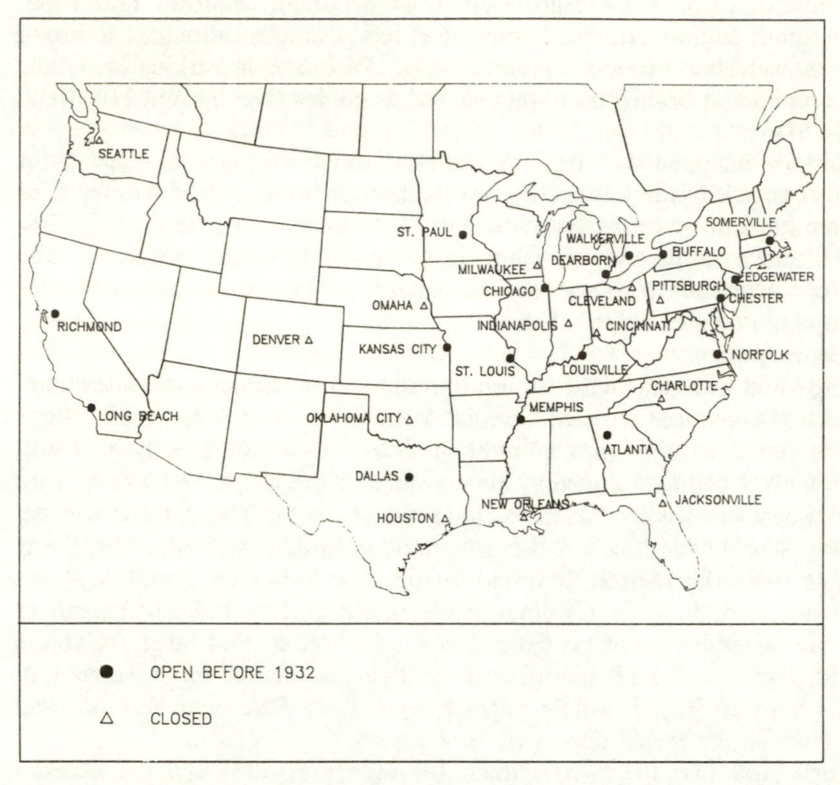

Figure 3.3 Ford Motor Company assembly plants closed in 1932. Ford actually closed all but eight assembly plants in 1932, but reopened several during the late 1930s, when sales increased again.

1.5 million in 1929 to 232,000 in 1932 – Ford could no longer afford to ignore the seriousness of the Depression. When production of 1932 models ended, Ford permanently closed thirteen of the twenty-nine assembly plants outside of Michigan, while six others were closed for two or more model years. The permanently closed plants included seven of the ten which were more than a decade old and had not been modernized during the 1920s, including Des Moines, Houston, Indianapolis, Milwaukee, Oklahoma City, Omaha, and Pittsburgh. Also closed were two older plants modernized during the 1920s – Cleveland and Denver – plus four new plants – Charlotte, Jacksonville, New Orleans, and Seattle. The Seattle plant shut only a few months after it was constructed earlier in 1932 (Figure 3.3).

The plants kept open through the Depression – or closed for less than two years – included only one which dated from prior to 1920, Ford's first branch assembly plant in Kansas City. The other ten were built in the 1920s, including a new one in Norfolk and replacement plants in Buffalo, Chester, Chicago, Dallas, Dearborn, Edgewater, Louisville, Richmond, and Somerville. Nine of the new plants, with the exception of Dallas, were located on deepwater ports. Five plants were closed in 1932 but reopened in 1935, including old plants in Cincinnati and St Louis and three new plants in Long Beach, Memphis, and St Paul. The new plant in Atlanta was reopened in 1937, giving Ford seventeen plants altogether for a short period in the late 1930s. By the end of the decade, the Cincinnati assembly plant was a warehouse. Thus, at the start of World War II, Ford had fifteen branch assembly plants in operation outside of Michigan. The thirteen permanently closed branch assembly plants became parts and sales centers during the 1930s.

The Atlanta branch was listed on the National Register of Historic Plants in 1984 and was renovated for apartments and boutiques. The Cleveland branch was bought by the Cleveland Institute of Arts in 1982. More commonly, the buildings, which were no longer suitable for manufacturing, were converted to warehousing and storage. The Kearney facility was sold to Western Electric in 1929. The Louisville plant was used for storage and leased as a WPA project during the 1930s. The Philadelphia branch was sold to Mack Warehouse Corporation.

4 General Motors builds branch assembly plants

William C. Durant was by all accounts a charming individual and an exceptional fundraiser. Walter Chrysler characterized Billy Durant most succinctly: 'I cannot hope to find words to express the charm of the man. He has the most winning personality of anyone I've ever known. He could coax a bird right down out of a tree' (Chrysler 1937: 143). Having made a fortune in Flint, Michigan, as the nation's largest wagon maker, Durant was largely responsible for the creation of General Motors in 1908.

Durant was more comfortable selling products than administering a large corporation's production schedule. According to Alfred E. Sloan, president and then chairman of the board of General Motors from 1923 until 1956, 'Mr. Durant was a man of charming personality. He was generous. He was always ready to do something for anybody that needed help. He was the salesman-promotor type. [But] he was not a scientific operator or a business administrator' (Sloan 1953: 2412). At a time when the Ford Motor Company was building a branch assembly plant network based on minimizing freight costs, Durant was operating largely through instinct. Under Durant, GM became the second largest producer of automobiles, but the company was run haphazardly. Durant grasped the logic of the branch assembly plant concept, but he never had the resources or administrative talent to structure GM's production sites in full accordance with neoclassical location theory.

Then, during the 1920s, new leaders took over General Motors and pushed the company's sales past Ford. Among the critical innovations was adoption of Ford's branch assembly plant concept. Like Ford, General Motors had begun as a Michigan-based company but added branch assembly plants elsewhere in the country to minimize distribution costs. After World War II, GM relied even more heavily on neoclassical location theory; an expanded network of branch plants helped the company capture half of the US market.

Even before he founded General Motors, Durant had perfected the skill of playing off one community against another in order to extract the most concessions. Shortly after taking over as manager of the struggling Buick Motor Company, in 1904, Durant moved assembly operations from

Flint to an empty building in Jackson, Michigan, owned by the Imperial Wheel Company, a subsidiary of the Durant-Dort Carriage Company. Engines, transmissions, and bodies were still built in Flint and shipped 100 km to Jackson for final assembly, clearly inefficient from a production standpoint.

Durant induced a rivalry between the bankers of the two cities in order to secure more capital for financially strapped Buick. The heads of Jackson's four major banks were informed that a suitable location had been found in Jackson to consolidate all of Buick operations, but Durant needed $100,000 from them in order to buy the building from the Smith Middling Purifier Company. The four bank managers were asked to purchase $25,000 each in preferred Buick stock to finance the relocation from Flint. A Jackson newspaper reporter who had learned about it vigorously promoted the idea, but the bankers turned it down (Durant 1944).

Meanwhile, Durant was negotiating with Flint bankers to keep the Buick operations there. Shortly after the collapse of the Jackson proposal, in April 1905, Durant secured $1.5 million from Flint investors, primarily the Genessee Bank, in exchange for an understanding that he would move Buick assembly back to Flint. Ground was broken in 1907 for a three-storey factory on the north side of Flint, the 90-hectare former William Hamilton farm, which Durant had purchased back in 1898. Over the next several years Buick's operations in Jackson and on the west side of Flint were consolidated at the Hamilton farm site. Frequently modernized and expanded over the years, the north Flint complex has been long identified as Buick's home plant and in the 1980s was renamed Buick City (May 1975: 212–213).

BIRTH OF GENERAL MOTORS

Recognizing the economic benefits resulting from large size, Durant pursued a policy of vertical integration, that is acquiring control of suppliers of the major components which went into automobiles, as shown in the next chapter. However, Durant's first priority was horizontal integration, that is control of a large percentage of automobile production. During the spring of 1908, he negotiated with other leading car makers to consolidate operations into a large holding company; even Henry Ford participated in the early discussions (Cray 1980: 58; Nevins 1954: 412–414). Unable to reach an agreement with the other major companies at the time, Durant established General Motors in 1908 as a holding company to acquire as many smaller firms as possible to supplement Buick.

GM's early acquisitions

Durant started with the Olds Motor Works in Lansing. Sales had plummeted after Ransom E. Olds left the company following a bitter argument with the Smith family, who owned two-thirds of the company's stock. When their strategy of building a luxury model proved a marketing failure, the Smiths were eager to sell the company to Durant, who was a family friend. Soon after purchasing Olds, Durant took his lowest priced Buick model, the Model 10 White Streak, to the Lansing plant, had it put on saw horses, and ordered the body cut lengthwise from front to rear and crosswise from side to side. The wider and longer version of the Model 10 immediately gave the Olds a more popularly-priced car to sell. Oldsmobile production remained concentrated in Lansing until recent years, although the division did not regularly turn a profit for General Motors until after World War II (Crabb 1969: 78–80, 246; Cray 1980: 64–66; Durant undated; May 1975: 139–142; Pound 1934: 62).

Durant bought the Pontiac-based Oakland Motor Car Company in early 1909. The firm had been started two years earlier by Edward M. Murphy, Pontiac's leading carriage manufacturer. Facing decreased demand for carriages and increased production costs, Murphy plunged into the automotive business by selling his carriage plants to Oakland. The largest financial stake in Oakland was held by James Dempsey, a wealthy retired lumberman from the western Michigan town of Manistee. Oakland under Murphy's leadership remained underfinanced and never produced more than a handful of cars. Durant, seeking to acquire Oakland, directed his legendary charm towards a reluctant Murphy, making frequent trips to Pontiac to praise his organizational talents. At the same time, Durant convinced Murphy's financial backers that their investment would be safer if they traded their shares in Oakland for General Motors. Murphy finally agreed to turn over his Oakland stock to Durant, a few days before he died (Cray 1980: 75; Durant undated; Pound 1934: 94).

As part of General Motors, Oakland performed well for a few years, but production difficulties during the 1920s hurt sales. The division introduced a new lower-priced model in 1926 called the Pontiac, immediately much more popular than the Oakland, and one year later expanded from its cramped facilities on Oakland Avenue into a large complex on the north side of Pontiac, near Harris Lake. The division's name was changed to Pontiac Motor Division in 1931, and the Oakland model was discontinued. The north side plant remained Pontiac's home production facility until the 1980s, when GM replaced it with a new facility immediately north of the city, near Route 24, the road to Lake Orion.

The Cadillac Motor Car Company already held a reputation as a leading high-quality automobile manufacturer when Durant acquired it in July 1909. A few months earlier Cadillac had been awarded the Dewar Trophy, regarded at the time as the highest award in the automotive industry 'in recognition of the most noteworthy performance of the year' for demonstrating the ability to standardize parts. On 29 February 1908, the contest's judges supervised the dismantling of three Cadillacs of different colors, scrambled some of the parts and replaced others, and watched the vehicles being completely rebuilt with only a wrench, hammer, screwdriver, and pliers. Cadillac's head Henry M. Leland and his son Wilfred agreed to continue managing Cadillac after GM's acquisition but insisted on complete independence. At the outbreak of World War I, the intensely patriotic Lelands sought to convert their plant to war production, but when Durant blocked the request, the Lelands resigned from GM and established the Lincoln Motor Car Company, initially to produce Liberty airplane engines (Crabb 1969: 114; Cray 1980: 155; May 1975: 256–258; Pound 1934: 106–107).

Two other Durant acquisitions became part of the GM Truck and Bus Group. The Rapid Motor Vehicle Company was established in Detroit by Max Grabowsky in 1902 and two years later moved to Rapid Street in Pontiac, the first plant in the United States built exclusively for the assembly of commercial vehicles. Rapid specialized in smaller, 0.5–3-ton trucks when General Motors acquired it in 1908. Early in 1909, GM bought the Reliance Motor Truck Company, which built larger 4 to 7.5-ton trucks. The products were both marketed as General Motors Trucks in 1912 and built together at Rapid's Pontiac plant beginning the following year. The former Reliance plant in Owosso, Michigan, was sold in 1914 to the American Malleables Company (Pound 1934: 134). A new truck plant, opened in 1927 in Pontiac, became the company's main production center for large trucks, until a number of plants elsewhere in the country began to assemble small pickups, vans, and sport utility vehicles.

Durant bought a number of other Michigan motor vehicle producers which did not survive. The Welch Motor Car Company, formed by brothers Fred and Allie Welch in 1904, was acquired by GM in June 1909, but within five months production of the high-priced model stopped. Durant convinced the Welches to build a second, less-expensive model, known as Welch-Detroit, in the former Olds plant in Detroit, but GM sold the operation to Peninsular Motor Company in 1914. GM acquired the near-bankrupt Rainier Motor Car Company, a Saginaw company founded in 1904 but in 1914 dropped the high-priced model, which was also sold to Peninsular. The Randolph Motor Car Company, another early acquisition, never actually produced automobiles. The Cartercar, founded in Detroit in 1905 by Byron Carter, held patents to propel vehicles by friction drive rather than transmission gears. GM took over the company which had moved to Pontiac, but ceased its production in 1916 and sold the plant.

Long after he was forced out of General Motors, Durant characterized his early policies in a 1940 letter to Alfred Sloan.

> The early history [of General Motors] reminds me of the following story: General Wheeler, who came up from the ranks, met Major Bloomfield, a West Pointer, at the Chickamaugua battlefield at Chattanooga. In speaking of the engagement General Wheeler said to Major Bloomfield, 'Right up on that hill there is where a company of infantry captured a troop of calvary.' Major Bloomfield said, 'Why, General, you know that couldn't be, infantry cannot capture calvary,' to which General Wheeler replied, 'but, you see, this infantry captain didn't have the disadvantage of a West Point education and he didn't know he couldn't do it, so he just went ahead and did it anyway.'
>
> (Durant 1940)

Durant's fall, return, and second fall

Unable to repay all of the loans he had secured to pay for the company's rapid expansion, Durant was forced out of GM management in 1910. Bankers agreed to loan money under the condition that for the five-year term of the loans a five-member committee appointed by them would run the company. Over the next five years, the company regained its profitability but lost most of its market share, from 21 per cent in 1910 to 8.5 per cent in 1915. As president of General Motors during the 1980s, Roger Smith made a similar decision; GM's market share fell from 46 per cent to 35 per cent under Smith's leadership, but the company recorded high profits during the period.

After leaving General Motors, Durant established two companies in Buick's abandoned west Flint plant – the Mason Motor Company organized in July 1911 to make four-cylinder engines and the Little Motor Car Company established three months later to make small cars. Meanwhile, in November 1911 Durant organized the Chevrolet Motor Company to produce a model based on a prototype developed by a famous race driver Louis Chevrolet. Durant rented a factory at 1145 West Grand Boulevard in Detroit that had formerly housed facilities of the Corcoran Lamp Company of Cincinnati (Dammann 1986: 6). The Little was well-designed but underpowered and poorly named, while the Chevrolet had a famous name but was high-priced and produced in limited numbers.

Durant relocated Chevrolet production to the Flint plant in 1913, but also transferred the Chevrolet name to the Little. Outraged at the association of his name with a small car unsuitable for racing, Louis Chevrolet quit the firm. Durant, ever the showman, bought a tract of land across Woodward Avenue from Ford assembly plant and erected a sign proclaiming that the Chevrolet Motor Company would soon erect a large completely modern factory on the site, but Chevrolet production was relocated to Flint, and the Highland Park land was sold to Ford in 1917. A new assembly plant built in 1917 in west Flint served as Chevrolet's home plant until 1947, when it was replaced by a facility in southwest Flint which was turned over to GM's truck assembly division during the 1970s (Crabb 1969: 323–327; Pound 1934: 144–148).

In one of the most remarkable events in American industrial history, Billy Durant, through his Chevrolet Motor Company, regained control of General Motors and became President in 1916. He again embarked on a rapid expansion program, including construction of new plants in Flint and Detroit and acquisition of two new automotive producers, the Scripps-Booth Company in Detroit and the Sheridan Motor Car Company in Muncie, Indiana.

Back in 1908, Durant had concluded a fifteen-year agreement to sell Buick engines and other parts to Robert S. McLaughlin, the son of a sleigh and carriage manufacturer in Oshawa, Ontario. McLaughlin remained loyal to Durant, first becoming a director of General Motors and then agreeing to build Chevrolets when it was an independent company. In 1918, with Durant again in charge, GM bought the McLaughlin Company. Robert McLaughlin ran the Canadian operation until he retired in 1924. His brother George stayed as president and a director of GM until 1967, when he retired at the age of 96 (Smith 1971: 101–103).

Once again, Durant overextended General Motors and was forced to resign. Control of the company passed to the duPont Company, which owned more than one-fourth of the stock. Pierre duPont became President in November 1920, and duPont officials moved into other top GM management positions. Rumors circulated that GM's Michigan factories would be closed and production relocated to plants closer to duPont's Delaware headquarters; real estate prices tumbled in Flint and Detroit. Denying the reports, Pierre duPont spent much of his time in Flint to be near the production facilities (Crabb 1969: 389). DuPont's refusal to move GM production to the east in the early 1920s marked the last occasion that Michigan's dominance of automotive production was seriously challenged.

Pierre duPont and especially Alfred Sloan, who became president in 1923, replaced Durant's instinctive, slipshod style of management, with a structured system, in which day-to-day decision-making was allocated to the operating divisions, while a central committee set broad corporate policies. Strict inventory controls were instituted, and production levels of particular models tied to sales performance. Years later, a reluctant Sloan was forced to testify in court about Durant.

> While it would be unfair to say that Mr. Durant did not believe in accounting, yet it would be fair to say that he didn't understand or believe in the wonderful possibilities of accounting in terms of indicating what ought to be done in the administration of the business.
>
> (Sloan 1953: 2412)

The testimony came during a trial in which the government ultimately proved that duPont's commanding position as a GM supplier was achieved not on competitive merit alone, but because its control of GM stock and management freed it from fair competition.

To prevent one passenger car division from cannibalizing the sales of another, GM created a distinct price niche for each. First-time customers were encouraged to purchase a low-priced Chevrolet and trade for higher-priced models. Sales were stimulated by changing the appearance of models each year, in contrast to Ford's preference for selling only black Model T's. Chevrolet, installed as GM's lowest-priced model, replaced Ford as the best-selling nameplate once Model T production ceased in 1927. GM's market share increased from one-third in 1929 to over 40 per cent between the 1930s and the 1980s.

As for Billy Durant, he went on to form another automotive firm, in 1922, financed through the sale of stock in the 1920s bull market to people who were attracted by his name. Durant Motors followed the familiar pattern of rapid growth through acquisition of ailing companies, including Locomobile in Bridgeport and Sheridan in Muncie, but the company collapsed during the Depression. In 1937, the former Durant Motors home plant on the south side of Flint, then used by GM's Fisher Body division to make Buick bodies, was the site of a famous sit-down strike which led to recognition of the United Auto Workers union. Durant lost a fortune in the 1929 Stock Market crash and declared personal bankruptcy in 1936. He died in poverty and obscurity in 1947, his brash and instinctive style of management something of an embarrassment to future GM executives. Durant did start to write an autobiography, which provides some information on GM's early years.

Durant embraces the branch assembly plant concept

Billy Durant embraced the branch assembly plant concept at an early date, although characteristically he left few explanatory records. The same year that the Ford Motor Company formally adopted a policy of building branch assembly plants, Durant announced that Republic Motors would establish factories in twelve cities, including Flint, to produce Little and Chevrolet automobiles and Mason and Sterling engines. The difference between Ford and Republic in 1912 was that Ford had already built a quarter of a million automobiles, whereas Durant could claim sales of only 2,999 for the Little and Chevrolet. However, from the outset of Republic production, Durant saw a network of branch assembly plants as central to a strategy of rapid growth.

Like Ford's Norval Hawkins, Durant wanted to assemble automobiles near consumers not to minimize transport costs, but because Durant believed that

> The problem today is not that of production, but of distribution. The enormous waste and extravagance in the marketing of automobiles, if continued, must result in the undoing of the industry. Regardless of high commissions, the majority of dealers are unable to make a profit. Under the plan outlined by us, the cost of distribution is materially reduced and each district is given the type and style of car best suited to its local requirements. Our trademark will be 'Built on the Spot.'
>
> (*Flint Journal*, 11 July 1912, quoted in Gustin 1973)

Looking for ways to maximize interest in his Republic automobiles, Durant hit on the idea of opening his first assembly plant outside of Michigan in the heart of New York City. Durant observed that the nation's largest city 'had no automobile plant, and it occurred to us that a factory located in the heart of the great city of New York, if it could be worked out successfully, would be of great advertising value' (Durant undated: 9). In July 1912, Durant purchased a building in Manhattan on Eleventh Avenue between 56th and 57th Streets, formerly owned by Rothschild Company, an early builder of automobile bodies. Components were shipped from Flint for final assembly of Little and Chevrolet models, some of which were then sent back to the Midwest for delivery to customers (Flink 1988: 67).

> When our good friends heard of our plan, they naturally, not knowing what we had in mind, took the position that we had lost our reason. Yes, we listened to the can'ts and don'ts, but we went right along with our well-advertised (supposedly foolish) plan.
>
> (Durant undated: 10)

New York City attracted Durant, because he considered that the key to selling the new vehicle was to create confidence in the product on the part of dealers, who – like consumers – were still concentrated in the northeast. By opening an assembly plant in New York, dealers as well as their customers could arrange visits more easily than if the nearest one were in Michigan.

> Grown-up people are very much like children in many respects, they like to see the wheels go round . . . . If for no other reason than for the convenience of our eastern dealers, we must have a plant in New York City.
>
> (Durant undated: 9)

Durant was well aware of the problems involved in locating a manufacturing plant in New York City, especially the high labor costs and inhospitable surroundings. In his memoirs, Durant recounted two stories that reflected on the cost of locating a plant in New York City. The first incident concerned children throwing 'live and dead trash' into the deep area designed to permit light and ventilation to reach the basement, where the blacksmith shop was housed.

> As the neighborhood was against us, we had to be careful not to offend, tact was needed. I had to have help and the help must come from the kids themselves.
>
> I spent a little time getting acquainted with the neighborhood kids and discovered a bright-eyed individual with whom I made a bargain to act as my representative, telling him the necessity of cleanliness and order for the benefit of the men working in the basement.
>
> (Durant undated: 10–11)

Durant located a portable merry-go-round and arranged with its owner to visit the plant twice a week from four until six in the afternoon. Children who had behaved properly were given free rides. This incentive appeared to solve the problem.

The second incident involved reaching an understanding with the neighborhood boss. Soon after Durant opened the plant, the owner of a saloon directly across the street visited him and informed him that he was the boss of the district. He told Durant that he was giving a party the next week 'and his organization was very much interested in its success. The tickets were $1 each, and he asked how many we would take.' Durant's reaction was to engage the saloonkeeper in conversation, learning his name, how long he had been in operation across the street, the condition of the business, and the nature of the organization. Durant asked 'how many tickets he could reserve for the "great event". He said ten, and he had that number with him. I told him that our boys like to have a good time, and asked if he could let us have twenty.' This reaction surprised and pleased the boss, who delivered the other ten tickets later. 'A week or two before our friend called upon me, our engineer, going to lunch, was terribly beaten up because he would not join the gang. After the ticket incident and a few other contributions, peace reigned' (Durant undated: 11–12).

Durant did not assemble vehicles in the middle of Manhattan for long. In June 1914, he bought the large plant in North Tarrytown, approximately 30 km north of Manhattan, formerly occupied by the Locomobile and most recently the Maxwell. The North Tarrytown site, between the Hudson River and the New York Central Railroad, north of the Tappan Zee Bridge, has served as a location for motor vehicle assembly longer than anywhere else in the United States (Pound 1934: 149).

The third branch assembly plant for Durant's company, by now renamed the Chevrolet Motor Company, was opened in St Louis in 1915. Locational theory suggested to Durant that other locations might be more efficient, but as always his choice was influenced by lack of cash. Russell E. Gardner, President of the Banner Buggy Company, one of the region's largest manufacturers, was an associate of Durant from their carriage building days. Interested in starting motor vehicle production, Gardner agreed to manufacture Chevrolets under a licensing arrangement, which meant that Chevrolet did not have to invest in the operation. The arrangement lasted for only a few years. Gardner decided to organize his own automotive company, the Gardner Motor Company, which produced vehicles in St Louis from 1919 until 1931. A new assembly plant was built in St Louis in 1920, soon after Chevrolet became a GM division (Durant 1915; Newmark 1936: 422).

Durant's fourth branch assembly plant was located at Fort Worth, Texas, beginning in 1916. By this time, he had regained control of General Motors but kept the Chevrolet Motor Company independent of GM until 1918. Production at the Ft Worth plant never matched the levels of the other assembly plants, and it was closed in 1921 (Newmark 1936b: 422). Later in 1916, assembly operations started in Oakland, California. As was the case in St Louis, the Oakland plant was built with the help of local capital. Norman deVaux, for many years Durant's chief lieutenant on the Pacific Coast, headed the company, and Durant's son Clifford, a race driver, was also involved in the California operation (Cray 1980: 131; Newmark 1936a: 380; 1936b: 422; 1936d: 756). Durant announced his intention to build other branch assembly plants for Chevrolet in Atlanta, Georgia; Kansas City, Kansas; and Minneapolis, Minnesota; but production cutbacks during World War I, shortage of capital, and reorganization problems prevented him from continuing the expansion program.

Average production at the early Chevrolet branch assembly plants matched the typical output at one of Ford's two dozen facilities at the time. In 1920, Durant's last year as head of GM, the five Chevrolet plants produced approximately 117,000 automobiles, including 33,000 each at Tarrytown and St Louis, 28,000 at Flint, 14,000 at Oakland, and 9,000 at Fort Worth. Chevrolet's branch plants had much larger capacities than Ford's, a characteristic which proved to be an asset as production expanded after World War I. The Tarrytown and St Louis plants were approximately 40,000 m^2, the Oakland facility was nearly 30,000 m^2, and the short-lived Fort Worth plant nearly 20,000 m^2, whereas the typical early Ford branch assembly plant was less than 20,000 m^2.

GM builds branch assembly plants

After Durant was forced out, the pace of new plant construction slowed at General Motors for a few years. The former duPont officials, now in charge of GM, felt compelled to impose emergency regulations limiting spending by the various divisions in order to return the corporation to a sound financial basis. The company adopted a policy of

> giving first consideration to coordinating and strengthening the large investments that have already been made in Plants, and restricting new expenditures to such items as will bring about greater efficiency and lower cost of operation, and that no expenditures be made for increasing our production facilities.
>
> (General Motors 1921)

According to the GM minutes, the suggestion came from J. Amory Haskell, a director and vice-president of duPont, who became GM's Finance Committee chairman and member of the Executive Committee in 1918. A rare glimpse into GM's decision-making process at this time can be found in the papers of John L. Pratt, who chaired a 1920 committee on the status of inventories and served as vice-president in charge of accessories divisions and member of the powerful Executive Committee at GM until his retirement in 1937; Pratt's papers are stored at the General Motors Institute Alumni Foundation Collection of Industrial History in Flint, Michigan.

Chevrolet's policy of expanding productive capacity resumed in 1922 with the hiring of William S. Knudsen, who had been in charge of setting up many of Ford's branch assembly plants a few years earlier. Knudsen joined General Motors in February 1922, as an assistant to Charles Mott, but within a month was made president in charge of production at Chevrolet. He became president of Chevrolet in 1924 and president of GM in 1937, succeeding Sloan who became chairman of the board. Two months after the Nazis had overrun his native Denmark, Knudsen left General Motors in 1940 to oversee production of military equipment, first as chairman of the National Advisory Defense Committee, then in January 1941 as codirector of the Office of Production Management, and finally in early 1942, after US entry in World War II, as director of war production in the War Department. He returned to GM after the war as a member of the board of directors. His son 'Bunkie' headed GM's Chevrolet and Pontiac divisions during the 1960s before being named president of the Ford Motor Company in 1967.

Bill Knudsen's first assignment at General Motors was to undertake long-range planning for Chevrolet's plant needs. Applying his knowledge of Ford's branch assembly plant concept, Knudsen fashioned the haphazard set of plants inherited from Durant into a network designed to minimize freight costs. During the 1920s, he increased the number of Chevrolet branch assembly plants from five to nine and rationalized their national distribution. Plants were opened in 1922 at Janesville, Wisconsin, in 1924 at Buffalo, New York, and Norwood, Ohio, in 1928 at Atlanta, Georgia, and in 1929 on Leeds Boulevard in Kansas City (Figure 4.1).

The North Tarrytown plant, a relic of the nineteenth century, was rebuilt and enlarged in 1930 to a daily production capacity of 900 automobiles, while the Oakland plant was replaced by one purchased from Durant

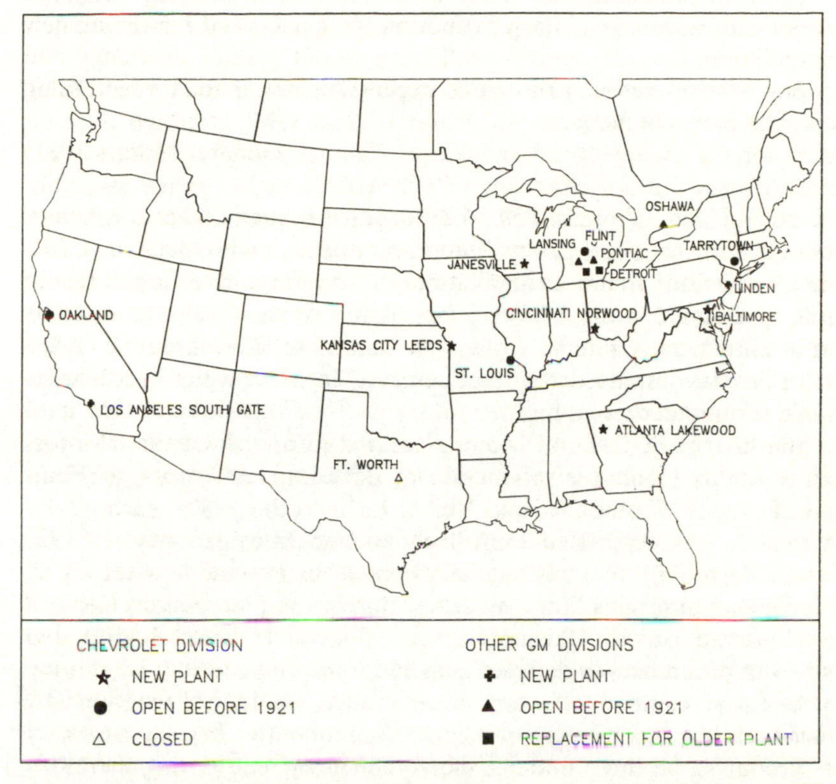

Figure 4.1 General Motors assembly plants opened and closed between 1921 and 1937. Baltimore, Linden, and Los Angeles South Gate were opened during the 1930s, the rest during the 1920s. Linden and Los Angeles South Gate assembled GM's mid-priced Pontiac, Oldsmobile and Buick models

Motors. Knudsen also closed the Fort Worth plant, which at the time was remote from major population concentrations.

The only location which represented a notable compromise with neoclassical theory was Janesville, 125 km northwest of downtown Chicago. A location closer to Chicago would have made more sense in order to minimize freight costs to the nation's second largest automobile market, but GM had inherited the Janesville plant from Durant's era. It had originally housed a farm implement firm acquired by Durant, who believed – like Henry Ford – that agricultural machinery was a lucrative market compatible with automotive production. Typically, Ford started his Fordson Tractor Company in Dearborn from scratch, while Durant spent $1 million in 1918 to acquire the locally-owned Janesville firm, which he renamed Samson after the biblical figure. Durant was attracted to Janesville, because it was 'strategically located [to] provide economical distribution [of tractors] to all points in the United States' (Newmark 1936c:

532). As tractor production was terminated at Janesville, using the plant for Chevrolet production was cheaper than building a new one from scratch closer to Chicago.

General Motors survived the 1930s depression better than Ford. Sales declined 56 per cent between 1930 and 1932 at GM, although over 60 per cent for the higher-priced Buick and Cadillac models. Unlike Ford, General Motors did not close any of Chevrolet's eight branch assembly plants. Alfred Sloan wrote in GM's 1935 Annual Report that the company 'believes that the soundest policy, both economically and socially, is to distribute [production] among as many different communities as is practically possible' (General Motors 1935). In fact, that year the Chevrolet division added a tenth branch assembly plant, in Baltimore, Maryland. By 1936, Chevrolet's sales surpassed the level achieved in 1929, while Ford's sales were still off by nearly one-half.

For the first time, General Motors decided during the 1930s to open branch assembly plants for production of the company's medium-priced models, Pontiac, Oldsmobile, and Buick. Prior to the 1930s, each of the three models was assembled exclusively in one Michigan city – Buick in Flint, Oldsmobile in Lansing, and Pontiac in Pontiac – a legacy of their independent origins. To save money during the Depression, General Motors ordered Buick, Oldsmobile, and Pontiac to share bodies and chassis, with differences in detailed trim and some engineering. By sharing many essential components, the three models could be economically produced on the same assembly line. Consequently, branch assembly plants producing all three middle-priced models opened during the 1930s on the east and west coasts. A plant managed by Buick opened in Linden, New Jersey, in 1935, for distribution to the northeast, and an assembly plant managed by Pontiac in South Gate, California, 12 km southeast of downtown Los Angeles, opened in 1937 for west coast distribution.

OTHER BRANCH ASSEMBLY PLANTS

General Motors and Ford were responsible for nearly all of the branch assembly plants built in the United States prior to World War II. The Chrysler Corporation joined General Motors and Ford as the third Michigan-based firm to dominate the US automotive industry. Although it passed Ford as the second leading US producer during the 1930s, Chrysler never developed an extensive network of branch assembly plants.

Chrysler Corporation

The roots of the Chrysler Corporation encompass a number of firms, some of which were based outside of Michigan, but by the time the Chrysler Corporation was established in the 1920s, Michigan's supremacy in automotive production was unchallenged, and there was no other logical location for a rapidly rising company to select. With few exceptions, Chrysler kept its production highly concentrated in the Detroit area.

Walter Chrysler was manager of the American Locomotive Company's Pittsburgh plant when one of the company's directors, James Starrow, at the time also a director of General Motors, introduced him to Charles Nash, then president of Buick. Fascinated with automobiles, Chrysler accepted a pay cut to work for Buick as Works Manager. When Durant regained control of GM, Chrysler was appointed Buick president to succeed Nash, who had left to start his own company. Chrysler resigned from Buick in 1919 but shortly thereafter was asked by the Chase National Bank to take charge of the reorganization of the Willys-Overland Company, which like many other automotive firms, faced financial problems during the recession of 1920–21. At the same time he was serving as executive vice-president at Willys-Overland, Chrysler was also asked by the banks to take the position of chairman of the Reorganization and Management committee of the Maxwell-Chalmers Company, another financially troubled company. In 1923 he became president of Maxwell and turned Chalmers into a subsidiary. A few months later, he incorporated the Chrysler Motor Corporation, which in April 1925 took control of Maxwell.

While at Willys-Overland, Chrysler had overseen development of a new model at the company's Elizabeth, New Jersey, plant, then the largest automotive factory in the United States. To reduce the company's debts, Willys-Overland sold the plant, but Maxwell-Chalmers was outbid by Durant Motors, which marketed the car developed at Elizabeth as the Flint. Chrysler enticed the engineering team to relocate from New Jersey to the Maxwell-Chalmers plant in Detroit to develop an improved version, which was introduced in 1924 as the first car named Chrysler. When prototypes were well received, the company was able to generate funds to enter production. The Chrysler car proved so successful that by 1926, the company dropped the Maxwell name altogether and introduced several Chrysler models. The former Chalmers plant on East Jefferson Avenue in Detroit became Chrysler's home plant.

Chrysler emulated GM's policy of building a variety of models, each targeted to a different price class. The most important step in the company's rapid growth was its acquisition of the Dodge Brothers in 1928. Chrysler had been forced to buy nearly all of its parts from outside suppliers, but the Dodge Main plant, located in the Detroit suburb of Hamtramck, provided Chrysler with a foundry and forge to make cast iron and other metal parts, as well as a large assembly plant. Chrysler completed its roster by introducing two new models in 1929, the low-priced Plymouth to compete with Chevrolet and Ford and the medium-priced DeSoto, positioned between Dodge and the high-priced Chrysler.

Chrysler operated a separate assembly plant in the Detroit area for each of the four models, as none of the four initially achieved sufficiently high sales to justify constructing branch plants elsewhere in the country. To supplement the Dodge plant in Hamtramck and the Chrysler plant on East Jefferson Avenue, the company opened two other assembly plants in Detroit, on Lynch Road in 1929 for Plymouth and on Wyoming Road in 1934 for DeSoto. The Jefferson Avenue plant was replaced by a new one next door in the early 1990s, while the other three Detroit-area plants all closed during the early 1980s.

Unlike Ford and GM, Chrysler increased sales during the Depression, from 375,000 in 1929 to 926,000 in 1936; its market share rose from 8 per cent in 1929 to 25 per cent in 1933. Chrysler became the second largest producer behind GM, a position it held until 1950, when it slipped to third behind a revitalized Ford.

During the 1930s, Chrysler took the first steps towards developing a national network of branch assembly plants outside of Michigan. First priority went to the west-coast market. Chrysler first assembled cars outside of the Detroit area in 1932, when a plant was acquired in Los Angeles to produce its low-priced Plymouth model for west-coast distribution. In 1939, the company opened a west-coast assembly plant for its medium-priced Dodge model in San Leandro, California, on the east side of San Francisco Bay, approximately 15 km south of downtown Oakland. The San Leandro plant also assembled Plymouths for distribution in the northwestern United States, while the Los Angeles reciprocated with assembly of Dodges for the southwest.

Prior to World War II, the other Chrysler assembly plant outside of Michigan or California was in the unlikely location of Evansville, Indiana, where the company inherited a plant when it acquired Dodge. The Evansville plant had been owned by Graham Brothers, which made bodies for Dodge trucks; Dodge gained control when it acquired Graham in 1925. Beginning in 1935, Chrysler used the plant to supply southeastern states with Plymouths, which were shipped by barge down the Ohio and Mississippi rivers.

Other companies

The only other pre-war branch assembly plants were built in Los Angeles by Willys-Overland and Studebaker. Willys opened its plant in the Maywood area, 10 km southeast of downtown Los Angeles; aided by its increased productive capacity, the company sold 315,000 vehicles, an all-time high, compared to 188,000 the previous year, more than any other US model, with the exception of Ford and Chevrolet, had ever achieved in a single year. However, sales plummeted to 69,000 just two years later and to under 8,000 in 1934. Willys accounted for less than 1 per cent of total automotive sales in the United States during the 1930s.

All production of automobiles for sale to civilians ceased in the United States by government order on 22 February 1942, eleven weeks after

Pearl Harbor. For more than two decades, the Ford Motor Company and General Motors had operated branch assembly plants in strategic locations around the country for production of their lowest-priced models, Ford and Chevrolet, respectively. When the US entered the war, Ford maintained sixteen assembly plants, in Atlanta, Buffalo, Chester, Chicago, Dallas, Edgewater, Kansas City, Long Beach, Louisville, Memphis, Norfolk, Richmond, St Louis, St Paul, and Somerville, plus the home plant in Dearborn. GM's Chevrolet division had ten, in Atlanta, Baltimore, Buffalo, Janesville, Kansas City, Norwood, Oakland, St Louis, and Tarrytown, plus the home plant in Flint. Otherwise, the branch assembly plant concept had been adopted but only on a small scale.

Automotive companies and individual plants within the corporations bid for contracts to produce military equipment, such as tanks and aircraft and marine engines. The military product with most enduring importance for the automotive industry was the four-wheel-drive General Purpose Vehicle, developed by Willys-Overland in response to an Army competition. The vehicle's name was successively shortened to its initials, GPV, GP, and eventually 'jeep'. After the war, with its passenger car business decimated, Willys stayed in business by selling Jeeps to veterans who remembered them fondly, the forerunner of the current craze for sport utility vehicles and light trucks.

POST WORLD WAR II: THE BRANCH ASSEMBLY PLANT AT ITS HEIGHT OF POPULARITY

Even while producing material for the war effort, the automobile companies were planning for the eventual return to civilian production. GM president Harlow H. Curtice testified a few years later that

> Before the end of World War II our research indicated that post-war demand for automobiles and all our products would rise substantially. This was at a time when many economists were predicting a post-war depression with at least eight million unemployed. We did not share this gloomy view and in 1943 announced a 500 million dollar post-war expansion program. Actually our capital investment expenditures for the period 1946–1953 came, not to 500 million, but to two billion dollars.
>
> (Curtice 1955: 35)

Following the end of World War II, US auto makers wasted as little time as possible in expanding productive capacity to meet long suppressed demand. More than at any other time, automotive producers explicitly embraced neoclassical location theory during the post-war era to determine where new assembly plants should be built. GM officials even appeared before a US Senate investigating committee in 1955 armed with maps depicting the market areas surrounding each of the company's branch assembly plants and financial data justifying why the locational model was optimal.

GM's post-war branch assembly plants

The map of GM's Chevrolet Division displayed ten assembly plants, each surrounded by a hinterland, irregularly shaped to conform to major transportation corridors. Chevrolet's ten assembly plants included nine in operation prior to World War II, with the exception of one at Buffalo, which was converted to production of components. Reflecting the changing distribution of the US population, the Buffalo assembly plant was replaced by a new one in the Van Nuys section of Los Angeles, 25 km northwest of downtown, in the San Fernando Valley. Production began at Van Nuys in September 1947, bringing Chevrolet back to ten branch assembly plants (Figure 4.2).

Chevrolet branch assembly plants served markets with mean areas of 873,000 km^2 and mean populations of eighteen million in 1960. Market areas varied between eleven and thirty-one million inhabitants and between 250,000 km^2 and 2.8 million km^2. The wide variation reflected the fact that population was not distributed uniformly across the country. Market areas were smaller but more populous in the northeast; Tarrytown, near New York City, encompassed the smallest service area, but the largest population. The two California plants contained the smallest number of inhabitants, as of 1960, although the differences would be less today if the branch plant system still existed, reflecting the fact that GM had accurately anticipated future migration trends within the United States.

Kansas City's market area was by far the largest, because it supplied much of the sparsely inhabited region between the Missouri River and the Rocky Mountains. Although the western part of the Kansas City market area was actually closer to the California plants, GM preferred to minimize 'back-hauling,' that is shipping components from Michigan all the way to the west coast before sending finished cars part of the way back towards the east.

Only three of the plants – Atlanta, Baltimore, and Flint – were located near the geometric center of their service areas. Oakland and Los Angeles were at the extreme west of their service areas, Norwood and St Louis at the extreme north, Janesville and Kansas City at the extreme east, and Tarrytown at the extreme south. However, most of the plants – not just Atlanta, Baltimore, and Flint – were located near the population center of their regions. A large percentage of the population served by the two California plants lived in the Los Angeles and San Francisco Bay areas. The Janesville and Tarrytown plants were located just north of Chicago and New York, respectively, where most of the people in the two service areas were clustered. Similarly, most of the people served by the Norwood plant lived in the northern part of the region. Only the Kansas City and

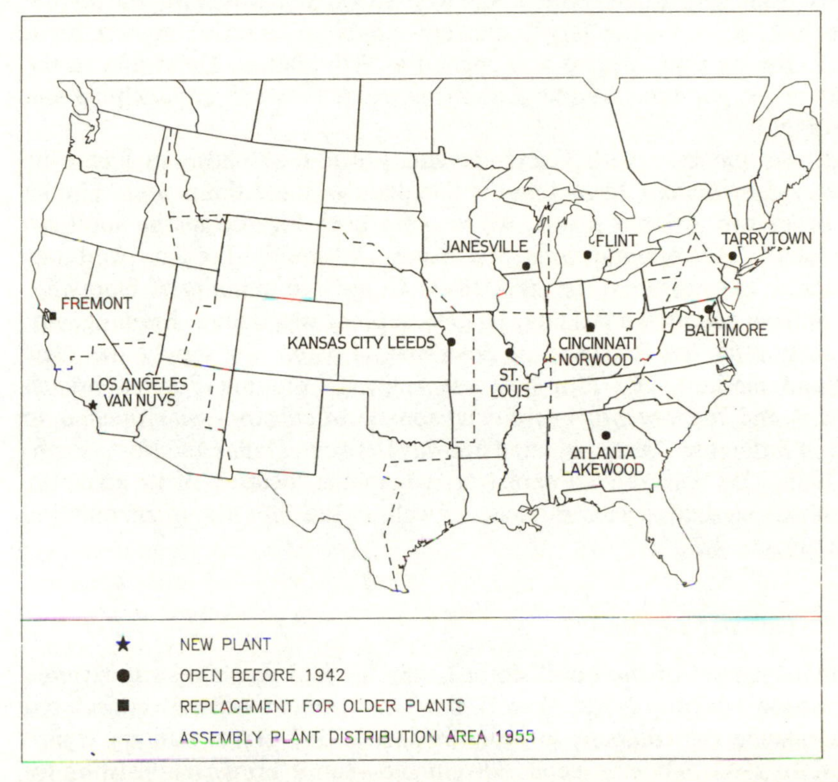

Figure 4.2 General Motors Chevrolet Division assembly plants opened between 1946 and 1959. This map, as well as Figure 4.3, was presented during testimony by General Motors officials at a 1956 US Senate hearing
Source: US Senate 1956: 895

St Louis plants were located some distance from the population, as well as the geometric, centers of their service areas.

General Motors also embraced the branch assembly plant concept for production of its medium-priced models after World War II. Two assembly plants for joint production of Pontiac, Oldsmobile, and Buick models had already been opened during the 1930s near the two largest car markets, including Linden, New Jersey, near New York City, and the South Gate section of Los Angeles. Not burdened by an inheritance of poorly located assembly plants, General Motors was able to create a network of assembly plants which conformed closely to neoclassical location theory, given the uneven distribution of the US population.

Sites for new combined Buick-Oldsmobile-Pontiac plants were identified before the end of the war, and within weeks of V-J Day construction started on four new plants. The first new post-war plant was opened in 1946 at Kansas City, Kansas, on Fairfax Avenue, 8 km north of Chevrolet's

Leeds Boulevard plant, which was located on the Missouri side of the state line, to serve the large, sparsely inhabited, interior market area. The following year, a plant was opened in Wilmington, Delaware, in the midst of the populous Middle Atlantic corridor between Philadelphia and Washington.

As was the case with Chevrolet and Ford, GM chose to locate its southeastern Buick-Oldsmobile-Pontiac plant in the Atlanta area. Unlike the other two Atlanta plants, which were both located in the southern part of the metropolitan area, the Buick-Oldsmobile-Pontiac plant was located in the northeast, in the DeKalb County community of Doraville, 20 km from downtown Atlanta. In 1948, a plant was built at Framingham, Massachusetts, 40 km west of downtown Boston, to supply the New England market. The fifth post-war Buick-Oldsmobile-Pontiac branch plant – and the seventh overall outside of Michigan – was opened in 1953 in Arlington, Texas, situated midway between Dallas and Fort Worth. By 1955, GM was able to depict on a map the location of its assembly plants for medium-priced models, as well as the hinterland surrounding each (Figure 4.3).

Ford's post-war recovery

Mismanagement of the Ford Motor Company had reached such extreme proportions during World War II that government officials considered nationalizing the company in order to ensure delivery of military equipment. In 1943, while a feeble 80-year-old Henry Ford was insisting on retaining the title of president, his son Edsel died of cancer at age 49, and an ill Charles Sorenson resigned as chief of production. The Navy released Edsel's son, 26-year-old Henry II, in order to take charge of the company. Henry I turned over the presidency to his grandson in 1945 only when his wife and his son's widow threatened to sell their company shares.

Henry II told an interviewer about the extent of the company's lack of administrative controls when he took over. 'In one department they figure their costs by weighing the pile of invoices on a scale' (Nevins and Hill 1962: 255). Henry II decided that the key to the company's survival was to replace his grandfather's casual decision-making approach with a decentralized structure patterned after General Motors.

The Ford Motor Company revived its branch assembly plant network to produce its best-selling Ford model after World War II. Fifteen of Ford's seventeen pre-war assembly plants reopened during late 1945 and 1946, as soon as they could be reconverted from war production. The other two pre-war assembly plants, at Atlanta and St Louis, had been sold to the government and were not reacquired (Figure 4.4). First to resume automobile production were assembly plants which had built tanks and jeeps during the war, including Edgewater and Louisville in July 1945 and Dallas, Chester, and Richmond in August. The assembly plants which had

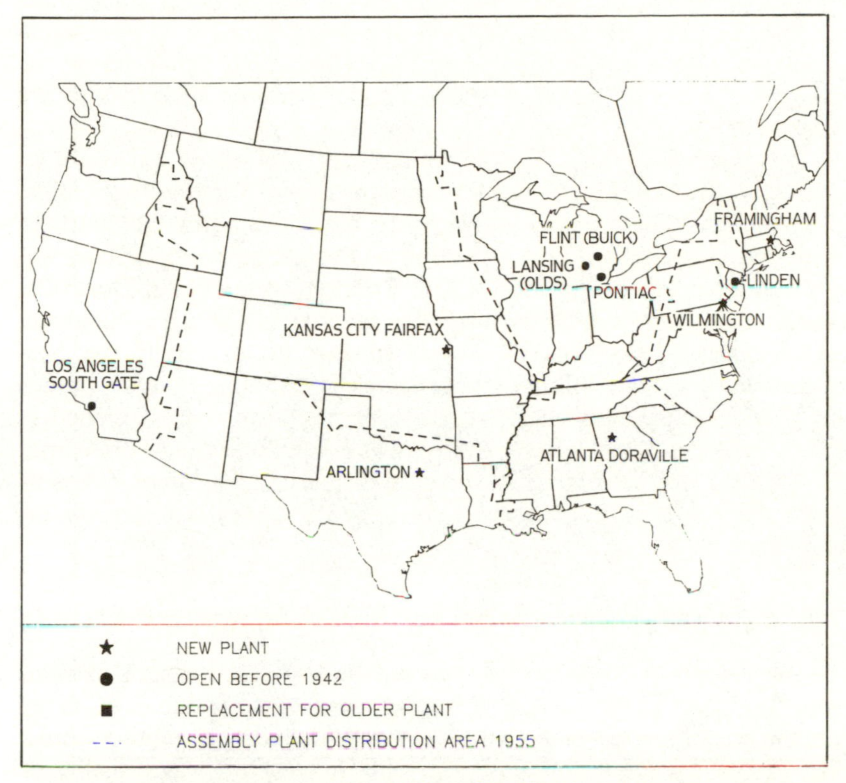

Figure 4.3 General Motors Buick-Oldsmobile-Pontiac divisions assembly plants opened between 1946 and 1959. Plants outside of Michigan assembled all three models, while each of the three 'home' plants in Michigan remained under the separate control of the three divisions
Source: US Senate 1956: 895

required more radical modifications to produce guns, armored carriers, and aircraft parts, opened later in 1945, including Somerville in October, Chicago and St Paul in November, and Kansas City and Memphis in December. The company also resumed automobile production by the end of 1945 at the Buffalo and Long Beach assembly plants, which had been leased to the government during the war, as well as at the Norfolk plant, which had been sold to the navy but reacquired following the war (Miller 1952).

Ford's branch assembly plants historically held responsibility for both sales and production within a region, but the two functions were separated, and new parts and distribution centers were built at different locations than the assembly plants. However, systematic modernization of the plants did not begin until the 1950s. Immediately after World War II, the only major change in the assembly plants for the low-priced Ford was construction of a replacement in the Atlanta area for the building sold to the government.

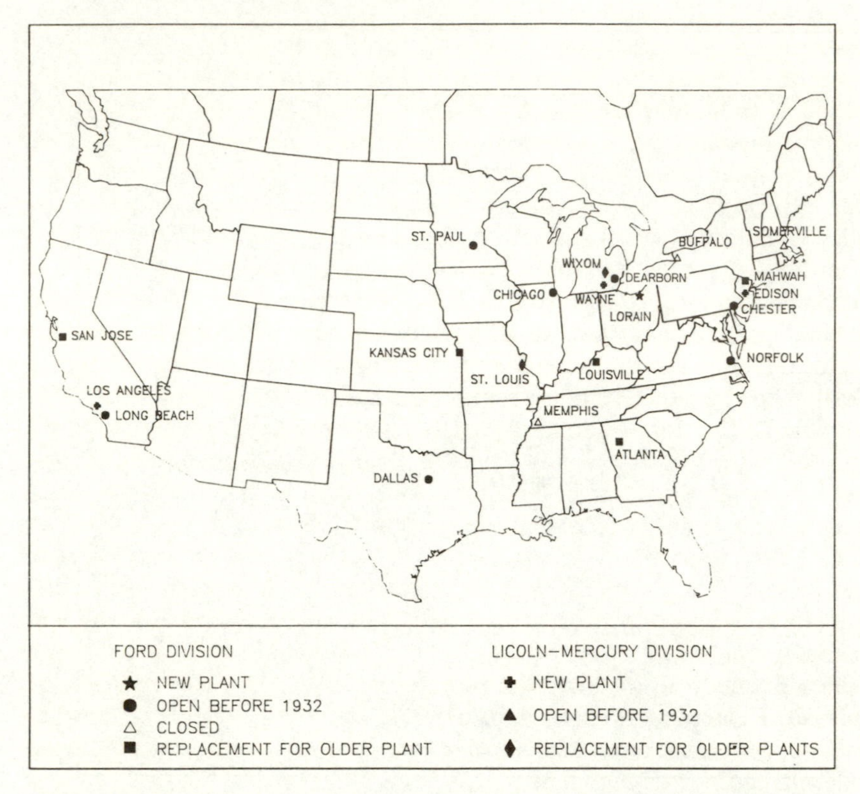

Figure 4.4 Ford Motor Company assembly plants opened and closed between 1946 and 1959

On 25 May 1945, the day that the War Production Board gave the automotive industry permission to start limited manufacturing of passenger cars, Ford announced the purchase of land for a new plant in the Atlanta suburb of Hapeville, 10 km south of downtown, near Hartsfield International Airport. Construction started in November 1945, and the plant opened one year later.

L.D. Crusoe, Ford's general manager and later vice-president for the newly created Ford Division, advocated a program of modernizing Ford's pre-war assembly plants, beginning with the Louisville plant, but finance officials, especially Robert S. McNamara, argued that the company couldn't afford the $1 billion price and that it was more profitable to maintain the existing plants. Finally, at a directors' meeting in February 1953, Henry Ford II authorized spending $500 million for plant modernization (Halberstam 1986: 237–244).

Construction was approved at the February 1953 meeting for replacement plants in Louisville and the New York and San Francisco areas. Delmar S. Harder, who had worked for a number of years at General

Motors and then was president of E.W. Bliss Company, which made power presses and packaging equipment, became Ford's vice-president of operations in December 1946, in charge of identifying and acquiring sites. According to Harder, sales forecasts were critical in deciding which plants would be modernized. 'Expansion plans were pretty nearly all predicated on a three year basis, though there was some thinking farther in advance. They did use the standard volume studies (estimates) determined by the Policy Committee' (Harder 1959; Nevins and Hill 1962: 340).

Harder was responsible for selecting the specific sites in the three areas. Critical factors included transportation of personnel to the plant, access to rail facilities, and availability of utilities. Conspicuously absent from the list of critical factors was access to deepwater. The locations reflected changing priorities in the site selection process. During the 1920s and early 1930s, plants had to be located near deepwater ports, because of Henry Ford's obsession with water transport. By the 1950s, land-based transport was more important and, for the first time, sites were selected with access to recently built limited access highways (Harder 1959).

The New York area plant was located in Mahwah, in extreme northern New Jersey, close to the New York State line, 30 km northwest of the company's Edgewater plant. The San Francisco Bay Area plant was located 6 km north of San Jose, near the junction of Route 17 and Capitol Avenue, on a 60 hectare plot acquired from the Western Pacific Railroad. Relocating the Bay Area plant 80 km south, from Richmond to San Jose, generated opposition from officials in Richmond, who advocated modernizing the existing plant.

Crusoe had to defend the San Jose site.

> The Richmond plant does not lend itself to the kind of expansion required to build the number of Ford cars and trucks to be sold in this area . . . . You can't 'splice on' to an assembly line. To increase capacity, each department within a plant must be expanded, and the building of an addition is not the simple solution . . . . [It would be] costly to the company to build additions and re-work the present plant. At a time when our present production cannot keep up with the demand for our product, we could not take a plant out of the system for the long period of time the job of virtually rebuilding it would require.
>
> (Miller 1952)

In addition to the three plants constructed between 1953 and 1955, the Ford Division also replaced its Kansas City facility. In 1949, Ford had bought land owned by the Weber Engine Company immediately south of the venerable Winchester Avenue plant, where the country's first branch assembly plant had opened in 1912, with the intention of modernizing the body and paint shops. Construction never proceeded, because the following year the company bought land along US Route 69 at Claycomo, in Clay County, 13 km northeast of downtown Kansas City. Ground was broken for a new plant in January 1951, but the plant was first used to manufacture B-52 bomber wings during the Korean War. Passenger car assembly began in January 1957. At 170,000 m^2, the assembly plant was ten times larger than the 40-year-old Winchester Avenue facility.

After replacing four of its fifteen pre-war plants, Ford decided to close five others. Branch assembly plants were closed in 1958 in Buffalo, Long Beach, Memphis, and Somerville, and in 1961 in Chester. Long Beach was apparently chosen because only one west-coast plant was needed, and the recently built San Jose plant was the logical one to retain. However, a new assembly plant was built in Lorain, Ohio, in 1958, bringing the total to eleven. The closures reduced Ford's number of branch assembly plants to only one more than its chief competitor Chevrolet; in 1957, the year before the closures, production averaged under 100,000 vehicles at Ford's assembly plants and under 50,000 at Somerville. Further, the distribution of Ford's plants after the closures closely matched that of Chevrolet. Eight of Ford's plants were now located within 50 km of Chevrolet's.

Further emulating General Motors, Ford placed responsibility for production, as well as design and purchasing, of the medium-priced Mercury and luxury Lincoln in a new division separate from the low-priced Ford model. To reinforce the distinct identity of the new division, four new branch assembly plants were built for Mercury production and one for Lincoln during the 1940s and 1950s. For the east-coast market, a plant was located in Metuchen, Edison Township, New Jersey, approximately, 30 km southwest of Manhattan, near the junction of the New Jersey Turnpike and the Garden State Parkway. A west-coast plant was located in the Pico Rivera suburb of Los Angeles, about 15 km east of downtown.

A third new branch assembly plant was located about 30 km south of downtown St Louis, near the banks of the Mississippi River, at the confluence of the Meramec River. The land had been acquired in 1937 to build a replacement for a Ford division branch assembly plant on Forest Park Boulevard which dated from 1914. When the plant was built after the war, it was used for assembly of Mercurys instead of Fords. All three new Mercury plants opened in 1948. Mercury's fourth new assembly plant was a new 'home' plant in the Detroit suburb of Wayne, Michigan, in 1952. Finally, Lincoln, which had never moved from its original home on Warren Avenue in Detroit, received a new 'home' assembly plant in 1957 at Wixom in Novi Township, outside of Detroit.

Other post-war branch assembly plants

The Chrysler Corporation was much less active than Ford and General Motors in establishing new branch assembly plants after World War II. The company maintained its three branch plants established during the 1930s in Los Angeles, Evansville, and San Leandro, plus its four 'home' plants in the Detroit area. But by failing to expand its productive capacity after World War II, Chrysler lost much of its market share and lost its position as the nation's second largest carmaker it had gained from Ford in the 1930s.

> During the eight years that followed V-J Day, Chrysler squandered the advantages so painstakingly accumulated during the Thirties. The contrast between what took place at [Ford headquarters in] Dearborn and what failed to take place at [Chrysler headquarters in] Highland Park explains much of Chrysler's later troubles.
>
> (Moritz and Seaman 1981: 46, 55)

The cars were unfashionable and poorly built, and the company's cost accounting was criticized by government reports.

Belatedly, Chrysler tried to catch up. The San Leandro Dodge plant was closed in 1955, and west-coast operations were consolidated at the Los Angeles facility. Chrysler finally opened an east-coast branch assembly plant in 1957, in a facility at Newark, Delaware, originally built in 1952 for construction of tanks during the Korean War. The antiquated Evansville assembly plant ceased production in 1959 and was replaced by a new facility in St Louis, which opened the following year. The company's medium-priced DeSoto model was terminated in 1960, but the plant where it was assembled, on Wyoming Avenue in Detroit, remained open until 1979 for production of export vehicles.

Two of the smaller automotive producers – Nash and Kaiser-Frazer – built branch assembly plants in the Los Angeles area after World War II. Nash, which had been assembled exclusively in Kenosha for fifty years, decided to launch a second branch plant in El Segundo, near the Pacific Ocean immediately south of Los Angeles International Airport. The company's sales increased from 118,621 in 1948, the year the El Segundo plant opened, to an all-time high of 189,543 in 1950. Kaiser-Frazer, which began production in 1946 in the former Ford aircraft plant at Willow Run, Michigan, added a second plant in Long Beach, California in 1949. The company also produced cars in Portland, Oregon, during 1950 and 1951.

With the opening of the Nash and Kaiser-Frazer plants in the late 1940s, the Los Angeles area had nine branch assembly plants. The other seven included two each owned by Ford and General Motors and one each by Chrysler, Studebaker, Willys-Overland. The concentration in the Los Angeles area reflected the region's rapid growth into the nation's second largest metropolitan area, combined with the increasing expense of shipping assembled vehicles more than 3,000 km from Midwestern plants (Figure 4.5). By the 1990s, all nine were closed.

The branch assembly plant system served Detroit well for fifty years. Shipping parts from Michigan for final assembly near consumers proved the most efficient spatial organization, consistent with neoclassical location theory. However, critics like David Halberstam saw in the operation of

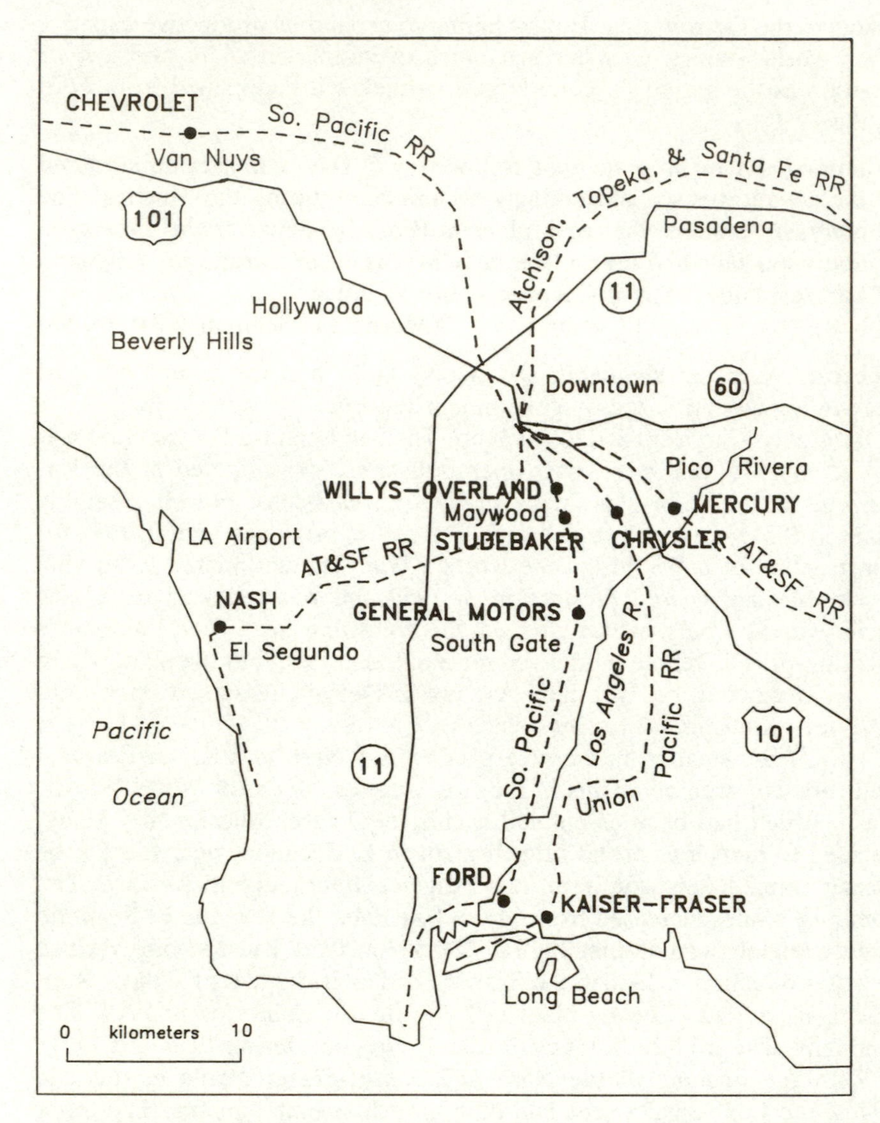

Figure 4.5 Assembly plants in the Los Angeles area, late 1940s. By 1983, only the Van Nuys facility still operated

the branch plants the seeds of the crisis which would hit the automotive industry in the 1970s. After McNamara became vice-president and general manager of the Ford Division in 1955, each plant was rated according to its efficiency, computed as production compared to costs of operation. Halberstam graphically portrayed the perspective of the Chester branch plant, apparently typical of the reaction to the cost accounting imposed by McNamara:

> [The world of branch assembly plant managers] was filled with secrets, and the name of the game was Screw Detroit . . . . They had learned to cheat Detroit as best they could in order to preserve the integrity of their own operation. They did this with admirable cunning . . . . Were there too many parts left over at the end of a model's life? Detroit hated that, so each year the plant people faithfully reported to the home office that they had only sixty-one of one part remaining and only forty-eight of another. Detroit, they were telling the home office, had been every bit as efficient as it hoped. Meanwhile [at the Chester plant] they dumped thousands and thousands of useless parts into the nearby Delaware River. Detroit loved how little waste there was, how well the numbers had matched out, and the people in Chester joked that you didn't have to swim the Delaware, you could walk across on the rusted parts of 1950 and 1951 Fords. It was the most critical part of the code: Tell Detroit what it wanted to hear and then do the best you could with the limited time and resources available to you . . . . Detroit had a system for demanding quality, and it periodically sent out inspectors to check the plants. But Chester could always rig the system, delivering to the visiting auditors precisely the cars it intended, cars of higher than normal quality . . . . The plant managers were giving them what they wanted, numbers, while paying lip service to quality. Years later in Vietnam some American officers, knowing McNamara's love of numbers, cleverly juggled the numbers and played games with body counts in order to make a stalemated war look more successful than it was . . . . In doing so they were the spiritual descendants of the Ford factory managers of the fifties.
>
> (Halberstam 1986: 220–222)

Stability during the 1960s and 1970s

During the 1960s and 1970s, US automotive producers opened and closed only a handful of assembly plants. After twenty-five new plants were built around the country during the first fifteen years after the end of World War II, only four were opened between 1960 and 1978. General Motors built a new plant during the 1960s at Lordstown, Ohio, and a replacement for the Oakland plant in nearby Fremont, California. Chrysler continued its effort to catch up with the other two large carmakers by opening two plants during the 1960s. The largest and most modern assembly plant in the country was built at Belvidere, Illinois, 120 km west of Chicago. The 200,000 m^2 plant began operations with the 1966 model year. The same year, an 88,000 m^2 Dodge Truck assembly plant was opened in Fenton, Missouri, 25 km west of St Louis (Moritz and Seaman 1981: 71).

Only four assembly plants closed between 1962 and 1978. General Motors closed its Oakland plant in 1964 when it opened a replacement in Fremont. In 1971, Chrysler withdrew from west-coast production, closing its Los Angeles assembly plant. Ford's only move between 1962 and 1980 was to close its Dallas assembly plant in 1970. Studebaker closed its South Bend, Indiana, plant in 1964. In addition, a Chevrolet plant in Bloomfield, New Jersey, which produced knocked-down kits for export, was closed in 1968, symbolizing the change in the United States from exporter to importer of motor vehicles.

By the late 1970s, massive changes were sweeping the US automotive industry. Sales of US-made cars declined, as Americans were attracted to foreign vehicles, especially Japanese, after the 1973–74 energy crisis. Chrysler reached the brink of bankruptcy, saved only by government financial guarantees. Through the 1970s, US auto makers held on to their plants, envisioning a return to prosperity in the 1980s. Most of the plants were modern, had been constructed or substantially renovated in the late 1940s and 1950s. Of the forty-six plants assembling automobiles in 1980, only eighteen continued to do so a decade later.

5 Components plants locate in the southern Great Lakes region

Manufacturers of automotive parts, like assemblers, owed their initial locational pattern to a combination of behavioral and neoclassical factors. However, parts suppliers did not follow the same spatial distribution as final assemblers. While producers relocated most final assembly operations to branch plants near large population concentrations around the country, parts suppliers remained clustered in southeastern Michigan and adjacent southern Great Lakes states.

The different spatial pattern stems partly from the behavior of the founders of parts suppliers, as well as Henry Ford, Billy Durant, and other officials at the large producers. The difference between the distribution of parts suppliers and assembly plants also derives from differences in applying neoclassical location theory. Assembly plants were located near customers primarily to minimize shipment of bulky products. Parts suppliers had to balance two concerns – access to customers, but also access to steel and other inputs.

Early automobile companies – little more than assemblers and distributors – had to obtain parts from other manufacturers. The automobile industry clustered in southeastern Michigan at the beginning of the twentieth century primarily because firms capable of building key components, such as engines and bodies, were already located there. By adapting components they had been making for other industries, existing firms successfully moved into automobile production, and new carmakers arrived to take advantage of the concentration of talented parts manufacturers which had rapidly formed in the region. If necessary, these suppliers could fill orders for complex components in accordance with precise specifications, but frequently they simply sold parts from existing stock which could be modified and bolted together at the assembly plant by skilled mechanics.

Depending on outside suppliers for major components soon proved embarrassing to producers. Parts did not always arrive when ordered, delaying final assembly. With demand for cars insatiable at the turn of the century, production delays resulted in lost sales during the spring buying season. More critically, financially-strapped carmakers, operating on credit extended by parts suppliers, couldn't settle their debts until they sold the cars and received full payment. Carmakers also began to make their own components in order to assure standardized size and composition. Consumers were convinced through advertisements that companies making their own parts produced better quality cars, because the components would fit together better.

Ford and General Motors owed much of their early leadership and long-term domination of US automobile production to aggressive policies of exercising increasing control over parts supplies. Half a century later, the term 'vertical integration' would be coined to apply to the strategy. Yet not even Ford and General Motors were ever able to achieve complete self-sufficiency and continued to purchase many of the thousands of parts which go into motor vehicles from independent suppliers. General Motors spent $30 billion and Ford $17 billion on outside suppliers in 1989 (Harney 1989: E32).

Nevertheless, the policies of Ford and GM to promote vertical integration caused a reversal in the original sequence of locational decisions. Whereas producers at the turn of the century located in the southern Great Lakes region to be near suppliers, in later years independent suppliers located in the region in part to be near corporate offices and distribution centers maintained by the large producers.

Ford started making its own parts because no independent supplier could match the volume of production to which the Model T was being pushed. GM's founder Billy Durant drew upon his experience building carriages, as noted in his memoirs.

> We started out [in the carriage industry] as assemblers with no advantage over our competitiors. We paid about the same prices for everything we purchased. We realized that we were making no progress and would not unless and until we manufactured practically every important part that we used.
>
> (Durant undated: 12)

By acquiring his suppliers Durant transformed his company into the nation's largest carriage producer. He achieved economies of scale through operating plants at capacity and eliminating the need for time-consuming and expensive negotiations with independent suppliers.

For both companies the first priority was to produce the engine, drivetrain, and chassis. Later, the other major modules or subassemblies which are brought together in the final assembly process, notably the body and electrical and nonelectrical accessories, were integrated into the operations of the large carmakers. By the late 1920s, Ford and GM had become largely self-sufficient in the production of these key subassemblies, at a time when competing carmakers still relied heavily on independent suppliers.

The principal modules or subassemblies in turn involved assembly of hundreds of components, some extremely complex in their own right.

The drivetrain, for example, requires bringing together components such as gears, axles, and differentials, each of which comprises hundreds of individual parts. Consistent with the principle of vertical integration, the large automobile producers have tried to make most of their own components, as well as subassemblies.

The hundreds of parts which go into components are produced through casting, stamping, molding, and forging metals, plastics, and nonmetallic materials into coils, tubes, sheets, blocks, and other shapes. Vertically integrated automobile producers perform some of these processes themselves, such as casting engine blocks at foundries and stamping body panels, but some metal and nonmetal parts are purchased from specialized independent suppliers. The transformation of raw materials, such as production of steel and plastics, is done by independent suppliers, although at one time Ford tried to control resource and raw material handling, as well.

Ford and General Motors followed different strategies to achieve vertical integration. Henry Ford preferred to initiate production himself, while Billy Durant acquired existing independent companies. Ford was more concerned than GM with handling materials, such as steel, glass, and wood, but GM acquired more independent producers of parts, such as bearings, springs, and wires.

As a result of these differences, Ford and GM did not adopt the same spatial distributions for their components plants. Ford followed the neoclassical theory to a logical conclusion: as southeastern Michigan was the point which minimized the aggregate cost of bringing in various inputs and shipping parts to assembly plants, then all parts production should be clustered in the area. On the other hand, GM's distribution of parts production was strongly influenced by Durant's behavior. While Durant recognized the benefits of centralizing parts production in Michigan, he maintained production in plants throughout the southern Great Lakes region acquired from other companies. Since World War II, Ford has remained highly clustered in southeastern Michigan and northern Ohio, whereas General Motors has shifted some parts production to southern states.

FORD'S COMPONENTS PLANTS

Ford's highly centralized locational model for parts production was initially implemented at the Highland Park plant, which opened on the first day of 1910, a year after Model T production had begun. Highland Park achieved the highest degree of vertical integration under one roof seen up to that time in the automotive industry.

Chassis components, engines, transmissions, and axles were made in a machine shop on the main floor of the Highland Park factory. The body compartment was welded together on the fourth floor, while bodies were painted and accessories such as lamps and trim were made on the third floor. Subassembly of the major modules and final assembly operations, such as attaching accessories and the body to the chassis, took place on the second floor. Engine blocks and other metal parts were cast in an adjacent foundry building. Moving lines were installed for subassembly of many of the parts and components, which were then brought together on the second floor for final assembly along a moving line (Nevins 1954: 453–456).

Ford's complex along the River Rouge in the Detroit suburb of Dearborn, which replaced Highland Park during the 1920s, was an even more centralized operation. Ultimately, 'the Rouge' (as it was widely termed) employed more than 75,000 workers in 93 structures, spread over 450 hectares, at the heart of extensive road and rail transport networks. In addition to producing most subassemblies and components, the Rouge also contained an integrated steel mill, including blast furnaces, an open hearth mill, a rolling mill, a pressed steel facility, and coke ovens. Not content with producing his own steel, Ford extended control of inputs to natural resources, as well, through acquisition of forests, coal mines, and iron ore fields. Barges built by Ford brought iron ore from Minnesota's Mesabi Range through Lake Superior, Lake Huron, and the Detroit River to a deep water port dredged on the River Rouge adjacent to the plant. Ford controlled forests on Michigan's Upper Peninsula and bought the Detroit, Toledo, and Ironton Railroad to facilitate hauling coal from Appalachia (Nevins and Hill 1957: 283–293).

Steel produced at the Rouge was relatively expensive in part because Henry Ford insisted that the plants be kept clean and serve as a laboratory for experiments, such as trying to eliminate the need for pig iron by transporting molten iron directly to the foundry (Nevins and Hill 1957: 289–292). In recent years, the steel mill, no longer limited to one customer, was first turned into an independent subsidiary of the Ford Motor Company and then sold in 1989 to Marico Acquisition Company, founded for the sole purpose of the acquisition, headed by Carl L. Valdiserri, retired chief operating officer of Weirton Steel Co. At the time of the sale, Rouge Steel's workforce included 3,300 hourly production and 750 salaried employees.

Prior to World War II, Ford maintained only two conventional components plants outside the Rouge, for production of glass in St Paul, Minnesota, and steering wheels in Hamilton, Ohio. The company also generated hydroelectric power at Green Island, New York, and operated lumber camps and woodworking facilities in several communities on Michigan's Upper Peninsula, including Alberta, Iron Mountain, L'Anse, and Pequaming.

In addition, Ford established eighteen 'village industries' for automotive production between 1918 and 1944 in small southern Michigan communities within 100 km of Dearborn. The village industries contributed to Henry Ford's vision of decentralizing production from urban to rural areas, taking advantage of low production costs, especially surplus farm

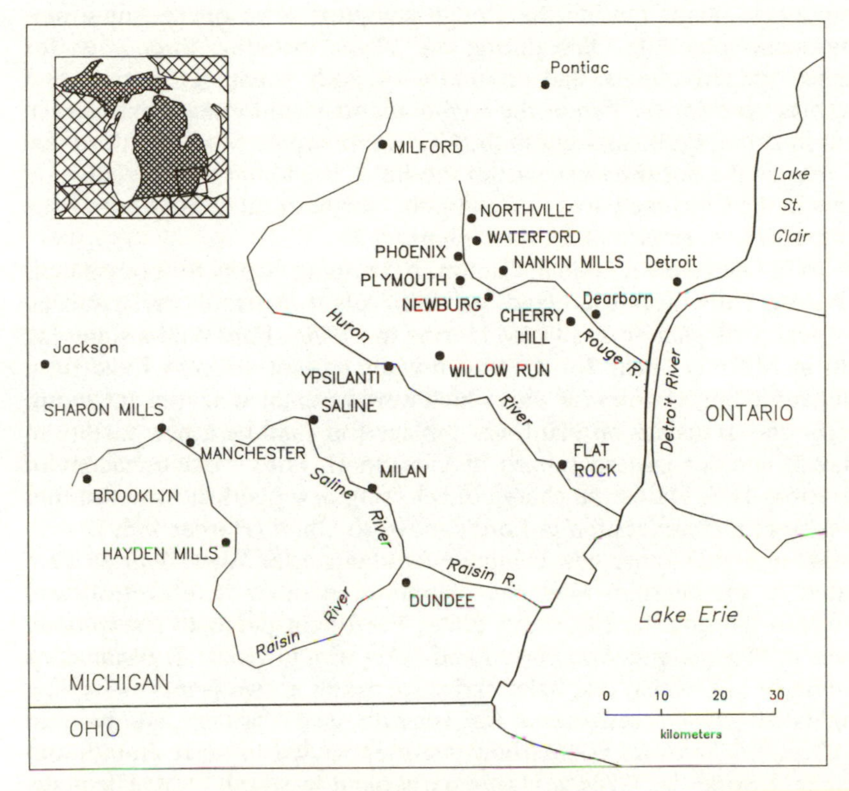

Figure 5.1 Henry Ford's village industries. Ford believed that manufacturing should be decentralized to small plants in rural settings
Source: adapted from Segal (1988)

laborers. The plants, which employed between 17 and 1,500 workers, produced electrical parts, such as ignition coils, gauges, lamps, and starters, plus small metal parts, such as valves, tools, and dies. Given Henry Ford's obsession with the benefits of hydroelectric power, all of the village industries were located alongside rivers, including seven near the Rouge, five near the Raisin, four near the Huron, and two near the Saline (Figure 5.1).

Most of the village industries were closed during the 1940s. The only survivor is Northville, which makes fuel tanks, although the Ypsilanti village industry was absorbed into a larger electrical parts complex after World War II. Ford built new plants near several other former village industries, including Flat Rock, Milan, and Plymouth. The closed structures have been converted into offices, garages, shops, and high-tech industries (Segal 1988: 183).

Once Henry Ford I departed from the scene in the late 1940s, the company for the first time embarked on a large-scale program of constructing components plants outside the Rouge complex. Ford opened nine new components plants in Ohio during the 1950s, including three each for engines and drivetrains, and one each for body stamping, casting, and electrical accessories. Two of the engine plants were located next to each other in Brook Park, adjacent to the Cleveland airport, while the third was in Lima, in the northwestern part of the state. Ford built two transmission plants in the Cincinnati area, including the northern suburb of Sharonville and the eastern suburb of Fairfax (Figure 5.2).

Also in Ohio were a forge in Canton; a stamping facility for body panels in Walton Hills, near Cleveland; a casting plant in Brook Park; and an electrical parts plant in Sandusky. During the 1970s, Ford built a stamping plant at Maumee, near Toledo, and bought a plant at Avon Lake from Fruehauf to finish bodies for vans which were assembled nearby at Lorain. The Fairfax transmission plant was replaced in 1980 by a new facility in Batavia, another eastern suburb of Cincinnati. Ford's vice-president of operations D.S. Harder, in charge of selecting new plant sites, stated that 'taxes were a consideration in Ford's move to Ohio' (Harder 1959).

Most of Ford's other new components plants after World War II were located in southeastern Michigan communities other than Detroit and Dearborn. During the 1950s and 1960s, Ford built plants to stamp body panels in Monroe and Woodhaven and body trim in Utica. Transmissions were made in Livonia, rear axles and suspensions at two plants in Sterling Heights, electrical accessories at Rawsonville, near Ypsilanti, and paint at Mt Clemens. Facilities at Dearborn were expanded to make frames and engines. During the 1970s, a plastic parts plant was built at Mt Clemens. These selections were part of a widespread movement in American industry to relocate production from the older cities to peripheral locations, where it was easier to assemble large tracts of low-cost land accessible to highways.

GENERAL MOTORS' PRE-WORLD WAR II COMPONENTS PLANTS

Like Ford, GM's components facilities at first were highly concentrated in southeastern Michigan. However, whereas Ford clustered production in one city – Detroit and later Dearborn – GM operations were always spread among several Michigan cities, including Detroit, Flint, Lansing, and Pontiac, a legacy of the company's origin as a holding company for a number of autonomous producers. Ford retained production of virtually all components in Michigan until after World War II, but GM soon expanded into other states, primarily through acquisition of independent suppliers.

Documenting the changing distribution of GM's components plants is formidable, because the organizational structure is perpetually changing, not just in recent years but since the company's birth. In periods of expansion and creation of new products, GM reorganizes so that each

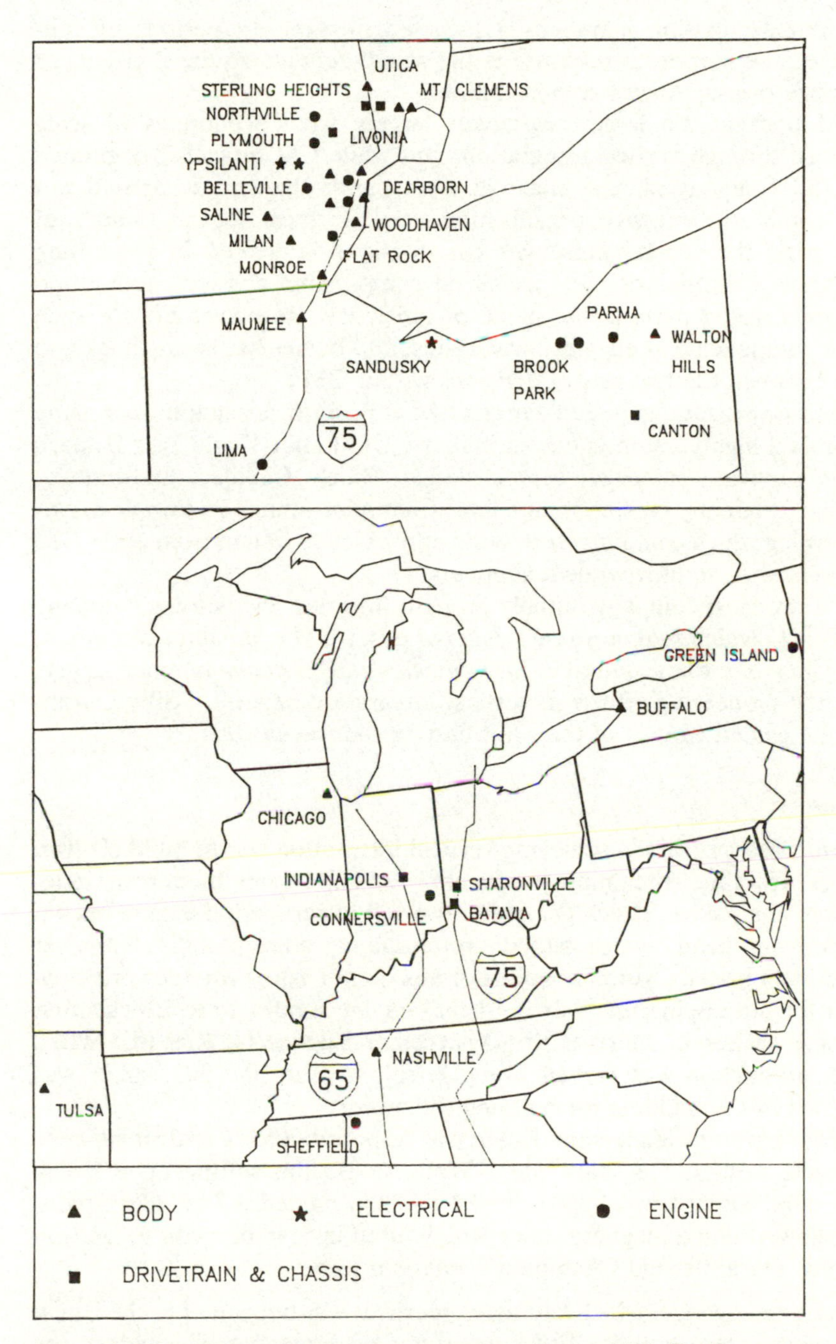

Figure 5.2 Ford Motor Company components plants, 1991. Ford's components production remains highly clustered in southeastern Michigan and northern Ohio

components division maintains a focussed mission. In periods of contraction, the corporate response is the same: reorganize the divisions to minimize overlapping of responsibilities.

GM became the largest carmaker largely from economies of scale achieved through vertical integration. But while GM officials determined that the company should make its own parts, they never figured out who within the massive organization was the most logical builder of each part: the several passenger car divisions interested in promoting distinctive identities or the specialized components divisions committed to standardizing parts in the name of efficiency. Responsibility for such major components as engines, drivetrains, and bodies has swung back and forth between the two sets of divisions within GM.

GM's organizational muddle originated at its birth as a holding company for several highly autonomous carmakers. Until after World War II, each of the surviving passenger car divisions – Buick, Cadillac, Oldsmobile, Pontiac (originally Oakland), and later Chevrolet – made its own decisions concerning which components to build and which to obtain from other GM divisions or from independent suppliers.

The strongest company initially brought into the GM holding company was Buick, which sold more cars than the other GM companies combined, until Chevrolet was acquired in 1918. Buick's early success stemmed largely from the pioneering efforts in vertical integration begun by Billy Durant after he gained control of the Flint-based producer in 1904.

Engines

Durant's first priority in achieving vertical integration was to build his own motors, at a time when most producers were still buying them from independent companies. Buick gambled on a still-experimental engine known as a 'valve-in-head,' which gained a reputation as efficient and reliable yet moderately priced. Automotive historians do not know whether principal credit for developing the 'valve-in-head' engine should go to Buick's first manager Walter L. Marr or to his successor Eugene C. Richard. When Buick production was moved from Detroit to Flint in 1903, space was allocated in the building for building the motor.

Under Durant's leadership, Buick was responsible for a critical advance in engine design, the spark plug. While in Boston setting up a Buick showroom, Durant was approached by a man named Albert Champion, who showed him a magneto, once widely used instead of a coil in ignition systems. Durant found Champion's magneto to be

> a very neat gadget which had much merit. It was not suited to the Buick because at that time the Buick was not a 4-cylinder car. The gadget was well designed and showed good taste. I thought that anyone who could produce that kind of a device might do other worth-while things as well.
>
> (Durant undated)

Durant recalled his conversation with Champion.

> 'Have you a factory?' I asked.
> 'No, just a shop.'
> 'What are you making?'
> 'Magnetoes and spark plugs.'
> 'We do not use magnetoes, but I am interested in spark plugs. Can you make a good one?'
> 'I have just started in that line, but I worked for a number of years with Mr. Renault of Paris, France, and am following his methods which have been most successful.'
>
> (Durant undated)

Champion accepted Durant's offer to relocate to Flint, but he turned out not to be the owner of the company which bore his name. The owner, a man named Stranahan, agreed to sell the company for $2,000, but he refused to part with the name Champion Ignition. Durant told Champion 'I was not interested in the name – I was interested in spark plugs.' But Champion replied, 'I am very much interested in the name. That is my name.' To avoid litigation, GM changed the name from Champion to AC Spark Plug Company, after Champion's initials. AC began in Flint in 1908, in a corner of the Buick building. AC moved to a separate building in Flint in 1912 and has continued production at the same site since then while expanding into other automotive products, such as speedometers and catalytic converters (Durant undated; Weisberger 1979: 140).

The founders of two other GM companies, Henry M. Leland, of Cadillac, and Ransom E. Olds were accomplished builders of gasoline motors for marine and agricultural applications. Cadillac built its own high-quality – although expensive – engines, but Oldsmobile and Oakland, as well as other GM companies, bought motors from independent suppliers. One of the leading engine builders, Northway Motor and Manufacturing Company, was acquired by General Motors and continued to supply Oldsmobile and Oakland until after World War I, when it switched to production of truck engines. Northway was dissolved in the mid-1920s, once GM's final assembly divisions were all able to build their own engines, and its plant in Detroit was transferred to Chevrolet (Pound 1934: 488).

One year after losing control of General Motors in 1910, Durant organized the Little Motor Car Company and the Chevrolet Motor Company to assemble and sell cars. He applied the same strategy of vertical integration he had followed at Buick to his new Chevrolet venture, again developing self-sufficiency in production of motors first and then in drivetrain. The Mason Motor Company was established in Flint to produce engines for both the Little and Chevrolet. In 1913, Durant transferred the name Chevrolet to the Little and placed a redesigned Mason engine in the new model. As a unit of the holding company known as Chevrolet Motor

Company of Michigan, the Mason Motor Company continued to make Chevrolet engines and expanded into axle production; the name Mason disappeared when Chevrolet was acquired by General Motors in 1918.

In addition to the 'home' plant cities, such as Flint and Pontiac, GM also established an important Michigan center at Saginaw for the production of components, especially motors. Saginaw had been home to one of Durant's first acquisitions, the Rainier Motor Company, which built the Marquette, expected to be GM's top-of-the-line luxury model. But when the Marquette proved unprofitable, production was halted, as part of GM's restructuring in 1911 to restore financial health. Rainier's Saginaw plant remained idle until 1917, when trench-mortar shells were built for the army. After the war, the plant built four-cylinder engines for Chevrolet until 1922 and for Oldsmobile until 1923, when the plant was closed (Pound 1934: 489).

When responsibility for engines was transferred to the passenger car and truck divisions during the 1920s, Saginaw remained a center for castings. In 1919, GM built the world's largest grey iron foundry at Saginaw to provide an adequate supply of castings for motors and acquired an independent company, the Saginaw Malleable Iron Company, established two years earlier to provide malleable castings to GM's Saginaw plants. Later that year, GM consolidated all of its Saginaw operations into the Saginaw Products Division, but as output increased and products became more complex, plants were allocated during the late 1920s to several newly created divisions. The Saginaw grey iron foundry was transferred to Chevrolet during the 1920s, while Pontiac, Buick, and Cadillac constructed foundries near their 'home' city plants during the period (Figure 5.3).

Drivetrain

Durant next turned his attention to acquiring suppliers of drivetrain components for Buick. Axles required substantial improvement over rigid ones used in carriages. So that the automobile could travel at high speeds, the axle had to carry power from the engine to the wheels – normally the rear ones in the early days – and permit the wheels to turn at different speeds. One of the leading axle producers was the Weston-Mott Company, which had previously manufactured wheels for bicycles, buggies, and pushcarts in Utica, New York, in the late 1890s, when orders started arriving from automobile companies, begining with the Olds Motor Works in 1900. Two years later, Olds switched to wood wheels, but orders arrived for axles as well as wheels from other companies, including Buick.

In 1904, Durant set out to woo Weston-Mott from Utica to Flint. The axle was proving Durant's biggest headache at Buick, because rail shipments from Utica were not arriving in Flint in a timely manner. Charles Mott, president and general manager of Weston-Mott was reluctant to

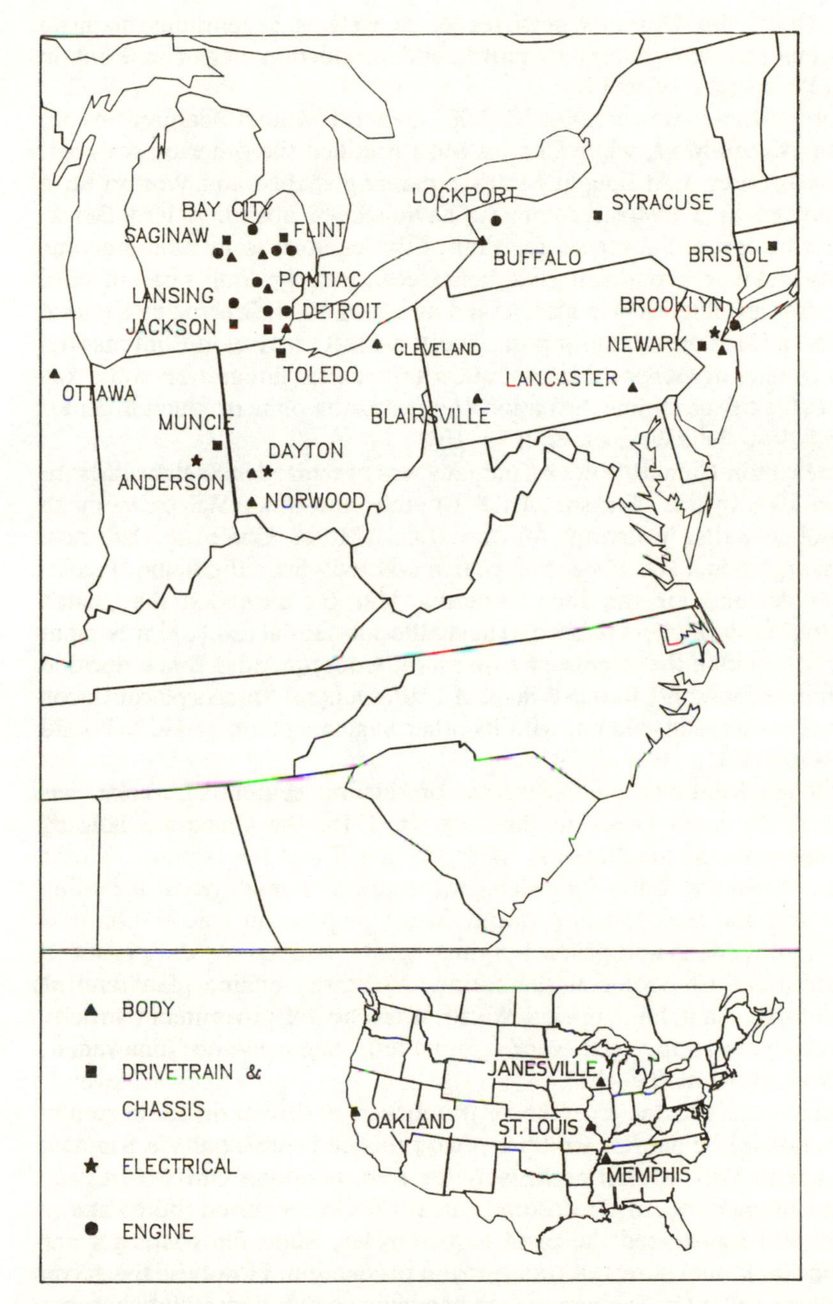

Figure 5.3 General Motors components plants, 1923. Plants were dispersed throughout the southern Great Lakes region, because most were acquired by GM rather than established from scratch
Source: adapted from General Motors annual reports

leave Utica, but Durant's persistence, as well as a generous financial arrangement, induced him to move, and production began in Flint in 1906 (Weisberger 1979: 102).

Shortly after it was created in 1908, General Motors acquired 49 per cent of Weston-Mott, while Charles Mott retained the other 51 per cent. Four years later, GM bought Mott's remaining shares, and Weston-Mott disappeared as a distinct company, having been absorbed into Buick. Despite his low initial regard for Flint, Charles Mott went on to become the city's mayor, prominent civic benefactor, and life-long resident until he died in his nineties. He also served as a long-time director of General Motors and vice-president for parts. Weston-Mott's arrival in Flint marked the beginning of General Motors' quest for vertical integration and a key event in the city becoming the nation's largest automobile production center (Pound 1934: 490; Weisberger 1979: 104).

The Jackson-Church-Wilcox Company was organized in Saginaw, Michigan, in 1908 by J.L. Jackson, E.D. Church, and M.L. Wilcox to make automobile parts, primarily for the Buick Motor Company. Its most successful product was a steering gear marketed under the name 'Jacox'. Buick's demand for the Jacox steering gear far exceeded the plant's capacity, but the owners refused to make the substantial level of investment required to make the necessary expansion. Consequently, Buick decided to purchase Jackson-Church-Wilcox in 1910. General Motors placed Jacox under common management with its other Saginaw plants in 1919 (Pound 1934: 478–479).

Although Flint became its largest production center, Chevrolet was not as clustered as Buick in that city. In 1916, the Chevrolet holding company acquired the National Cycle Company and the National Motor Truck Company in Bay City, Michigan, to produce small parts, including carburetors and dies, and the Warner Gear Company, in Toledo, Ohio, to make transmissions and gears. Forgings, gears, and axles were produced in Detroit for Chevrolet at the former Northway engine plant and at a drop-forge plant built during World War I to fill government orders. Chevrolet powertrain parts were also made in Saginaw and Tonawanda, New York (Pound 1934: 151).

Saginaw also became a center for production of drivetrain components. The National Engineering Company had produced crankshafts in Saginaw beginning in 1907, at first exclusively for Reo, although GM's Northway Motor Company became a customer in 1913. GM acquired the company in 1919 and transferred the plant to Chevrolet, when the passenger car divisions took full control of transmission production. However, the Jacox steering gear plant in Saginaw, which had been acquired by Buick, became a supplier to other GM divisions, as well. When Saginaw Products Division was split up in the late 1920s, the former Jacox facilities formed the nucleus of the Saginaw Steering Gear – and more recently simply Saginaw – Division (Pound 1934: 478–479, 488–489).

After General Motors was transformed from a holding company to a corporation with operating divisions in 1918, most of the newly acquired firms supplied components to more than one of the passenger car divisions. An exception was the Armstrong Spring Company, a Buick supplier located in Flint, where the company originated as a producer of carriage and wagon springs. GM bought Armstrong in 1923 and maintained it as a separate division until 1932, when it became part of the Buick Motor Division (Pound 1934: 484).

More typical was GM's acquisition during the 1920s of the T.W. Warner Company, which was renamed Muncie Products Division, the name of the city in Indiana where the plant was located. Warner, a major supplier of transmissions and steering gears to Chevrolet since 1915, also made transmissions for Oakland, Oldsmobile, and GM Truck. The Muncie plant was taken over by Chevrolet during the 1930s and transferred to the Hydra-matic Division after World War II. In 1990, the plant became part of New Venture Gear, established by GM and Chrysler, the first joint venture among the Big Three producers in US history (Pound 1934: 487–488; Connelly 1990a: 6).

Similarly, GM acquired Brown-Lite-Chapin Company, which produced differentials for automobile gears in Syracuse, New York. The firm originally built bicycle gears but started to supply differentials in 1900. Weston-Mott, which soon became the largest customer for Brown-Lite-Chapin's differentials, acquired a financial interest in the company. Brown-Lite-Chapin passed into GM's hands as part of the Weston-Mott acquisition and became a GM division in 1923. During the 1930s, the Guide Lamp Division took over the plant for production of hubcabs and bumper guards (Pound 1934: 484–485). More recently, the division, known as Inland Fisher Guide has made molded plastic parts at the Syracuse plant.

United Motors acquisition

The United Motors Corporation constituted by far GM's single most important acquisition of an independent components supplier. Billy Durant, never content with only one iron in the fire, established United Motors as a holding company for several components producers in 1916, at the same time he was maneuvering to regain control of General Motors. United Motors functioned as an independent company until 1918, when it became a group within GM (Sloan 1941: 101–104). Heretofore confined almost exclusively to Michigan, GM became a major producer of components in half-a-dozen other states, from Indiana to Connecticut, thanks to the United acquisition.

United Motors expanded GM's self-sufficiency in production of engine and drivetrain components. Major additions to GM included Harrison Radiator Corporation of Lockport, New York, Hyatt Roller Bearing Company of Harrison, New Jersey, and the New Departure Manufacturing

Company of Bristol, Connecticut. Herbert C. Harrison established the company bearing his name in 1910 to build radiators, which minimized overheating, a heretofore serious problem with gasoline-powered motors. Harrison built several plants in Lockport and nearby Buffalo, New York, and during the 1920s operated a facility in Detroit, as well. After GM sold its Frigidaire Refrigerator Division in the early 1980s, Harrison took over production of automobile air conditioners at two plants in the Dayton, Ohio, area.

Hyatt and New Departure brought General Motors control of production of ball and roller bearings, both essential for efficient operation of the drivetrain and chassis. Bearings permit parts to turn freely without rubbing against other ones, thus minimizing friction and maintenance costs. Bearings comprise cages which house either hollow cylinders (in the case of roller bearings) or loose steel balls. Hyatt and New Departure so dominated production of bearings for wheels, steering gears, transmissions, axles, and other components, that they continued to be the most important suppliers to other auto makers even after GM acquired them.

The Hyatt Roller Bearing Company was founded in 1892 by John Wesley Hyatt, who had long held a reputation in other fields, most notably the development of celluloid during the 1860s. Hyatt initially invented roller bearings during the 1880s to install in a mill for crushing sugar cane; production of bearings began in 1896 for early automobiles, including early Olds and Ford models. Production was concentrated until the late 1960s at Harrison, New Jersey, near Newark.

Hyatt was a better inventor than financier, and the company struggled until the father of one of the draftsmen provided fresh operating capital with the proviso that his son become general manager. The young general manager – Alfred P. Sloan – became president of United Motors in 1917, when it took over Hyatt, and came into GM as a vice-president as part of the United merger (Cray 1980: 149–150).

The New Departure Manufacturing Company was originally established in 1888 to produce a bicycle bell, which was so radically different in design that the inventors, Edward D. and Albert F. Rockwell, called it the 'New Departure Bell.' The Rockwell brothers also built hand brakes for bicycles before applying themselves to ball bearings for automobile axles beginning in 1907. In addition to the plant at Bristol, production took place at Meriden, also in Connecticut (Pound 1934: 474–476). New Departure and Hyatt were consolidated into one division in 1965, when the Meriden and Harrison plants were closed. New Departure Hyatt in turn merged with Delco-Moraine, the brake-making division, in 1989.

United Motors gave GM another drivetrain supplier, the Jaxon Steel Products Company, which made wheels and wheel rims. The company, originally called Jackson Rim, was established by a relative of Charles Mott in Jackson, Michigan, and four years after production had begun, in 1917, it was taken over by the Perlman Rim Corporation of New York.

Bucking the trend towards vertical integration, GM decided to sell Jaxon to the Kelsey-Hayes Wheel Corporation in 1930 (Pound 1934: 486).

Electrical accessories

Purchasers of early automobiles received little more than chassis, engine, transmission, and seat, unless they paid extra for such accessories as tires, a horn, gauges, lamps, and a top. While these accessories soon became standard features on all automobiles, other options were introduced, typically only on luxury models at first. The acquisition of United Motors propelled GM's pursuit of vertical integration into production of accessories, especially electrical ones. Two makers of electrical parts joined GM through United Motors – the Dayton Engineering Laboratories Company, and the Remy Electric Company. From these two companies can be traced a complex web of GM accessory divisions, perpetually being reorganized as demand for particular products increases or wanes.

The electric self-starting ignition, one of the first important accessories, was largely responsible for making Dayton, Ohio, the largest center for production of automobile components outside of Michigan. Before the invention of the self-starting ignition, starting a motor was difficult, requiring 'the strength of Ajax' to turn the hand crank on the front of the car, 'the cunning of Ulysses' to set the controls which advanced the spark and regulated the flow of fuel, and 'the speed of Hermes' to run back and forth between the crank and the controls (Pound 1934: 271–272). Turning the crank was too hard for most women and caused many injuries.

The Dayton Engineering Laboratories Company, organized in 1909, eased the difficulty of starting cars. The Laboratories' electric self-starter was first installed as standard equipment in 1912 on Cadillacs and within the next few years on most cars, with the notable exception of the Ford Model T. The Laboratories – by then generally known by its acronym Delco – was sold to United Motors in 1916 and joined GM two years later. Delco's director Charles F. Kettering was appointed head of the newly established General Motors Research Laboratories in 1920.

Anderson, Indiana, ranked second to Dayton as GM's most important production center outside of Michigan, thanks to Remy, which had been founded by B.P. and Frank Remy in 1895 to make magnetos for a number of early producers, including Buick. Remy's business suffered in 1912, when Buick and other producers switched to Delco's new self-starting ignition system. Even after Remy and Delco both joined United Motors and later GM the two companies engaged in marketing battles, with Delco holding the edge because it ran more aggressive advertisements and built more attractive-looking parts.

Remy prospered during the 1920s under the leadership of general manager Charles E. ('Engine Charlie') Wilson, later president of General Motors and Secretary of Defense under Eisenhower (and the man who coined the phrase 'what was good for our country was good for General Motors'). GM had bought the Klaxon Company – originally the Lovell-McConnell Manufacturing – of Newark, New Jersey, the nation's leading producer of automobile horns. In fact, the name 'Klaxon' – derived from the Greek for 'shriek' – had become synonymous with the word 'horn' in the United States. After GM acquired the company in 1920, production was moved from Newark to nearby Bloomfield, New Jersey (Pound 1934: 465). Then, in 1924, Charlie Wilson convinced GM officials that Remy could make Klaxons for less money; the New Jersey plant was closed, and production was transferred to Anderson. Further expanding Anderson's importance, production of Delco's ignitions was transferred there in 1927, and the division was renamed Delco-Remy ('Anderson tales: From Perry Remy to Delco Remy', *Automotive News*, 1983: 206–207).

Delco-Remy opened other battery-making plants at New Brunswick, New Jersey, and Muncie, once the home of the Interstate Motor Company and GM's short-lived Sheridan automobile model built in 1920 and 1921. Delco-Remy was able to make a profit building Delco batteries for installation in new GM cars at a time when competitors were selling them at a loss to other new car producers so that customers who needed replacements would be attracted to their brand. Delco-Remy also temporarily ran a plant which made radios in Kokomo, Indiana, acquired by General Motors in 1936 from the Crosley Manufacturing Company. The plant was then turned over to the newly created Delco Radio Division; after World War II, the division was known as the Delco Electronics.

GM officials were so concerned that Anderson's labor force and infrastructure could not cope with Remy's rapid growth, that they came down from Detroit to inspect the city. Wilson organized an elaborate ceremony to alleviate their fear of further expanding in Anderson ('Anderson tales: from Perry Remy to Delco-Remy', *Automotive News* 1983: 207). In 1929, a year after GM acquired Guide Lamp, production of headlamps was relocated from Cleveland to Anderson. The same year, GM bought the Bu-Nite Piston Company and moved its operations from Indianapolis to Anderson.

In 1930, production of starting, lighting, and ignition systems made by the North East Electric Company was transferred to Anderson from Rochester, New York. The previous year GM had acquired North East, which was founded by Edward A. Halbleib in 1908 as the Rochester Coil Company. The Rochester operation, which was renamed the Delco Appliance Division, continued to build car-heating blowers, as well as nonautomotive products transferred from Delco-Light in Dayton. A second GM division, Rochester Products, was established in the city immediately before World War II to build a variety of automobile accessories, including instrument panels, speedometers, generators, and starters, for distribution to GM's eastern assembly plants (Pound 1934: 461–462; Rowand 1983: 191).

Dayton hardly suffered when production of ignitions was transferred to Anderson. Typical of GM's attitude towards controlling supplies of parts, Delco-Remy was asked to design new shock absorbers. When it was determined that the design infringed on a patent held by the Lovejoy Shock Absorber Company of Boston, Massachusetts, GM simply bought Lovejoy so that Delco-Remy could produce the shocks legally ('Anderson tales: from Perry Remy to Delco-Remy', *Automotive News* 1983: 204). When production of ignitions was moved to Anderson, the Dayton plant made Lovejoy shock absorbers instead and was renamed the Delco Products Division. Delco Products soon acquired a second product, when GM bought the Sunlight Electrical Company in 1933 and relocated production of its small motors from Warren, Ohio (Rowand 1983: 191).

Delco Products was responsible for further expansion in Dayton. Production of hydraulic brakes, developed by the division in 1934, was transferred to a separate plant, and the Delco Brake Division was established in 1936. The following year, a plant was built next door for Moraine Products, which had been organized by Kettering in 1923 to make a self-oiling power metal bearing, which was marketed under the name Durex. In 1942, Delco Brake was merged into Moraine Products, but in 1960 the name was changed to Delco Moraine in order to regain the trade name for the brakes.

Meanwhile, back in 1916, the Dayton Laboratories had launched a second successful product, the Delco Farm Light System, marketed as Delco-Light. Delco-Light provided a 32-volt home-lighting system, purchased mainly by farmers living in rural areas not yet electrified. Again, production was moved from Dayton, this time to Rochester, in 1930, as part of the new Delco Appliance Division. However, another early Delco-Light product was an electric refrigerator called Frigidaire. Frigidaires had been originally built by the Guardian Frigerator Company, organized in Detroit in 1916 by the Murray Body Company, one of the leading builders of automobile bodies. Billy Durant bought Guardian for General Motors, renamed it Frigidaire, and relocated production from Detroit to the Delco-Light plant in Dayton, because Kettering had taken an interest in it. Frigidaire sales increased so rapidly that it was spun off as a separate division in 1926; Frigidaire production remained in Dayton, while Delco-Light moved to Rochester (Pound 1934: 304–309; Rowand 1983: 183).

Kettering was closely associated with other prominent Dayton inventors, notably the Wright Brothers, who although staged their first successful flight at Kitty Hawk, North Carolina, did most of their research and construction in their native Dayton. The Dayton-Wright Airplane Company, founded by Kettering and the Wright Brothers, was acquired by General Motors, and several of its buildings were used for manufacturing of various automotive components. Sales of one of the components – wood steering wheels – grew so rapidly in the early 1920s, that a separate division was organized to manufacture them in 1923, known as the Inland Manufacturing Company. As products made from wood were becoming obsolete by the late 1920s, Inland began to use rubber instead for steering wheels and expanded into other rubber-based accessories, such as running boards and battery containers (Pound 1934: 472). A second Inland plant was opened in 1938 in Clark Township, New Jersey, to make rubber products, taking advantage of a port location to use imports.

GM's last major acquisition of a parts supplier before World War II was Packard Electric in 1932. J.W. and W.D. Packard had started making incandescent lamps and transformers in Warren, Ohio, in 1890, and introduced the Packard automobile in 1898. The company started to make wire cables because they were unable to find a reliable supplier. Unwilling to relocate to Detroit where the automotive industry was clustering, the Packards sold their auto production venture to Henry Joy and their lamp department to General Electric, while selling wires to a number of auto makers (Pound 1934: 477–478).

GM built its powertrains in Canada for its Canadian cars during the 1920s. The company constructed a plant at Walkerville to make engines, front axles, and other parts. McKinnon Industries, in St Catherines, was acquired in 1929 to make axles for GM's Canadian vehicles. For a time, McKinnon also operated a plant.

Bodies

The body was the last major component to be controlled by the major auto makers. In early automobiles passengers rode in open seats exposed to the elements, but after 1910 coach builders began to design closed bodies, like other accessories initially for luxury models. With the introduction of mass-production techniques, even the low-priced Ford Model T had a closed body by the 1920s. Higher-priced models were then distinguished by their customized hand-crafted bodies. However, by the late 1930s, only Cadillac and Chrysler's Imperial model were still fitted with custom bodies. The adoption of the all-steel body during the 1930s hastened the decline of independent body builders, because welded construction protected against deterioration and squeaks as effectively as expensive custom bodies. Lower demand for luxury cars during the Depression also contributed to the decline of customized body builders (Pfau 1971: 144).

Amesbury, Massachusetts, a nineteenth-century carriage production center, attracted several automobile body builders, especially suppliers of customized bodies for luxury models, many of which were assembled in northeastern states until the 1930s. Biddle & Smart Company, a long-established carriage-builder, became Amesbury's leading body supplier, by selling to Hudson, the best-selling medium-priced model during the 1920s, as well as to a number of high-priced limited-volume producers. However, Amesbury's body-building industry collapsed during the Depression, when many luxury models ceased production and Hudson switched to low-cost bodies built in Detroit (Pfau 1971: 146).

Like other suppliers, the largest body builders soon clustered in southern Michigan. During the 1920s, the three largest – Fisher Body Company, Briggs Manufacturing Company, and Murray Corporation – were all based in Detroit; the fourth largest, Hayes Body Company, was located in Grand Rapids. Ford was the first large car maker to build most of its own bodies, prior to World War I. Detroit-based C.R. Wilson Body Company had been the largest body producer on the strength of sales to Ford. However, when Ford began to build its own bodies, Wilson's sales declined, and the company was taken over in the 1920s by Murray. Murray soon ceased body production but continued to supply other automobile components (Pfau 1971: 146).

In 1910, Cadillac decided to place a large order for closed bodies with a 2-year-old firm established by two brothers who had worked for Wilson, Fred and Charles Fisher. GM acquired three-fifths of the Fisher Body Company in 1919 and agreed to buy all of its bodies from the company for ten years. The six Fisher brothers – four others had joined Fred and Charles – continued to manage the division even after GM purchased the remaining shares in 1926. The phrase 'Body by Fisher' became a prominent feature of GM's advertising and was affixed to all of the company's cars.

Fisher acquired the Fleetwood Metal Body Corporation, based in Fleetwood, Pennsylvania, in 1925. Fleetwood, which specialized in customized bodies for higher-priced models, moved to a plant on Ford Street in Detroit during the 1930s in order to supply Cadillac's Clark Avenue assembly plant, 4 km away. The Fleetwood and Clark plants were closed in 1987. The name 'Fleetwood' survives as the name of a Cadillac model.

Briggs continued in business as an independent company even after World War II, because Chrysler bought all of its bodies from Briggs instead of making its own. LeBaron Incorporated, acquired in 1926 to obtain the services of the customized builder's design staff, was retained as a division of Briggs. In recent years, the name has been applied to some of Chrysler's higher-priced models. Briggs workers earned a reputation for untrammeled militancy and were known as 'the Dead-End Kids' (Moritz and Seaman 1981: 58).

Briggs also supplied bodies to another Detroit-based car maker after World War II, Hudson, until it merged with Nash to form American Motors and production was moved to Kenosha. Nash, for its part, back in 1936 had acquired its major body supplier, the Seaman Body Company, located in nearby Milwaukee, Wisconsin (Pfau 1971: 146). The Milwaukee plant provided stampings for AMC's Kenosha assembly plant until 1988, when both facilities were closed after Chrysler's takeover.

Fisher Body was the first major supplier to adopt a policy of deliberately locating large new components plants outside the southern Great Lakes region; in fact, prior to World War II, Fisher was the only notable exception to the highly clustered spatial pattern. As bodies were very bulky, Fisher found that hauling them by truck to distant assembly plants was more costly than piling steel, glass, aluminum, rubber, copper, plastics, fabrics, and other materials into rail cars. Just as important, eliminating long-haul shipments across the nation's poorly graded highway system minimized wear-and-tear and potential for damage to the bodies (Kuhn 1986: 149).

Consequently, Fisher constructed stamping and body plants near GM's final assembly plants, preferably on adjacent sites. By the late 1920s, Fisher was building bodies in every North American city where GM was assembling automobiles. Fisher's plant in Pontiac, built in the late 1920s, was a showcase: the first to be connected by overhead closed bridge to the final assembly plant, thus eliminating the need to haul bodies by truck even short distances, still necessary everywhere else (Fleming 1983: 116).

Fisher Body developed its own set of suppliers of parts and raw materials. To assure supplies of glass, in 1920 Fisher bought a controlling interest in the National Plate Glass Company, which in turn had acquired three large and well-known glass factories dating from approximately 1900, including Columbia Plate Glass Company in Blairsville, Pennsylvania, Federal Plate Glass Company in Ottawa, Illinois, and Saginaw Plate Glass Company in Saginaw (Pound 1934: 292).

Fisher controlled 25,000 hectares of timberlands in northern Michigan and 65,000 in the southern states of Louisiana and Arkansas, plus an additional amount in the Pacific Northwest, and operated woodworking plants in Memphis, Tennessee, and Seattle, Washington, and sawmills in Ferriday and Wisner, Louisiana. Fisher also operated plants for stampings in Cleveland, Ohio, and Grand Rapids, Michigan, and trim in Ionia, Michigan. Ternstedt Manufacturing Company, which made body parts, such as window cranks, at a plant in Detroit, was acquired by Fisher. A second Ternstedt plant opened in Trenton, New Jersey, during the 1930s to make body hardware (Pound 1934: 295).

General Motors was involved in processing other raw materials, as well, although never as intensively as Ford. The Lancaster Steel Products Corporation, in Lancaster, Pennsylvania, produced steel products for stamping, wiring, and special shapes, primarily to General Motors. GM acquired the plant in 1919 but sold it to the Armstrong Cork Company in 1927 (Pound 1934: 487).

GM's distribution prior to World War II

At the outbreak of World War II, General Motors facilities for producing components were still highly clustered in the southern Great Lakes region. Engine components were the most highly centralized, with most production at the 'home' plants of the passenger car divisions in Flint, Detroit, Pontiac, and Lansing. Drivetrain and chassis components were more dispersed across the northeast, as a result of acquisition of independent suppliers. Ohio and Indiana were centers for production of electrical accessories, again because of the growth of once-independent suppliers. Body plants were the most dispersed, primarily as a result of Fisher's policy of locating near branch assembly plants.

GM's leading production centers in Michigan included the 'home' plant cities of Flint, Detroit, Pontiac, and Lansing, plus the Saginaw/Bay City area. Outside of Michigan the leading centers were Dayton, Ohio; Anderson and Indianapolis, Indiana; upstate New York, including Buffalo, Lockport, Rochester, and Syracuse; northern New Jersey, including Clark Township, Harrison, New Brunswick, and Trenton; and Bristol and Meriden, Connecticut (Figure 5.4).

Other centers for GM components production included Muncie, Indiana; Grand Rapids and Ionia, Michigan; and Cleveland, Toledo, and Warren, Ohio; Outside the southern Great Lakes region, Fisher Body operated plants in Memphis, Tennessee, and Seattle, Washington, plus facilities adjacent to final assembly plants in Oakland, California; Atlanta, Georgia; Baltimore, Maryland; Kansas City and St Louis, Missouri; Tarrytown, New York; Norwood, Ohio; and Janesville, Wisconsin.

Durant started GM's dispersion around the southern Great Lakes through his policy of acquiring independent producers, although he did 'overload' the Flint area with too many components divisions compared to the available transport facilities and labor force (Kuhn 1986: 147–148). Under Sloan's leadership, GM turned the dispersed locational pattern inadvertently created by Durant into policy. Sloan stated in the 1935 GM annual report to stockholders that 'the soundest policy, both economically and socially, is to distribute [production] among as many different communities as is practically possible' (General Motors Corporation 1935).

Dispersing plants minimized problems of labor unrest and transport congestion. The development of a production center at Buffalo in the late 1930s was especially influenced by GM's desire to reduce its dependency on plants in southeastern Michigan, especially Flint (Kuhn 1986: 148). By striking at a handful of key components plants, GM workers were able to halt the company's entire output of automobiles in 1937.

GM's components plants after World War II

GM components production expanded rapidly after World War II, but in contrast to earlier periods, most of the growth came through construction of plants in new cities by existing components divisions rather than through acquisition of independent suppliers. One GM habit carried over from before the war – the constant reorganization of the components divisions.

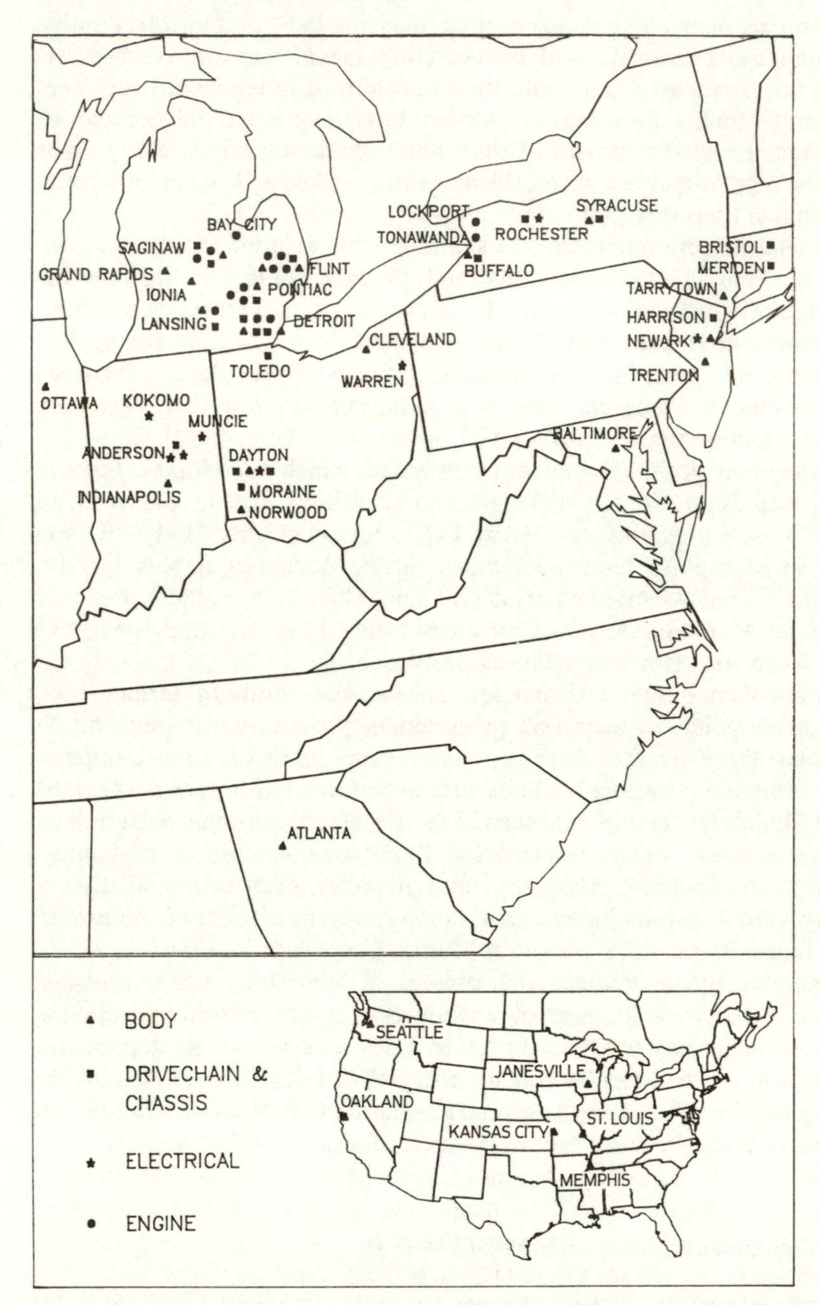

Figure 5.4 General Motors components plants, 1939. The distribution was more dispersed than in 1923, primarily because Fisher Body operated facilities near branch assembly plants
Source: adapted from Sinsabaugh (1940)

The most massive reorganization came in 1965, when functions previously performed by different components and passenger car divisions were combined into new divisions. By the late 1960s, a new locational model had emerged for the production of GM components, reflecting for the first time a spatial division of labor within the United States.

Before World War II, most GM parts were produced in the southern Great Lakes; body parts, the most notable exception, were stamped near the branch assembly plants to save shipping – not labor – costs. After the war, production of GM engine and drivetrain components, which required relatively skilled workers, remained clustered in the southern Great Lakes region. However, much of GM's electrical components manufacturing was relocated to the south, because the tasks could be performed by low-skill workers willing to work for lower wages than in the north. On the other hand, at the same time, production of body parts was recentralized in the southern Great Lakes, because most assembly plant operations returned to the area.

Diffusion within the southern Great Lakes region

Production of engines has remained highly clustered in the southern Great Lakes, despite reorganizations which transferred responsibility for building engines from the individual passenger car divisions. Instead of each passenger car division producing all of the engines at its 'home' plant, each engine plant produced a particular size of engine to be shared by more than one car model. In addition to the former 'home' engine plants originally operated by Buick in Flint, Oldsmobile in Lansing, Pontiac in Pontiac, and Chevrolet in Flint, new facilities were built at Livonia (originally for Cadillac) and Delta Township near Lansing. Outside of Michigan, the only passenger car engines were built at a former Chevrolet facility in Tonawanda, New York, and at St Catherines, Ontario (Figure 5.5).

The biggest change in drivetrain components after World War II was the diffusion of automatic transmissions. The 1940 Oldsmobile introduced an automatic transmission called Hydra-matic, described as 'the last major innovation in automotive technology' (Flink 1988: 208). The Hydra-matic was built by a newly created Detroit Transmission Division, operating out of a former Fisher body plant in Detroit. When automobile production resumed after the war, Detroit Transmission constructed a new plant in the Detroit suburb of Livonia, but the facility burned shortly after it opened in 1953. Transmission production was quickly relocated to a plant in nearby Willow Run, built by Ford in 1940 for aircraft production and occupied for a time after the war by Kaiser.

Typically, Buick developed its own automatic transmissions, called Dyna-flow, while Chevrolet had its own, called Powerglide. During the 1960s, transmission production was consolidated into a new division, called Hydra-matic, which enjoyed a high name recognition. The new division

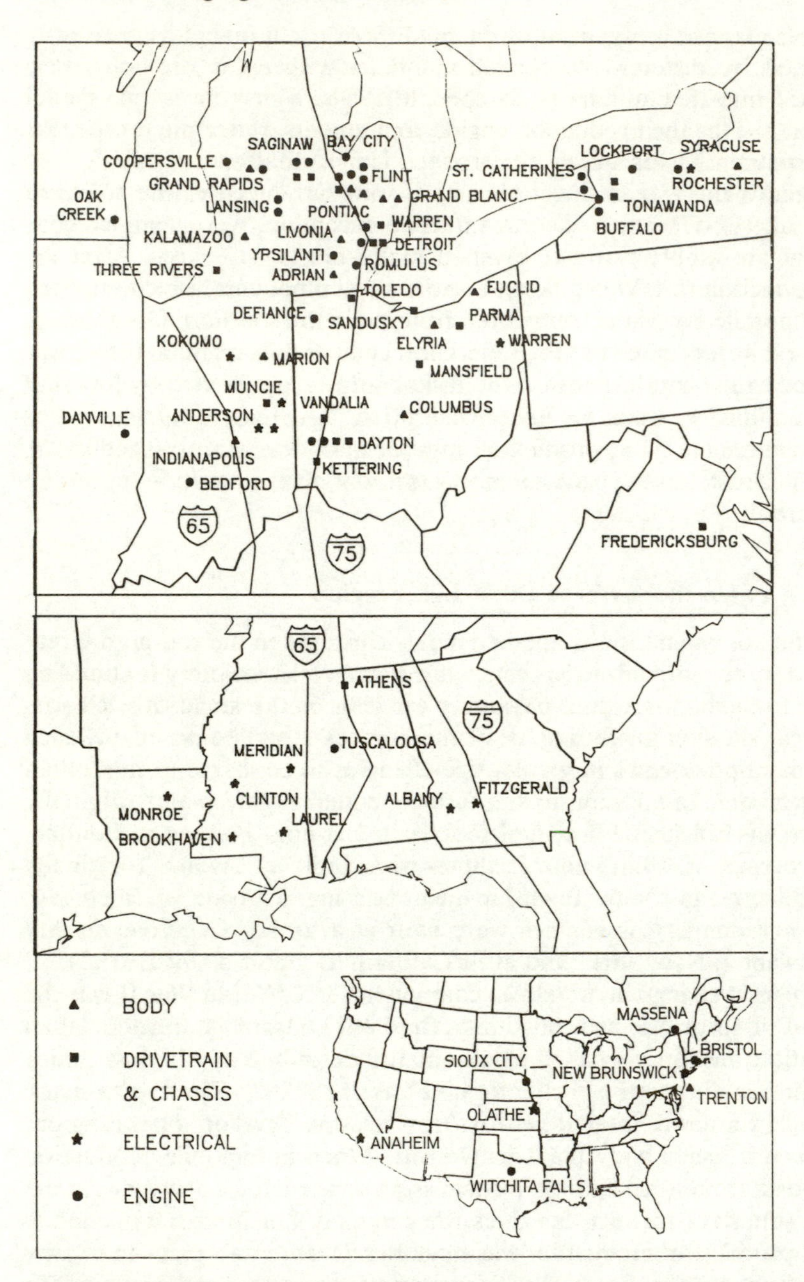

Figure 5.5 General Motors components plants, 1991. Production of engine and drivetrain components has remained highly clustered in and near southeastern Michigan. GM's electrical components are more likely to be produced in the south, whereas production of body parts, such as stamping operations, have been centralized in the southern Great Lakes

took over transmission plants run by Buick in Flint and Chevrolet in Toledo, as well as new plants built in 1961 at Warren, Michigan, and in 1977 at Three Rivers, near St Josephs, Michigan.

Given the concentration of engine and drivetrain production in the southern Great Lakes region, the production of aluminum and iron castings for engine blocks, transmissions, and chassis systems similarly clustered there. In 1946, GM created the Central Foundry Division, which included the Saginaw Malleable Iron plant, a malleable iron plant in Danville, Illinois, built during World War II, and a small nineteenth-century facility in Lockport, New York, transferred from the Harrison Radiator Division but closed in 1948. Central Foundry built a new facility at Defiance, Ohio, to supply grey iron castings to Oldsmobile, and in 1958 took over an aluminum casting plant in Bedford, Indiana, originally built in 1943 to supply aluminum cylinder-head castings for aircraft engines assembled by GM's Delco-Remy Division during World War II.

Meanwhile, Chevrolet was building its own foundries after World War II, including a second grey iron foundry at Tonawanda, New York, to supplement its Saginaw facility, and an aluminum foundry at Massena, New York, to take advantage of low-cost hydroelectric power along the St Lawrence River. But as the passenger car divisions now share engines, they have given up responsibility for producing their own castings. Cadillac's Detroit foundry closed in 1963, Buick's in Flint in the late 1970s. Central Foundry took over the foundry at Pontiac in 1977, at Massena in 1978, and at Saginaw and Tonawanda in 1983. Faced with excessive capacity, Central Foundry closed the Tonawanda and Pontiac plants during the 1980s.

Other powertrain work remained predominantly in the southern Great Lakes, as well. GM's various bearings divisions were merged, beginning in 1965, when the two venerable east-coast-based New Departure and Hyatt divisions combined to form New Departure Hyatt. Headquarters and a plant for the new division were built in Sandusky, Ohio, closer than the east-coast plants to the main customers, GM's other drivetrain divisions. Two of the three northeastern plants – Harrison and Meriden – were closed, leaving only the facility at Bristol, in addition to Sandusky. In 1989, New Departure Hyatt combined with Delco-Moraine, which also made bearings, as well as brakes, to form Delco Moraine NDH. Before the merger with New Departure Hyatt, Delco-Moraine had expanded brake production from its Dayton plant to Saginaw.

Harrison Radiator, which makes the company's heating and air conditioning products, remained for the most part in the southern Great Lakes region, as well. The major change was to take over the production of air conditioner compressors from Frigidaire, when GM sold the division in 1981. Two former Frigidaire plants in Dayton were taken over by Harrison to make the compressors.

Changes at other powertrain divisions were more modest. The Saginaw Division took over axle plants operated by Chevrolet in Detroit since the 1920s and in Buffalo since 1937. Rochester Products, which began to specialize in carburetors and fuel systems, expanded production from Rochester to Grand Rapids, Michigan. AC moved into similar products in the 1970s, building catalytic converters at a plant opened in 1973 at Oak Creek, Wisconsin, near Milwaukee. The two divisions were combined to form AC Rochester in 1989.

Delco Appliances was combined with Delco Products in 1965 to reduce costs. The division began to specialize in chassis rather than electrical components, especially suspension systems, although the division continued to make electrical parts, such as wipers, at Rochester. Bumpers and suspension systems were made at plant originally built by Chevrolet in Livonia and Flint, while the Dayton plant made shock absorbers. In 1989, the division took over two former Inland Division plants in the Dayton area which had made rubber hoses.

Southern plants

Meanwhile, the rapidly expanding electrical divisions broke the long-standing concentration of components production in the southern Great Lakes region. Attracted by the availability of unskilled workers willing to shun unions and work for relatively low wages, the electrical accessory divisions opened ten plants in the south and west, primarily during the 1970s. Delco Remy built batteries at Anaheim, California; Olathe, Kansas; Fitzgerald, Georgia; and Laurel, Mississippi; and starting motors and other electrical components at Meridian, Mississippi, and Albany, Georgia. Packard assembled wire harnesses in Brookhaven and Clinton, Mississippi, while Guide Lamp – which became part of the Inland Fisher Guide Division in 1989 – made headlamps at Monroe, Louisiana.

Powertrain divisions also dabbled with relocating production to the south during the 1970s. Plants were opened to make gears by Saginaw at Athens, Alabama, air filters by AC at Wichita Falls, Texas, clutches by Delco-Moraine at Fredericksburg, Virginia, and carburetors by Rochester Products at Tuscaloosa, Alabama. (The Tuscaloosa plant was transferred to Harrison Radiator in 1989.) However, these plants did not represent the 'wave of the future' in powertrain production, as most engine, transaxle, and chassis production remained firmly entrenched in the southern Great Lakes region.

Reinforcing the fact that GM was relocating only selected low-skill tasks to the south, body operations reconcentrated in the southern Great Lakes. Traditionally, the Fisher Body division erected body plants around the country and hauled finished bodies by truck to nearby branch assembly plants, where they were bolted onto the chassis. However, the adoption of the unitized body frame modified the process. The unitized frame was built at the assembly plant and door, hood, and other panels were hung on it. With the integration of body and final assembly operations, Fisher Body lost its responsibility for building bodies near the final assembly plants to the Assembly Division.

As Fisher Body was now in the business of supplying the assembly plants with less bulky body panels, interior trim, and other body parts, rather than complete bodies, production facilities were logically concentrated in the southern Great Lakes to minimize shipping costs. Fisher Body opened a half-dozen stamping facilities to make door and other body panels in the southern Great Lakes region in the late 1940s and 1950s, including Marion, Indiana; Willow Springs, near Chicago, Illinois; Grand Blanc, Michigan; Hamilton and Mansfield, Ohio; and McKeesport, near Pittsburgh, Pennsylvania. During the same period, Fisher opened interior trim plants in Grand Rapids and Livonia, Michigan, and Euclid, Ohio.

Fisher Body lost its distinct identity during the 1980s, first merging with Guide Lamp and then with Inland to form the Inland Fisher Guide Division. Guide Lamp had also built new plants for production of body hardware during the late 1940s and 1950s in the southern Great Lakes region, including Columbus and Elyria, Ohio, as well as Flint. Inland, for its part had opened three new plants in Michigan, in 1952 at Grand Rapids to make soft trim, in 1954 at Livonia to make interior trim, and in 1977 at Adrian to make plastic parts.

CHRYSLER'S COMPONENTS PLANTS

Chrysler never came close to the level of vertical integration achieved by GM and Ford. Even after World War II, most of Chrysler's components came from outside suppliers. During the 1960s, the company substantially increased its ability to produce parts, but many of these plants were closed in the late 1970s and early 1980s to help stave off bankruptcy.

Chrysler's initial strength came from the quality of its engineering. The first Chrysler model was powered by a moderately priced high-compression six-cylinder engine developed by Fred Zeder, Owen Skelton, and Carl Breer. When Walter Chrysler was in charge of Willys-Overland in the early 1920s, he hired the three engineers to develop a new engine at a Willys plant in Elizabeth, New Jersey. But in 1922, the plant was sold to Durant, and the three engineers set up a consulting firm. Walter Chrysler, now in charge of Chalmers, lured the three to Detroit to build their engine for a new model which became known as a Chrysler rather than a Chalmers.

At first, Chrysler had to purchase nearly all of its components from outside suppliers, but the acquisition of Dodge in 1928 gave Chrysler the large Dodge Main complex in Hamtramck, which included a forge, foundry, and other parts-making operations, in addition to a final assembly line. Prior to World War II, most of the in-house parts for all of Chrysler's models came from Dodge Main, although stamping operations were also performed at a plant on Kercheval Avenue and at the Wyoming Avenue DeSoto assembly plant. Chrysler's only major components plant outside of Detroit prior to World War II was at Kokomo, Indiana, where transmissions were produced. Small parts were also made during the 1930s at a plant in New Castle, Indiana, where suspension components are now manufactured.

Bodies were purchased primarily from Briggs Manufacturing Company, which was acquired in 1953 for $75 million (Moritz and Seaman 1981: 56). The Briggs 'home' plant on Mack Avenue in Detroit provided stampings for Chrysler's nearby Jefferson Avenue assembly plant, while a second former Briggs facility, at Eight Mile Road and Outer Drive, supplied stampings to the Plymouth assembly plant on Lynch Road. Stamping operations were terminated at both the Mack Avenue and the Eight Mile Road plants in the early 1980s, during Chrysler's first brush with bankruptcy.

Chrysler's most famous body prior to World War II was the 1934 Airflow installed on DeSoto and Chrysler models. The Airflow body provided more structural integrity and a roomier, more comfortable passenger compartment. Most prominently, the rounded aerodynamic shape marked a dramatic contrast to the boxy styles which predominated at the time. The Airflow proved to be unpopular with consumers in the mid-1930s, but by the early 1940s other manufacturers had successfully introduced rounded bodies. Fifty years later, after cars had become boxy again, Ford built much of its success on being the first US firm to reintroduce aerodynamic bodies.

During the 1960s, Chrysler substantially expanded its components production. The company opened engine plants at Mound Road in Detroit and Trenton, Michigan, an aluminum casting plant at Kokomo, and a foundry at Indianapolis; an engine plant at Kenosha was also inherited from American Motors (Figure 5.6).

INDEPENDENT SUPPLIERS

Mapping the spatial distribution of parts suppliers at one point in time, let alone changes, is a formidable task. Secondary sources of data provide limited assistance and in fact can offer a distorted view of the actual distribution. The alternative method is to generate names of suppliers, but the number of firms supplying the automotive industry is staggering: General Motors alone receives bills from 40,000 vendors.

Standard Industrial Classification

At first glance, it would seem that the changing distribution of automotive suppliers could be determined through Standard Industrial Classification

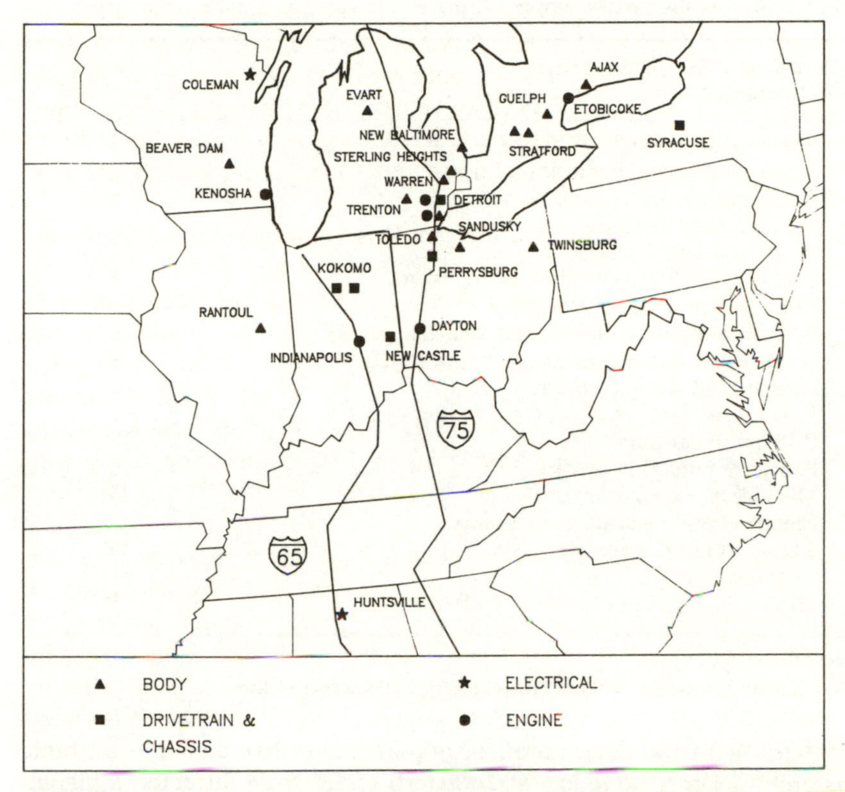

Figure 5.6 **Chrysler Corporation components plants, 1991. Chrysler's production facilities were more clustered than those of the other US automakers**

(SIC) codes. SIC 3714, titled 'Motor Vehicle Parts and Supplies,' would appear to be the logical classification for most manufacturers of automotive components. The US Bureau of the Census publishes *County Business Patterns*, which show the numbers of employees and firms for four-digit SIC codes, including 3714, for every state and county in the United States. Because *County Business Patterns* are published annually, changes in the distribution of SIC 3714 can be tracked.

In reality, SIC 3714 offers an incomplete and inaccurate impression of the distribution of automotive suppliers. One problem is that SIC 3714 includes two types of producers. One group makes components that are installed on new vehicles by the car makers; these suppliers are known as original equipment manufacturers (OEM). However, SIC 3714 also includes producers of replacement parts for older vehicles, known as aftermarket sales. More than half of the firms designated SIC 3714 supply the aftermarket rather than OEM.

Aftermarket and OEM suppliers do not have the same spatial distributions. Aftermarket suppliers are distributed around the country in

Table 5.1 Automobile parts supplier firms by two-digit SIC code, early 1980s

Standard Industrial Classification Code	*Description*	*Per cent of all suppliers*
22	Textile mill products	1
23	Textile products made from fabrics	1
24	Lumber and wood products	a
25	Furniture and fixtures	a
26	Paper and allied products	1
27	Printing, publishing, and allied industries	1
28	Chemicals and allied products	5
29	Petroleum refining and related industries	1
30	Rubber and miscellaneous plastics products	9
31	Leather and leather products	a
32	Stone, clay, glass, and concrete products	2
33	Primary metal industries	19
34	Fabricated metal industries	16
35	Machinery, except electrical	23
36	Electrical and electronic machinery	5
37	Transportation equipment	14
38	Instruments	1
39	Miscellaneous manufacturing industries	1

a = less than 0.5 per cent

Source: adapted from Kamath and Wilson (1983) and surveys of firms

accordance with the distribution of population rather than the location of assembly plants. Southern and western states have attracted a higher percentage of aftermarket suppliers in recent years, because population has increased relatively rapidly in these areas.

Even if allowance were made for the large number of aftermarket suppliers, SIC 3714 does not adequately reflect the diversity of OEM suppliers. According to a University of Michigan study (Kamath and Wilson 1983), car makers buy supplies from firms represented in eighteen of the twenty two-digit SIC codes assigned to manufacturing (SIC 20 through 39). SIC 20 – food – and SIC 21 – tobacco – are the two exceptions (beverage holders and ash trays appear under other codes).

Transportation equipment (SIC 37) might be expected to contain the largest percentage of automobile suppliers, but barely one-seventh are allocated to it. According to Kamath and Wilson (1983), more suppliers are classified under nonelectrical machinery (SIC 35), primary metal industries (SIC 33), and fabricated metal products (SIC 34). SIC codes 33, 34, 35, and 37 together account for approximately 70 per cent of all purchases (Table 5.1).

The percentage of suppliers in SIC 3714 is even smaller according to a study in the 1950s that took into account suppliers of raw materials, such as steel, as well as fabricated components. Firms in four of the two-digit SIC codes were responsible for over 80 per cent of the sales of components to carmakers (Table 5.2). The largest SIC code in terms

Table 5.2 Firms supplying components to car makers by standard industrial classification, 1957

Standard Industrial Classification Code	*Description*		*Per cent (value of sales)*
22	Textile mill products		7.5
	tire cord	5.8	
	other (carpets, pads)	1.7	
23	Textile products made from fabrics		3.5
	trim	2.6	
	seat covers	0.9	
26	Paper and allied products (felt)		0.1
30	Rubber and miscellaneous plastics products		29.5
	tires	25.5	
	other	4.0	
32	Stone, clay, glass, and concrete products		4.3
	glass	3.1	
	other (asbestos, linings, facings)	1.3	
34	Fabricated metal industries		20.4
	stampings	8.2	
	hardware	6.6	
	lighting	2.3	
	steel springs	1.6	
	seat and back springs	1.5	
	other	0.2	
35	Nonelectrical machinery (carburetors, pistons)		14.4
36	Electrical and electronic machinery		16.3
	batteries	4.3	
	radios	2.2	
	battery charging generators and regulators	1.7	
	spark plugs	1.4	
	cranking motors	1.4	
	wiring and cables	1.1	
	motor generators	1.1	
	distributors	0.6	
	switches	0.5	
	lamps	0.5	
	other (coils, condensers)	1.5	
37	Transportation equipment		2.4
38	Instruments (instrument panel indicators)		1.6

Source: Motor Vehicle Manufacturers Association 1957: 40-41

of value of sales of components to carmakers, was rubber (SIC 30), which accounted for nearly 30 per cent of all sales, primarily from tires. Second was fabricated metal products, such as stampings, which totaled approximately 20 per cent of the value of purchases. Firms allocated to the SIC codes for electrical and nonelectrical machinery (SICs 36 and 35) accounted for approximately 17 per cent and 14 per cent of total sales, respectively (Motor Vehicle Manufacturers Association 1957: 40–41).

The placement of rubber products at the top of the list of sales to car makers reflects the fact that the tire industry has remained dominated by independent firms. Three firms accounted for over half of the tire sales in the early 1990s: the US-based Goodyear, the French-owned Michelin, and the Japanese-owned Bridgestone-Firestone. Another 30 per cent of the market was held by five other firms, including Sumitomo-Dunlop, Pirelli-Armstrong, Continental-General, Uniroyal Goodrich, and Yokohama. Smaller firms accounted for the remaining one-sixth of sales.

The names of the leading tire producers reflect the globalization of the tire industry through mergers and acquisitions. During the late 1980s, Japanese-based Sumitomo Rubber acquired the British firm of Dunlop, General was sold to Continental, a German company, Uniroyal and Goodrich, two American companies, merged, Italian-based Pirelli took over Armstrong, a British firm, and US-based Firestone was taken over by Bridgestone. The difference between the two studies reflects the fact that car makers purchase large quantities of steel and other metal products for fabricating their own parts.

Although not all manufacturers classified as SIC 3714 are OEM suppliers, at least they are producing motor vehicle parts. In contrast, most of the firms in other two- and four-digit SIC codes are not producing motor vehicle parts. Consequently, conclusions about the changing distribution of motor vehicle suppliers cannot be made from data for SIC codes other than 3714.

Surveying individual firms

The other approach to learning about the spatial distribution of suppliers is to compile lists of individual firms and document their addresses. The principal difficulty with this alternative is the large number of firms involved in production of motor vehicle parts. GM classifies its suppliers into six categories based on the type of service provided, including fabricated components, raw materials, indirect materials, transportation, capital equipment, and insurance. Only the first two types of suppliers provide parts and materials that actually go into motor vehicles, while the remaining four types provide support services to the company. Yet, even when nonmaterial suppliers are excluded, several thousand firms make parts for US car makers.

Aside from the large number of firms involved, sorting out the distribution of components suppliers to a company such as General Motors is difficult, because the addresses on the bills that General Motors receives for purchasing parts may not correspond to the place where manufacturing occurred. A list of billing addresses would not distinguish the value of the shipment. Even if such a study were undertaken, the distribution of suppliers in the past would be impossible to reconstruct.

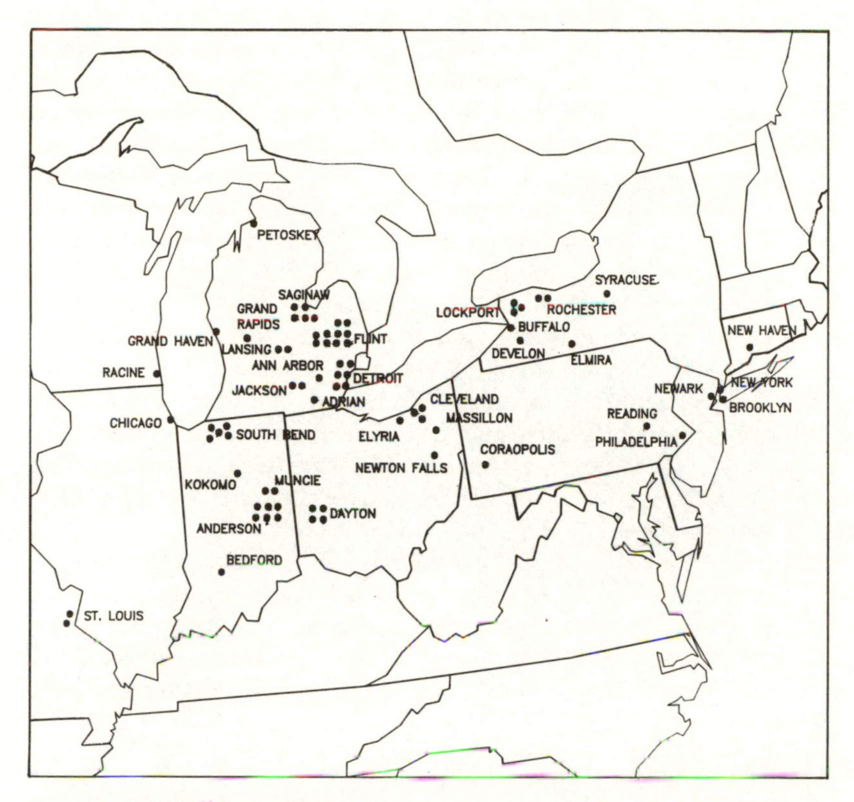

Figure 5.7 Suppliers to GM's Buick Division assembly plant in Flint, 1951
Source: Henrickson (1951)

A study by Henrickson documented the distribution of suppliers of parts and raw materials to GM's Buick Division assembly plant in Flint prior to World War II and in the immediate post-war period. Utilizing information supplied by Buick's purchasing department, Henrickson found that most suppliers were located in the southern Great Lakes manufacturing region during both periods (Figure 5.7). However, during the post-war period, suppliers were more likely to cluster in the heart of the southern Great Lakes region, especially southeastern Michigan and northern Indiana and Ohio, rather than in peripheral states within the region such as Illinois, New York, Pennsylvania, and Wisconsin (Henrickson 1951: 48).

The distribution of OEM suppliers has not changed drastically since the 1950s. Based on a 1982 US Department of Transportation study, Glassmeier and McCloskey concluded that two-thirds of the OEM suppliers were located in the Midwest in 1982. Michigan, Ohio, and Indiana remained the three most important states in terms of the number of firms and number of employees involved in the manufacturing of automotive parts (Glassmeier and McCloskey 1988: 148).

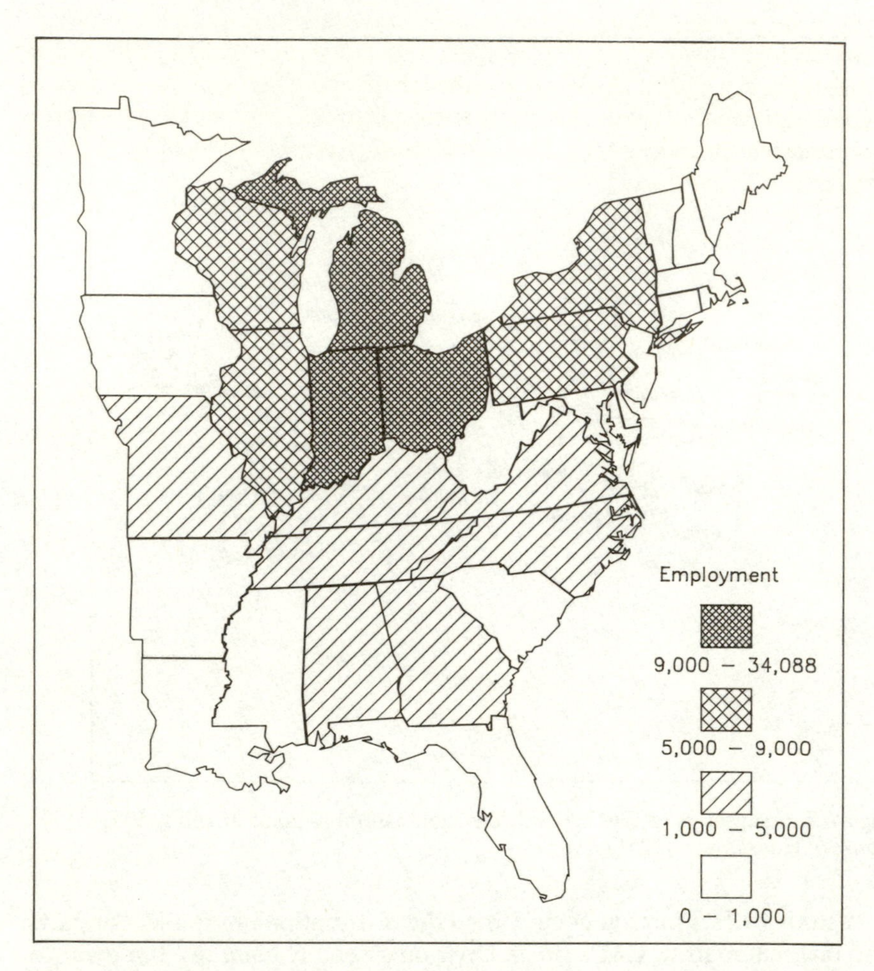

Figure 5.8 Number of employees in manufacturers of automotive parts, 1982
Source: Glassmeier and McCloskey (1988)

While most suppliers were still clustered in the southern Great Lakes region, Glassmeier and McCloskey did find a significant movement of suppliers towards the south during the 1970s. The percentage of firms clustered in the Midwest had dropped from 74 per cent in 1972 to 65 per cent in 1982. Most of the growth was in southeastern states (Figure 5.8). The southern movement coincided with similar policies on the part of the large carmakers, especially General Motors, to relocate some production away from the industry's traditional heartland in the southern Great Lakes (see Chapter 9).

Thus, in the late 1970s and early 1980s, when the restructuring of the automotive industry began, the predominant spatial tendency was relocation of some production from the north to the south. Because the principal reason for the movement of production to the south was a search for a lower cost, nonunionized labor force, changes in the location of suppliers appeared consistent with structuralist theories concerning the spatial division of labor. However, subsequent changes in the structure of automotive production in the United States have not encouraged further decentralization of suppliers.

Part II

Reasons for recent locational changes: regional scale

The long-term locational pattern of US automotive production rested heavily on Weberian-style identification of least-cost transport locations. The economic crisis of the 1970s forced auto makers to restructure much of their operations. Included in this restructuring was a re-evaluation of the optimal locations for assembly and components plants.

Each auto maker was forced to undertake this as a result of the 1970s economic crisis. In the late 1970s and early 1980s, Ford and Chrysler closed many of their assembly and components plants to bring productive capacity in line with lower sales forecasts. Chrysler faced a second round of plant closures to shed excess capacity after its 1987 acquisition of American Motors. In contrast to Ford and Chrysler, GM opened a number of new plants in the early 1980s. When sales declined in the late 1980s, GM was forced to close a large number of facilities. Meanwhile, foreign companies, beginning with Volkswagen, and then the Japanese companies, opened a dozen assembly plants in the United States and Canada during the period. Despite differences in approaches, the various auto makers reached a similar conclusion: coastal plants should be closed and new ones should be built in the interior.

Chapter 6 explains that assembly plants have returned to the interior of the country primarily because of changes in the North American motor vehicle market. Prior to 1960, all cars manufactured in North America were roughly the same size, and branch assembly plants built these models for distribution within a relatively compact market area. Since 1960, auto makers have added models of varying size, including intermediate, compact, and subcompact, as well as several types of trucks.

With the fragmentation of the market, auto makers now need only one or perhaps two assembly plants to meet demand for each of their models throughout the United States and Canada. Consequently, assembly plants have been converted from production of identical models for regional distribution to production of one or two models for sale throughout North America. In general, if the critical factor of production is minimizing distribution costs from the plant to consumers, and the output of one plant is sufficient to meet national demand for that product, then the optimal location for the plant is in the interior of the country.

The increasingly competitive environment of North American motor vehicle production has pushed companies towards making more precise calculations of the least-cost locations for new plant sites, as Chapter 7 explains. At the same time, companies are able to take advantage of sophisticated computer models to undertake Weberian-style analysis with more precision, despite frequent changes in freight rates as a result of the deregulation of the rail and trucking industries.

The distribution of components suppliers has also changed in recent years as a result of changing transport patterns. The diffusion of just-in-time delivery patterns has forced many suppliers to locate near their customer, a final assembly plant. However, Chapter 7 demonstrates that rapid delivery of parts is only one element of the restructuring of producer–supplier relationships embodied in the just-in-time concept.

6 Market fragmentation

For more than half a century, the spatial distribution of automotive production within the United States followed the regional branch concept. The two dominant producers – first Ford and later General Motors' Chevrolet Division – maintained branch plants in the largest metropolitan areas, where consumers clustered, such as Atlanta, Boston, Dallas, Los Angeles, New York, Philadelphia, and San Francisco. As bulky products fabricated from thousands of parts, automobiles were assembled as close as possible to consumers in order to minimize aggregate freight costs. However, the large coastal metropolitan areas have lost nearly all of their automotive plants in recent years. Minimizing the costs of transporting assembled vehicles to consumers remains a critical factor, but the optimal location for plants has switched to the interior of the country.

GM'S SATURN PLANT SITE SELECTION

General Motors' highly publicized process of selecting a site for its Saturn plant during the 1980s illustrates the logic underlying recent preferences for interior locations. According to William Hoaglund, Saturn's president at the time, the most critical factor in the calculation of costs was the freight charge for delivering assembled vehicles from the factory to consumers. Freight charges accounted for 80 per cent of the formula (Forney 1987).

Saturn officials knew that GM's national market center was north of Fort Wayne, Indiana. In other words, producing automobiles at that spot would minimize the cost of delivering them to consumers throughout North America, given the existing configuration of highways and rail lines and the cost per km of shipping various distances by truck and rail. The market center was further north and east than the North American population center, located in central Illinois, because foreign cars had captured a higher percentage of sales in the south and west, especially in southern California. The Chrysler Corporation's market center was even further east, in western Pennsylvania; this explains the company's decision to start an assembly plant at New Stanton which it never occupied, although Volkswagen completed it and assembled vehicles there between 1978 and 1988.

GM's marketing plan for Saturn called for higher market penetration compared to the company's other vehicles in the Great Lakes states, Texas, and the east and west coasts. Consequently, Saturn's market centroid was pulled 200 km southeast of Fort Wayne, near the town of Washington Court House, Ohio, 60 km southwest of Columbus and 70 km east of Dayton.

General Motors then concentrated on sites within a one-day driving range of the national market centroid, an irregularly shaped 1.2 million km^2 area with a radius of roughly 500 km around Washington Court House, depending on the position of highways. The area extended from southern Michigan to northern Alabama and from western Illinois to eastern Virginia. Locations outside the circle were computed to add between $400 and $500 per vehicle in freight charges, in part because truck drivers would be required to stop overnight more often, as well as cover longer distances. A few sites outside the circle were seriously considered, notably in Texas, but none had advantages which could overcome the higher freight costs.

The search was limited to sites less than 50 km from a metropolitan area of at least one-quarter million inhabitants containing a major university and airport. The site also had to be near two major long-distance interstate highways and a main or at least secondary main rail line. Within the 1.2 million km^2 area, eleven cities met these constraints – Chicago, Illinois; Cincinnati, Cleveland, Columbus, and Dayton, Ohio; Detroit, Michigan; Indianapolis, Indiana; Louisville, Kentucky; Nashville, Tennessee; and Pittsburgh, Pennsylvania (Figure 6.1).

Unsuitable locations within these eleven metropolitan areas could be eliminated quickly. Because of the large size of the needed parcel – roughly 400 hectares – only sites outside built-up areas needed to be examined. The search was further confined within each metropolitan area to a handful of major transportation corridors which contained a rail line and interstate highway. Hilly sites were also eliminated.

Essential to the narrowing process were topographic sheets published by the US Geological Survey at the 1:24000 scale. Topographic maps display elements of the human-built environment, such as roads, rail lines, dams, and buildings, as well as physical features, such as lakes, rivers, and forests. Topographic maps are also useful in the industrial site selection process because they contain contour lines, which connect points of equal elevation. Contour lines permit determination of whether the site is flat or hilly, through computation of relief (the difference in elevation between two points) and slope (relief divided by distance between the two points).

Additional information was collected for sites which passed the initial screening on topographic sheets, including suitability of soil for constructing a large plant, neighboring land use activities, number of parcels of land

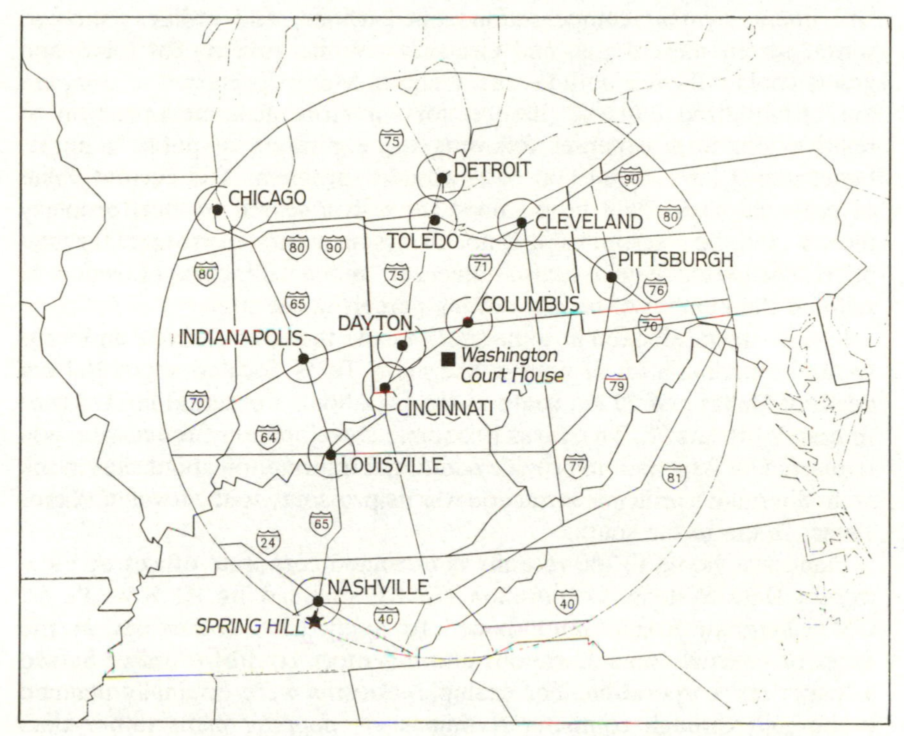

Figure 6.1 Large metropolitan areas within 500 km of Washington Court House, Ohio, the center of the anticipated market area for GM's Saturn. The circles drawn drawn around the metropolitan areas have 50-km radii

within the site, and the names of the property owners. Sites that appeared to be suitable from the topographic maps were visited in person by a Saturn official. On-site visits exposed problems not identified on topographic maps. For example, the maps showed the location of buildings but not the type of activity for which they were used.

Officials were limited in what they could learn at the sites, because they conducted the process in secret to avoid generating rumors about the company's intentions. Reservations for flights and lodging were made under pseudonyms, to prevent tracing the movements of Saturn officials. To minimize leaks in the tight world of Michigan automotive executives, Saturn officials didn't even tell their spouses where they were going. Telephones calls were routed through an intermediary to prevent tracing. Air reservations were booked on circuitous routes involving changes rather than direct flights to the desired destinations. Team members worked alone in the field.

Sites that appeared to be suitable from the topographic sheets and secret visits were turned over to Argonaut Realty to determine the costs of doing business at each. Critical factors included tax rates, including property and unemployment compensation contributions, and utilities, such as water, sewer, natural gas, and electricity. While government loans and grants could influence both factors, General Motors preferred to compare the unsubsidized costs of these factors, a more accurate reflection of relative long-term expenses following the expiration of public support. Determining land acquisition costs posed a problem. The current value of potential sites could be obtained through a search of local property records, but the assessed value is not always an accurate surrogate for final price. The purchase price is also influenced by the willingness of owners to sell and the number of people owning parcels in the site.

Saturn officials looked at some sites outside the interior circle and were spotted spending a lot of time in Sherman, Texas, located about 100 km north of Dallas and 25 km south of the Oklahoma border, along US (not Interstate) Route 75. The Texas Economic Development Commission was requested by Argonaut to provide additional information about sites in the area. Sherman's principal attraction was its proximity to the town of Plano, Texas, 70 km to the south.

Plano was home to the recently constructed corporate offices of Electronics Data Systems Corporation (EDS), founded by H. Ross Perot. GM Chairman Roger Smith hoped to apply EDS technology in the areas of internal communications and inventory control to make Saturn a 'paperless' corporation. For example, Saturns were originally planned to be sold through computer terminals at shopping malls rather than free-standing dealers. A customer would place an order for a specific model, color, and set of options. The order would trigger production of components, which would be shipped quickly to the factory for final assembly, and the finished vehicle would be delivered to the customer within 48 hours. Enthralled with the marriage between computer technology and automotive production, Smith induced Perot to sell EDS to General Motors in 1984.

In the end, Saturns would be sold through dealers in the conventional manner, and the advantage of proximity to EDS headquarters could not overcome the substantially higher freight costs in Texas. Perot, now GM's largest stockholder, also became the company's leading critic. On 1 December 1986, less than a month after announcing it would close eleven plants, GM decided to acquire Perot's shares for $742.8 million; eighteen months later, Perot set up his own electronics firm and hired many of the best EDS employees (Levin 1989a).

General Motors' decision to produce the Saturn vehicles represents one example of the current industry strategy of marketing a wider variety of models. Historically, companies sold a large number of models, but they were all placed on a handful of platforms, and therefore they could be built on the same assembly line. In recent years, the variation among companies' models has become much more substantial, especially the introduction of several sizes of platforms.

As assembly of more than one platform on the same line is difficult, assembly plants that had produced the same platform for distribution within one region of the country have been converted to specialized ones producing distinctive models for the North American market. If only one plant is sufficient to meet demand throughout North America, and the critical cost in the production process is minimizing the costs of distributing the product from the plant to the market, then the optimal plant location is in the interior of the United States.

DIVERSITY IN EARLY AUTOMOTIVE PRODUCTS

Early automobile manufacturers believed that they needed to produce three or four distinctive models, each built on a different chassis or platform. Models based on entirely new platforms had to be introduced every year or two, so that if one model failed to generate public interest the company could still survive. The policy of offering several new platforms each year proved to be extremely unprofitable, because producers faced the expense of making small batches of replacement parts each year for the constantly growing number of discontinued older models (Parlin and Youker 1914: 266).

The Ford Motor Company followed the conventional wisdom at first, selling three models in 1904–1905, two in 1905–1906, and four in 1906–1907. But Henry Ford was committed to concentrating all of his resources on one low-priced model, the Model T, beginning in 1909. 'If Ford's single model policy had failed he would have to scrap all his machinery' (Parlin and Youker 1914: 176). Mass production of one model encouraged Ford to establish branch assembly plants in order to minimize freight costs.

Many of the early motor vehicle producers distinguished their different models by letter of the alphabet rather than by name. The first model sold by a company was often called the Model A, and each time a major improvement was made – a frequent occurrence in the early years of automotive production – the new model was designated by another letter, frequently though not invariably the next letter in the alphabet. Ford, for example, ran through nearly a dozen letters in six years, until the Model T proved so successful when it was introduced in 1909 that no new models were produced until 1927. The Model A, which replaced the Model T in 1927, was the last Ford to carry a letter.

GM's 'car for every purse'

In contrast to Ford's one model policy, General Motors, from its inception in 1908, was committed to increasing sales by maximizing diversity, a legacy of the company's origin through the merger of several independent producers. During Billy Durant's two terms as its head, General

Motors acquired a number of producers which retained considerable autonomy within the corporation. The strategy of acquiring a number of companies also protected General Motors against uncertainty in the industry's formative years, before engineers had not settled on the optimal types of engines, transmissions, electrical systems, and other elements of automotive technology.

Alfred P. Sloan, Jr., president of General Motors from 1923 to 1937 and then chairman of the board until 1956, was instrumental in rationalizing the company's products so that they did not compete against each other. Five self-contained automobile production divisions were retained, each with complete responsibility for engineering, purchasing, production, marketing, and finances. Sloan assigned a specific price bracket to each of the five automobile manufacturing divisions, a strategy he called 'a car for every purse and purpose.'

In 1921 Chevrolet occupied the lowest price bracket, followed by Oakland, Buick, Oldsmobile, and Cadillac. Over the next few years, Buick moved to the second highest price bracket, while Oldsmobile became the second lowest-cost division for a while, before settling into the middle position. In 1926, Oakland introduced a lower-cost model named Pontiac, which captured three-fourths of the division's sales; the Oakland name was dropped altogether in 1932 and the division's name changed to Pontiac. Several generations of Americans then learned the names of General Motors' five automobiles in ascending order of price and social status – Chevrolet, Pontiac, Oldsmobile, Buick, and Cadillac. Consumers buying automobiles for the first time were encouraged to get Chevrolets and as their income increased 'trade up' to more expensive models produced by GM's other divisions (Sloan 1964: 59–60).

General Motors' marketing strategy generated a mixed pattern in the location of production. The low-priced Chevrolets, which accounted for half of the company's sales, were assembled at branch assembly plants near population concentrations. The other four divisions clustered in southern Michigan, producing all of the automobiles they needed at one assembly plant each, Oakland in Pontiac, Oldsmobile in Lansing, Buick in Flint, and Cadillac in Detroit.

After he acquired Dodge Brothers in 1928, Walter Chrysler emulated Sloan's policy of a car for every purse. The Chrysler model was moved to a higher-priced bracket to eliminate competition with Dodge, the low-priced Plymouth was introduced to compete with Ford and Chevrolet, and a medium-priced DeSoto was positioned between Dodge and Chrysler. Each of the four divisions was produced at a separate Detroit-area assembly plant, Plymouth at Lynch Road, Dodge at Hamtramck, DeSoto at Wyoming Avenue, and Chrysler at Jefferson Avenue.

Henry Ford stubbornly stuck to his policy of producing one model through the 1920s and 1930s. The only exception was made for the luxury-priced, low-volume Lincoln, which was acquired in 1921. Lincoln had been founded by Henry M. Leland and his son Wilfred in 1917, initially to make Liberty airplane engines. Leland had started in the automotive industry as a manufacturer of engines and went on to manage Cadillac before leaving in a dispute with Durant. In 1920, the Lelands began to manufacture the Lincoln automobile at their plant on Warren Avenue in Detroit, but running into financial difficulties, agreed to sell the firm to Ford in late 1921. The Lelands continued to manage the division for a couple of years but resigned in protest at the introduction of new manufacturing and sales techniques by Ford officials.

To fill the substantial gap between the low-priced Ford and the luxury Lincoln, Ford finally brought out the Mercury in 1939, although it never sold as briskly as its competitors in the other companies. Ford's second attempt to market a medium-priced model, in the 1950s, was poorly designed and marketed, and the model's name, Edsel, became a synonym for failure in the American language.

Designing many bodies for a single platform

Each auto maker achieved diversity within its handful of marketing divisions by offering a variety of body styles and accessory packages placed atop a single platform. Early automobiles had open bodies, like horse-drawn wagons, but by the 1920s passengers generally sat in an enclosed space. During the 1930s, producers began to design a greater variety of bodies to fit over a particular platform. Some bodies had two doors while others had four; some had a rear seat, while others had only a front seat. The shape of some bodies resembled a box, while others were rounded. Sedans had a rear trunk – sometimes containing an extra seat known as a rumble seat – set at approximately the same height as the front hood, while station wagons continued the roof line of the passenger compartment to the rear of the car.

More than one marketing division within a company sometimes shared bodies to reduce costs, especially during the 1930s Depression. By sharing bodies, GM's three medium-priced nameplates had a sufficiently high combined output to justify establishing branch assembly plants near New York City and Los Angeles before World War II. After the war, auto makers introduced new body styles, such as hardtops, which lacked center posts between the two sets of front and rear side windows. The arrangement provided a wider glass area but reduced the car's structural integrity, and the design was abandoned in the 1970s. Among the more bizarre body styles during the 1950s was Ford's 'hardtop-covertible', in which the hard top of the car retracted into the trunk.

The number of distinctive bodies reached an all-time high around 1970 and has declined since then, primarily because hardtops were eliminated for safety reasons and consumers have shunned convertibles and station wagons. Today, most passenger car bodies are variations of the traditional two-door and four-door sedans; the principal distinction is between a notchback, which has a separate trunk space, and a hatchback, which has a door at the rear and a folding rear seat to create a larger storage area.

The sales divisions also increased differences among mechanically similar automobiles through cosmetic treatments. The bodies of higher-priced models received additional metal trim, unusual paint colors, or decals. Interiors were made of more expensive materials, such as wood, leather, and cloth, or, increasingly, plastic simulated to resemble more expensive materials. Higher-priced models contained standard accessories which are otherwise optional or unavailable. Each generation has redefined what is standard and what is optional. In various decades, windshield wipers, mirrors, self-starters, and radios were introduced on luxury models but soon became available on all models.

Auto makers find the practice of outfitting a particular model with more elaborate trim to be profitable, because the additional cost of installing the material is less than the additional price which consumers are willing to pay. In the 1930s, moving the gear shift from the floor to the steering column was considered a technological innovation for which consumers were willing to pay extra. Three decades later, moving the gear shift from the steering column back to the floor was considered a sporty touch for which consumers were willing to pay extra, even though it was a less expensive and technologically simpler arrangement for auto makers to manufacture.

Chevrolet's 1957 models reveal the variety which carmakers could achieve with one platform. Chevrolet offered three levels of trim and accessories – Bel Air, 210, and 150 – plus nine bodies, including two- and four-door sedans, two- and four-door hardtops, two- and three-seat four-door station wagons, two types of two-seat two-door station wagons, and a convertible. The more elaborately trimmed Bel Air model was priced several hundred dollars more than a plainer 210 or 150 model for the same body type (Table 6.1).

The marketing of a large number of models all built on the same platform encouraged expansion of branch assembly plants. Each branch assembly plant could produce models based on the identical platform intended primarily for distribution within the immediately surrounding region. Dealers in some regions of the country received deliveries from more distant branch assembly plants when customers in the region placed an especially heavy demand on a particular body type or trim style.

RESTRUCTURING OF PRODUCTS DURING THE 1960s

The restructuring of the geography of North American motor vehicle production began during the 1960s, when producers adopted two strategies. First, auto makers introduced different sizes of platforms. Second,

Table 6.1 Body styles offered by Chevrolet, 1957

Body style	*List price $*	*Percent of production**
2-door sedan		18.7
Bel Air	2,338	3.9
210	2,222	10.0
150	2,096	4.4
150 utility	1,985	0.5
4-door hardtop		17.0
Bel Air	2,464	8.5
210	2,370	8.5
2-door hardtop		13.2
Bel Air	2,399	10.3
210 Sport	2,304	1.4
210 Delray	2,262	1.6
4-door station wagon		10.9
Bel Air Townsman 2-seat	2,680	1.7
210 Beauville 3-seat	2,663	1.3
210 Handyman 2-seat	2,556	7.9
2-door station wagon		2.4
Bel Air Nomad 2-seat	2,857	0.4
210 Handyman 2-seat	2,502	1.1
150 Handyman 2-seat	2,407	0.9
Convertible		2.9
Bel Air	2,611	2.9

Note: * Total Chevrolet production in 1957 was 1,623,152

Source: adapted from Dammann 986

companies expanded the practice of corporate twins, that is the sale of mechanically identical models under different names.

New platforms

Until the 1960 model year, virtually all of the Big Three's domestically produced cars exceeded 5.3 m in length; the only exception consisted of a few limited-production sports cars. But for 1960, each of the Big Three introduced compacts, 1 m shorter than the conventional full-sized models, including GM's Chevrolet Corvair, Ford's Falcon, and Chrysler's Plymouth Valiant. When the Corvair, with its air-cooled rear engine, proved disappointing in sales and quality, General Motors added a second compact model in 1963, first known as the Chevy II and later as the Nova.

The US compact cars responded to increasing sales during the late 1950s for smaller cars. European producers, led by Volkswagen, increased their

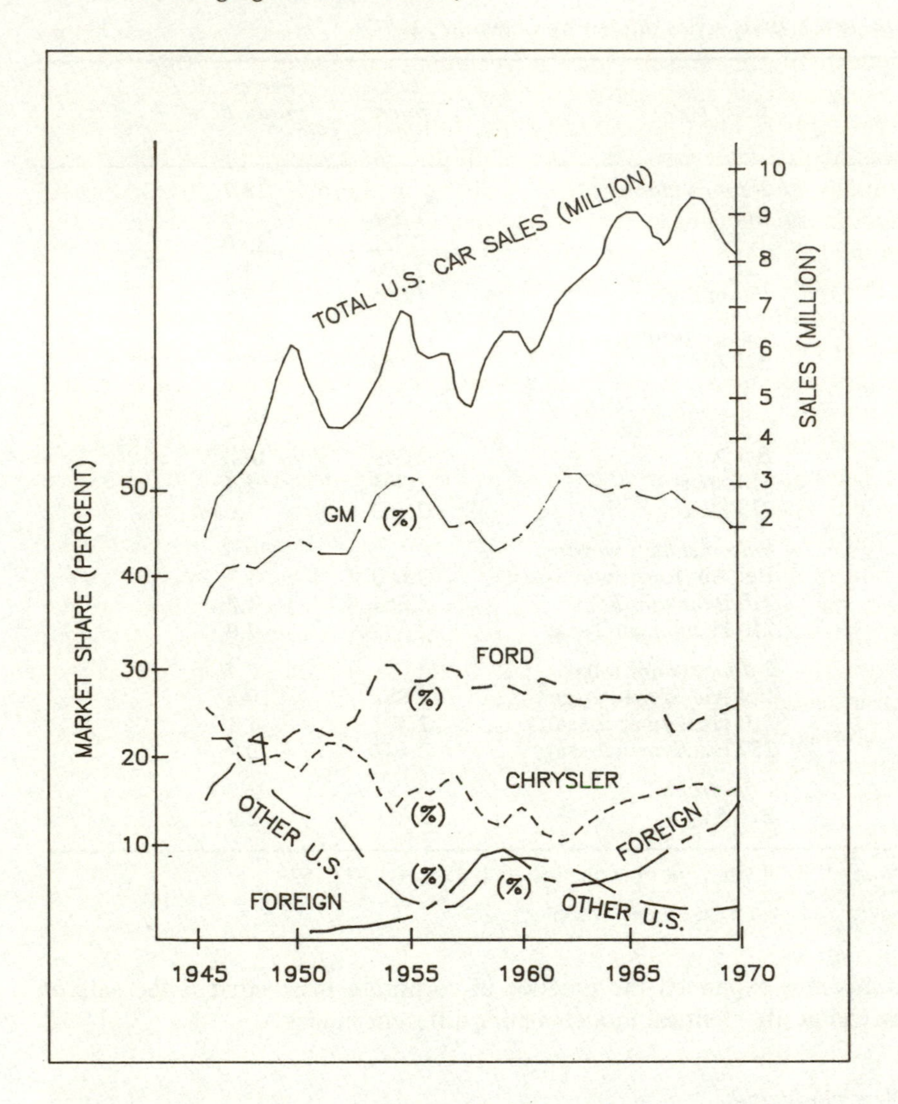

Figure 6.2 Share of the US market held by US and foreign companies, 1945–1970
Source: L. J. White, *The Automobile Industry Since 1945* and Automotive News, *1989 Marketing Data Book*

share of the US market from less than 1 per cent in 1955 to more than 10 per cent in 1959 (Figure 6.2). American Motors, which had been created in 1954 through the merger of Hudson and Nash, also achieved all-time high market shares of more than 6 per cent during the late 1950s and early 1960s by selling a Rambler model that was several inches shorter than any of the Big Three's offerings. After introduction of the domestic compacts, imports fell back to under 5 per cent in 1962 and American Motors began a steady decline in sales until the company disappeared in 1987.

Intermediate-sized automobiles were introduced by General Motors in 1961 and by Ford and Chrysler the following year. Ford's intermediate was approximately 5.1 m long, while the General Motors models were initially under 4.8 m but grew by 30 cm in the mid-1960s. The compact and intermediate models stemmed the tide of imports for a few years, but in the late 1960s the percentage of foreign car sales began to rise again, to 9 per cent in 1967 and 15 per cent in 1970. Again, the American companies responded by introducing subcompacts for the 1971 model year, less than 4.3 m long, including the Ford Pinto and GM's Chevrolet Vega. Thus, within a decade, American manufacturers had expanded their offerings from only full-sized models to four basic sizes – full-sized (over 5.4 m), intermediate (5 m), compact (4.7 m), and subcompact (under 4.3 m).

The number of distinctive models also expanded in the 1960s because of introduction of sporty models aimed at carving out market niches. Typically, these specialty automobiles had responsive engines, two-door bodies, and limited – if any – rear seating. Small sporty automobiles – less than 4.7 m – included both moderately-priced models such as the Ford Mustang and luxury models such as GM's Chevrolet Corvette. Similarly, both moderately-priced and luxury models were offered as intermediate-sized sports cars, close to 5.1 m in length.

During the 1950s, Ford and General Motors had both begun producing two-passenger luxury sports cars, known as the Thunderbird and Corvette, respectively. The Corvette, known at GM as the Y-body, remains an expensive sports car, with annual sales rarely exceeding 30,000. In 1981, the first year a fiberglass body was used, Corvette production was moved from St Louis to a new plant in Bowling Green, Kentucky. After a few years of production, Ford repositioned the Thunderbird as an intermediate-sized model, with a rear seat and an overall length exceeding 5.1 m most years. Thunderbird's annual sales have fluctuated wildly, from under 40,000 in the 1950s to over 350,000 in 1978. Between 1958 and 1976 Thunderbirds were assembled at Wixom, along with luxury Lincoln models. Production moved to Chicago between 1977 and 1980 and to Lorain during the 1980s. In addition, Thunderbirds were assembled along with other Ford models at Los Angeles between 1968 and 1979 for the west-coast market.

The first moderately-priced compact sports model designed for large-scale production was the Ford Mustang, which sold 120,000 in 1964, its first year, and 600,000 the following year. The Mustang expanded or contracted in length on several occasions but generally remained under 4.7 m. Production at the Dearborn assembly plant has been devoted almost exclusively to Mustang production since the model's introduction in 1964. In keeping with the branch plant concept, Ford also assembled Mustangs at San Jose between 1964 and 1970 and again between 1978 and 1981 for west-coast distribution and at Edison between 1965 and 1971 for east coast distribution. Since 1981, Mustangs have been assembled exclusively at Dearborn.

General Motors hustled to introduce a competing model in 1967, known as the F-platform and marketed as the Chevrolet Camaro and Pontiac Firebird. The F-platform lengthened from 4.7 m to 5.1 m by the late 1970s before being shortened in the early 1980s. For two decades, Camaro-Firebird production was allocated to two plants, Norwood for the eastern market and Van Nuys for the western. After Norwood was closed in 1987, Van Nuys continued as the sole assembly site for several years, until production moved to Ste Thérèse, Quebec, for 1993 models.

Intermediate-sized higher-priced sporty cars were introduced by General Motors in 1962, beginning with the Buick Riviera, and Ford soon enlarged the Thunderbird to fit this class. GM's Cadillac and Ford's Lincoln Continental nameplates normally offered high-priced sporty models as well, which rank among the most expensive made in North America. These models are typically assembled at only one location, usually in Michigan, such as GM's Hamtramck plant for Cadillacs and Ford's Wixom plant for Lincolns.

In the 1950s, every General Motors automobile (with the exception of the limited-production Corvette sports car) was built on three platforms that were all approximately the same size, but by 1970, the company produced nine distinctive platforms. Similar policies by Ford, Chrysler, and American Motors brought the total number of American-made platforms to twenty-eight in 1970, ranging in length from under 4.3 m to over 5.6 m (Table 6.2).

Corporate twins

The proliferation of new models was reduced somewhat by the practice of producing corporate twins. Back in the 1930s, General Motors had sold essentially identical models with different nameplates in order to save money during the Depression and justify the establishment of branch assembly plants for the company's medium-priced models. However, until the 1960s, the various marketing divisions within the large companies normally retained individual responsibility for manufacturing their own vehicles, even if they shared some components with other divisions.

In 1960, the Ford Motor Company's new compact model was given two names, the Falcon, sold by the Ford Division, and the Comet, sold by the Mercury Division. The Comet had more elaborate trim, justifying its somewhat higher price than the mechanically identical Falcon. The following year, General Motors and Chrysler adopted similar practices. GM's new intermediate model was simultaneously introduced as the Pontiac Tempest, Oldsmobile F-85, and Buick Skylark. A twin of Chrysler's

Table 6.2 Platforms built in the US and Canada by size, 1991

	Subcompact	*Compact*	*Intermediate*	*Full-sized*	*Specialty*	*Total*
Chrysler						
1970	0	1	1	2	1	5
1980	1	1	1	1	1	5
1990	1	1	2	0	2	6
Ford						
1970	0	1	1	2	4	8
1980	1	1	1	1	3	7
1990	1	1	1	1	4	8
General Motors						
1970	0	1	1	3	5	10
1980	1	2	1	2	5	11
1990	0	3	2	3	6	14
*Other**						
1970	1	1	1	1	1	5
1980	3	1	0	0	0	4
1990	5	2	1	0	2	10
Total						
1970	1	4	4	8	11	28
1980	6	5	3	4	9	27
1990	7	7	6	4	14	38

*Note:** Other includes American Motors in 1970 and 1980; Volkswagen in 1980; Cami, Diamond-Star, Honda, Hyundai, Nissan, Nummi, and Toyota in 1990

compact Plymouth Valiant was sold in 1961 as the Dodge Lancer and in 1963 as the Dodge Dart. By the end of the 1960s, most platforms were shared by corporate twins.

In the 1950s, General Motors' five automobile divisions all utilized similarly sized platforms and shared some bodies. By 1970, Chevrolets were built on six different platforms, Pontiacs, Oldsmobiles, and Buicks four each, and Cadillacs two. However, because of corporate twins, General Motors produced ten rather than twenty distinctive platforms in 1970. Overall, the company offered one compact platform, one intermediate, three full-sized, two small sport, and three large sport. Chevrolet's platforms included one compact, one intermediate, one full-sized, and three sport; Pontiac built one intermediate, one full-sized, and two sport; Oldsmobile and Buick both sold one intermediate, two full-sized, and one sport; while Cadillac sold one full-sized and one sport.

General Motors would pay a heavy price during the 1970s for its standardization program in the 'Chevymobile' scandal. When an owner took his recently purchased Oldsmobile Delta 88 for service in Chicago, the mechanic encountered a curious problem: Oldsmobile's fan belt and oil filter wouldn't fit the engine. The reason was that the engine had been built by Chevrolet rather than by Oldsmobile. Outraged, the owner complained to the Illinois attorney general's consumer fraud office, which then sued General Motors on the grounds of deception. Eventually, seventy suits were filed, including suits by the attorneys general of all fifty states.

The day after the first suit was filed, GM president Pete Estes revealed that not only had Chevrolet built engines for Oldsmobile, the practice of substituting components was widespread. Buick engines were powering Pontiac and Oldsmobile models, and transmissions were being interchanged. General Motors argued that the practice was not deceptive, because all parts were made by the same company, and they were of comparable quality. As an act of good faith, GM offered an owner of a Pontiac, Oldsmobile, or Buick, with a Chevrolet engine either a discount on buying a new car or a long-term warranty on the existing car.

GM's offer did not satisfy most litigants. For decades, GM had justfied charging higher prices for Pontiacs, Oldsmobiles, and Buicks, compared to Chevrolets, on the basis of additional features, such as the Oldsmobile engine. If Chevrolets and Oldsmobiles were identical, why charge more for Oldsmobiles? The suits were settled in 1979, when GM agreed to pay 132,000 affected owners $200, in addition to an extended warranty, a settlement which cost the company an estimated $40 million (Cray 1980: 509).

New truck models

The number of truck models has also increased because North American consumers increasingly regard trucks as alternatives to passenger cars rather than reserved for commercial and industrial uses. Producers offer three types of light trucks – pickups, vans, and sport utility vehicles. Pickup trucks may be less expensive than passenger cars, while vans and sport utility vehicles have substituted for station wagons. Truck production in North America nearly doubled during the 1970s, accounting for 30 per cent of total production in 1978, compared to 20 per cent in 1971. Following a sharp decline during the late 1970s and early 1980s, when consumers temporarily switched to more fuel-efficient cars, truck sales again increased relatively rapidly, to roughly 40 per cent of the total motor vehicle market by the 1990s.

For a number of years, full-sized pickup trucks were offered in three basic sizes – 4.8 m length with 2.9 m wheelbase, 5.3 m length with 3.3 m wheelbase, and 5.8 m length with 3.8 m wheelbase; further diversity was achieved by offering each of the three sizes with extended passenger cabs and chassis. Full-sized pickups were sold by General Motors as the Chevrolet C/K and GMC Sierra, by Ford as the F-series, and by Chrysler as the Dodge Dakota. During the 1980s, manufacturers added a compact pickup, with a wheelbase between 2.7 m and 2.8 m and length between 4.4 m and 4.6 m. GM called its compact pickups the Chevrolet S-10 and GMC S-15, while Ford's were known as the Ranger and Chrysler's as the Dodge Ram.

Vans and sport utility vehicles have accounted for much of the recent increase in light truck sales. Volkswagen began during the 1950s to import a van, which enjoyed popularity as an alternative to domestically built station wagons. Vans were first produced in North America a few years later primarily for commercial use, but some individual consumers bought them to haul groceries and children. These vans ranged in size from a wheelbase of 2.8 m and overall length of less than 4.6 m to a wheelbase of nearly 3.5 m and overall length over 5.1 m. The smallest of these vans was still substantially wider and taller than passenger cars.

During the 1980s, producers introduced minivans, comparable in length to the smallest version of the older vans, but closer in width and height to passenger cars. Minivans are 1.7–1.8 m tall, compared to 2 m for other vans and 1.2–1.5 m for passenger cars. At 1.8 m wide, minivans are 30 cm narrower than other vans, but near the average for passenger cars. In 1990, GM introduced a second minivan, which featured a plastic body; to muddle further the difference between passenger cars and trucks, GM created twins for the new minivan to Oldsmobile and Pontiac rather than GMC, in addition to Chevrolet, previously the only nameplate to sell both cars and trucks. Ford added a second minivan in 1992, part of a joint venture with Nissan.

Sport utility vehicles trace their origins to the lingering popularity of Jeeps for civilian transport. General Motors, Ford, and American Motors (later Chrysler) began to produce sport utility vehicles in the 1970s with wheelbase of approximately 2.7 m and overall length of approximately 4.6 m, sold as the Chevrolet Blazer, GMC Jimmy, Ford Bronco, and Jeep Wagoneer. These models were all substantially larger than the original military Jeep, which was approximately 3.8 m long. During the 1980s, mini-sport utility vehicles were introduced, with wheelbases of approximately 2.5 m and overall length of approximately 4.3 m. These were marketed as the Chevrolet S-10 Blazer, GMC S-15 Jimmy, Jeep Cherokee, and Ford Explorer (initially Bronco II); Mazda sold the Navajo, a twin of the Ford Explorer.

Jeep, which invented the market for sport utility vehicles, also had models comparable to the two sizes sold by the other companies, the smaller Cherokee and the larger Wagoneer. However, the Jeep models were somewhat larger, more luxurious, and more expensive than the competing full-sized and mini-sport utility vehicle models. To retain its market share against increased competition, Chrysler developed smaller sport utility vehicles designed to compete more precisely against the models of the other companies.

Organizational changes during the 1960s

Motor vehicle producers initially reacted to the proliferation of models by trying to build more than one-sized platform on the same assembly lines, especially in California, to minimize shipping cars from distant assembly plants. However, the freight savings were outweighed by additional production costs, such as line down time for retooling and complex materials management procedures, necessitated by mixing platforms in the same plant. Therefore, producers returned to the branch assembly plant concept traditionally used to allocate production of full-sized models. By the early 1970s, most assembly plants again produced only one size of model for regional distribution.

The proliferation of models induced the auto makers to reduce the autonomy of the traditional marketing divisions. Why should both the Oldsmobile and Buick divisions hold separate responsibility for building virtually identical cars? In 1965, over the objections of the five marketing divisions, General Motors transferred six former Chevrolet and the former Buick-Oldsmobile-Pontiac branch assembly plants, plus the paired Fisher Body plants, to a newly created Assembly Division (GMAD). The remaining Chevrolet factories were transferred in 1968 and 1971. Chevrolet's former home plant in Flint was converted to exclusively truck production in 1970, but the other four marketing divisions were allowed to retain control of their home plants in Michigan until the 1980s. Ford and Chrysler similarly transferred control of production from the divisions responsible for marketing the various nameplates to production-oriented managers.

Under GMAD management, assembly plants entered a period of turmoil. As GM increasingly standardized parts among the five marketing divisions, products were shuffled annually among the different plants, resulting in the need for frequent retooling. Only two plants – Janesville and Tarrytown – assembled the same product throughout the 1960s, in both cases full-sized Chevrolets. Only two other plants were even able to concentrate exclusively on one size of automobile during the decade; Willow Run produced only compact models and Doraville only full-sized models. GMAD also induced unrest by restructuring work assignments following merger of the labor forces at the assembly plants and the former companion Fisher Body plants. New managers were installed who had reputations for being hard-nosed supervisors. Strikes against the harsh discipline and faster line speeds frequently crippled GM production during the late 1960s and 1970s (Cray 1980: 448).

GMAD still allocated production among assembly plants according to the regional branch concept during the 1970s. For the 1973 model year, full-sized models were built at Linden, Tarrytown, and Wilmington for the east-coast market, at Fairfax, Janesville, and St Louis for the interior, at Doraville for the southeast, and at South Gate for the west. GM distributed intermediate and similar-sized sports models to the west from Fremont, to the southwest from Arlington, to the southeast from Lakewood, to the interior from Leeds, to the mid-Atlantic from Baltimore, and to the northeast from Framingham. GM built compacts at Van Nuys for the

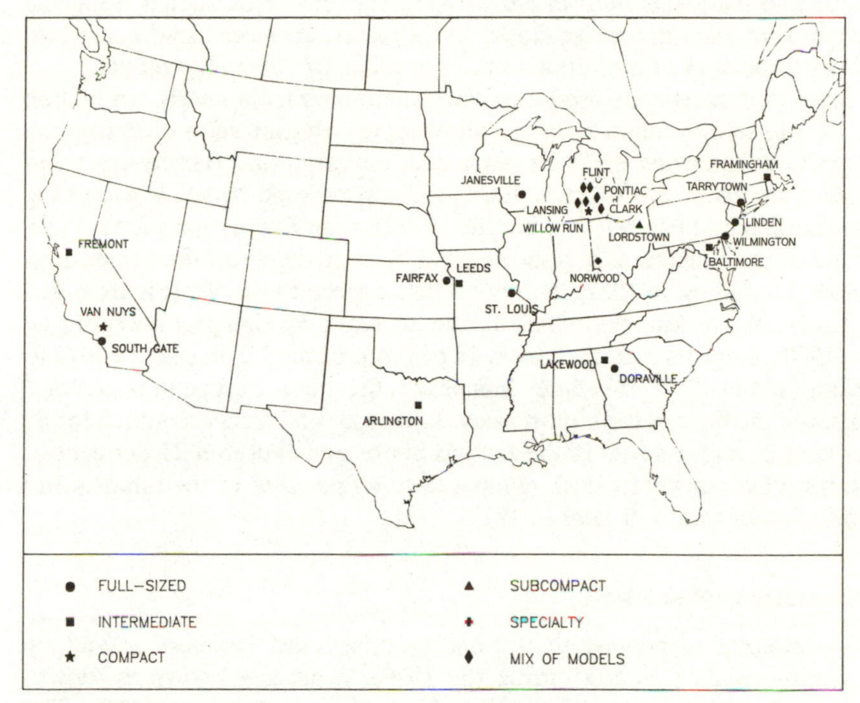

Figure 6.3 General Motors automobile assembly plants by type of model produced, 1973. The concept of a network of branch assembly plants for regional distribution still influenced GM's allocation of products

west and at Willow Run for the east. Two Ohio plants – Lordstown and Norwood – built the subcompact Vega and sporty Camaro and Firebird twins, respectively, for national distribution. The four 'home' plants in Michigan – Detroit, Flint, Lansing, and Pontiac – built a variety of luxury and specialty models for national distribution, as well as more popular models for regional distribution (Figure 6.3).

US FIRMS SCRAP THE BRANCH PLANTS

The geography of motor vehicle production within the United States recovered from the disruption caused by the introduction of different-sized platforms, and in 1973, the all-time peak sales year for US-owned companies, the branch assembly plant system was still strongly entrenched. The structural crisis of the 1970s faced by the automotive industry delivered the fatal blow to the branch assembly plant system of allocating production.

The assault on the concept of regional distribution came from two directions. On the one hand, US producers closed a number of automobile assembly plants in the face of lower sales triggered by increasing petroleum prices and Japanese competition. On the other hand, the number of distinctive platforms built in North America further expanded, reducing demand for each particular model. As a result, no model sold enough to justify allocation of more than one or at most two assembly plants.

Sales of domestically produced cars plummeted from nearly ten million in 1973 to seven million in 1975, following the Organization of Petroleum Exporting Countries' 1973–74 petroleum embargo and subsequent rapid price rises. Sales of domestic automobiles recovered to levels exceeding nine million in the late 1970s, but the 1979 Iranian Revolution set off a new round of price increases and shortages. US car makers saw their sales drop below 5.5 million in 1982, the lowest figure since the 1950s (Figure 6.4).

Americans seeking fuel-efficient cars turned to foreign producers during the 1970s. Imports increased from 15 per cent of the US market in 1972 to 27 per cent in 1980. This time, Japanese rather than European companies captured the bulk of the import sales. Japanese producers accounted for 80 per cent of imports sold in the United States and captured 21 per cent of the total US market in 1980, compared to 40 per cent of the imports and 6 per cent of total US sales in 1972.

Downsizing automobiles

US producers responded to the energy crises and Japanese assault by designing smaller models during the 1970s, a process known as 'downsizing.' Automobiles were shrunk to their sizes in the early 1960s. The length of full-sized models was reduced from 5.5 m first to 5.3 m and then to 5 m, intermediates from 5.3 m to 4.8 m, and compacts from 5 m to 4.6 m; subcompacts remained approximately 4.3 m. The downsizing process hastened the demise of the branch assembly plant system. At first, US producers rushed downsized products into whatever assembly plant could be adapted most efficiently, cheaply, and quickly, regardless of regional balance.

Downsizing took a decade to accomplish. General Motors, with the most financial resources, accomplished the task first, reducing the size of its full-sized models beginning with the 1976 model year, compacts in 1980, and intermediates in 1982. Ford shrunk its full-sized models in the late 1970s but did not introduce compacts until 1984 and intermediates two years later. Chrysler, near bankruptcy, did not begin downsizing until the late 1970s, beginning with its compacts, and did not introduce downsized intermediate models until 1988; lacking the resources to downsize its full-sized models, the company abandoned that market.

The staggered introduction dates for downsized models meant that for several years recently redesigned replacements for full-sized models were actually smaller than older intermediate-sized models, presenting a formidable marketing challenge in a country that traditionally associated larger size with luxury and expense. Ford tried to make a virtue of its slower pace of downsizing by advertising that it offered consumers a choice of larger

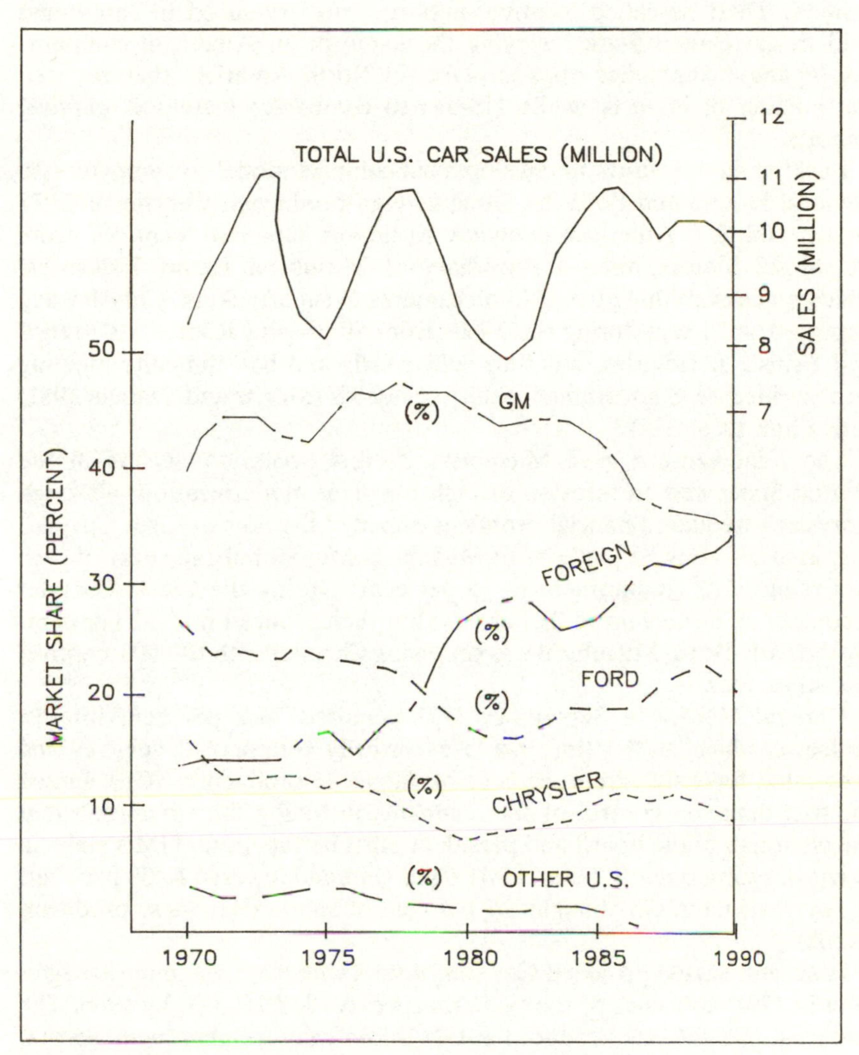

Figure 6.4 Share of the US market held by US and foreign companies, 1970–1991
Source: L. J. White, *The Automobile Industry Since 1945* and Automotive News, *1989 Marketing Data Book*

and smaller models, while GM only had smaller models. Consumers were not impressed: Ford's sales plunged to their lowest level since the 1950s.

'Captive' import models

Unable to compete with the Japanese in producing small fuel-efficient vehicles, US-owned companies began to import models from Japan partners instead. These so-called 'captive' imports were produced in Japan and sold in the United States carrying the name of an American company. As Japanese companies opened plants in North America, they reduced the number of imports, while US-owned companies increased 'captive' imports.

Lacking the resources to develop a subcompact model to compete with General Motors and Ford, let alone foreign producers, Chrysler in 1971 became the first American company to import Japanese 'captives' from Mitsubishi Motors, then a subsidiary of Mitsubishi Heavy Industries, which made everything from Nikon cameras to supertankers. Chrysler had imported small cars during the 1960s from Simca and Rootes, its French and British subsidiaries, but they sold poorly and had difficulty meeting more stringent US government safety standards (Moritz and Seaman 1981: 116; Flink 1988: 332).

The joint venture gave Mitsubishi its first access to dealers in the United States and an infusion of cash for plant modernization, although Chrysler's frequent financial problems impeded the flow of cash. Chrysler first agreed to buy 35 per cent of Mitsubishi Motors but ran short of cash and reduced its commitment to 15 per cent. During the 1980s, Chrysler increased its ownership to 24 per cent but then reduced it to 12 per cent. By the early 1980s, Mitsubishi was supplying Chrysler with 100,000 'captive' imports a year.

General Motors in September 1971 acquired 34.2 per cent interest in Isuzu, which at the time made exclusively commercial vehicles and accounted for only three per cent of Japanese production. GM agreed to strict limits on control of the company, including the stipulation that the chairman of the board and president must be Japanese. GM's stake in Isuzu was subsequently raised to 41.6 per cent and lowered to 38 per cent. A few years later, GM bought 5.3 per cent of another Japanese producer, Suzuki.

Isuzu and Suzuki provided General Motors with 'captive' imports beginning in 1984, and sales of the two models exceeded 150,000 by 1986. The two cars, plus the one produced at GM's joint venture plant with Toyota, were designed to give Chevrolet dealers a stable of small cars, consistent with the division's traditional corporate role as the lowest-price entry-level nameplate. When sales proved disappointing, GM changed the names of the 'captive' imports from Chevrolet to Geo, a strategy that proved successful.

Ford began negotiations to buy 20 per cent of Toyo Kogyo in 1972, but negotiations collapsed, because the companies couldn't agree on a price or on whether Ford would have the right to increase its holdings in the future. At the time, Toyo Kogyo had achieved success selling Mazdas equipped with a rotary engine, which provided faster acceleration than conventional ones. However, because the rotary engine got only eleven miles per gallon in city driving, Mazda sales dropped after the 1973–74 petroleum crisis by nearly one-half world-wide and more than one-third in the United States. Rescued from near bankruptcy by Japanese banks, Toyo Kogyo cut costs and eliminated production of the rotary engine. The company was again profitable by 1979, when Ford reopened negotiations and this time purchased a one-fourth interest.

Ford began to import 'captive' imports from Mazda in 1988, but the vehicles were not assembled in Japan. Instead, Mazda shipped parts to a Ford plant in Hermosillo, Mexico, where the vehicles were assembled and sold in the United States under Ford nameplates. In the 1990s, Ford had effectively muddled the concept of 'captive' imports: more than three-fourths of the parts used at Hermosillo were shipped from the United States, but the vehicle was designed in Japan, based on a Mazda model.

As production costs in Japan increased during the 1980s, US firms turned to Korea to expand 'captive' imports, further entangling a web of global interaction. General Motors acquired a 50 per cent interest in Daewoo Motor Company, which then built a 'captive' model for sale in North America based on GM's German Opel. Ford sold a 'captive' subcompact model in the United States built by Kia Motors Corporation, in which Ford had acquired a 10 per cent interest; further, Mazda, one-fourth owned by Ford, purchased 8 per cent of Kia and assisted in the design of the 'captive' model. Chrysler obtained a 'captive' model from Hyundai, 15 per cent owned by Mitsubishi, in which Chrysler owned a stake. 'Captive' imports from Japan and Korea accounted for 3 per cent of the North American market in the early 1990s.

Ford closes its coastal plants

Among US producers, Ford's changing distribution of assembly plants during the 1980s most clearly displayed the impact of new products. During the early 1970s, Ford retained the branch plant concept as much as possible in allocating products among assembly plants. By 1990, the company had concentrated most of its assembly operations in the interior of the country and had closed its coastal facilities.

In 1973, Ford built subcompacts at San Jose for the west and at Edison for the east and intermediates at Atlanta for the south and at Lorain for the north. Full-sized Fords were assembled at seven plants, including Los Angeles for the west, Mahwah for northeast, Norfolk for the southeast, and Chicago, Louisville, St Paul, and Wayne for the interior. Products with limited sales were distributed nationally from one plant, including compacts from Kansas City, full-sized Mercurys from St Louis, Mustangs from Dearborn, and luxury models from Wixom. Five of the fifteen automobile assembly plants were located in coastal states.

After the closures of the early 1980s, Georgia was the only coastal state containing a Ford assembly plant, although the plant is actually located at

Atlanta, 400 km inland. During the early 1990s, Ford's subcompacts were produced at Wayne, compacts at Kansas City and St Louis, intermediates at Atlanta and Chicago, full-sized Ford and Mercury models at St Thomas, luxury Lincolns at Wixom, Thunderbirds at Lorain, and Mustangs at Dearborn.

Ford reduced its number of automobile assembly plants in the United States and Canada during the 1980s from fifteen to nine. Three automobile plants were closed, two in 1980 and a third in 1983, while production was shifted from automobiles to light trucks at three others. Consistent with neoclassical locational theory, freight costs were clearly critical in Ford's choice of three plants to close. All three closures were located near large east- and west-coast population concentrations, including Los Angeles and San Jose, California, and Mahwah, New Jersey, near New York City. One of the three North American plants converted from car to truck production was also a coastal location, Edison, New Jersey. The other two conversions were at Louisville, Kentucky, in the I-65 corridor, and Oakville, in Canada. Ford also opened a new van assembly plant in Avon Lake, Ohio (Figure 6.5).

Ford's announcement of the San Jose closure in 1983 revealed that transport costs figured prominently in the decision. According to William E. Scollard, Ford vice-president for body and assembly operations,

> We considered every possibility to keep the plant open in order to maintain employment but we could find no way in the present economy and with today's California market to make the plant viable. Backshipping units from San Jose to markets east of the Rocky Mountains would be uneconomical and would simply cost the jobs of other Ford employees at plants outside California.
>
> ('Ford blames imports for plant shutdown' *Automotive News* 1982: 9)

The 200,000-m^2 San Jose plant was sold in 1983 to the Mariani Financial Company of nearby Cupertino, California, for development of a high-tech industrial park ('Last Ford plant in Calif. sold for high-tech park', *Automotive News* 1983: 6).

GM restructures its assembly plants

General Motors struggled to find a balance between two approaches to maximizing profits. On the one hand, the company tried to sell more automobiles – and therefore increase revenues – by producing a diversity of products. On the other hand, the company wanted to reduce costs by standardizing components and producing more twin models.

GM's traditional strength came from Sloan's strategy of marketing five nameplates clearly differentiated by price and image. By the 1970s, GM's policy of standardizing components was criticized for producing

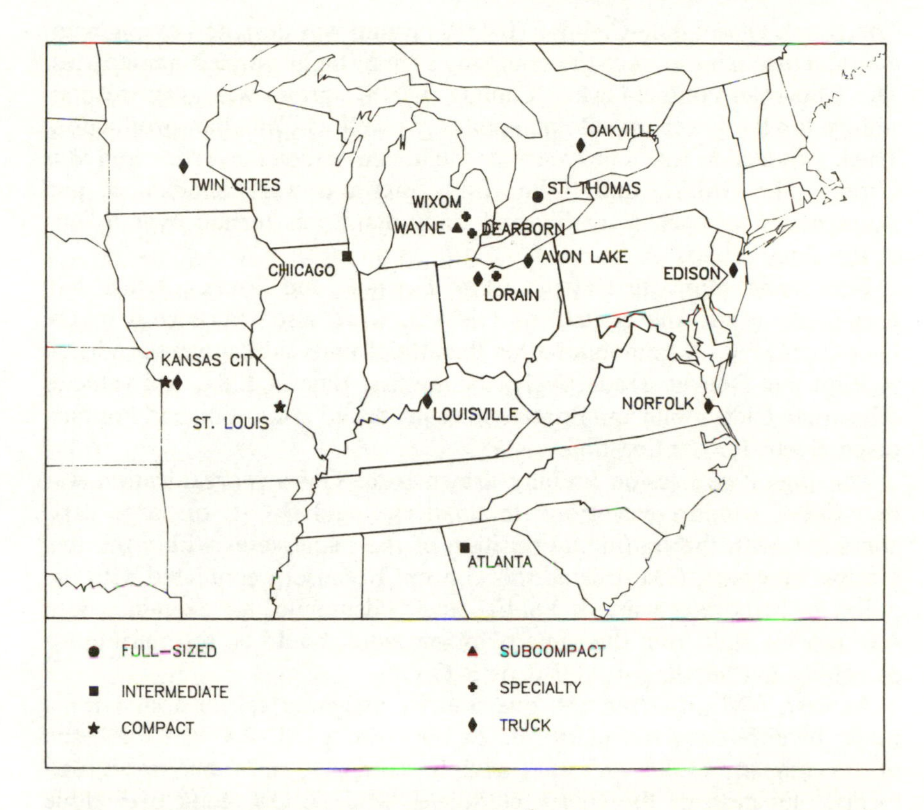

Figure 6.5 Ford Motor Company assembly plants by type of model produced, 1991. Ford concentrated its US automotive production in a handful of interior states

look-alike corporate twins, and one division's increased sales were obtained by cannibalizing those of another. To increase diversity, GM tried to find a compromise – give the marketing divisions more latitude in designing distinctive bodies and trim but further standardize the chassis, engine, transmission, and other platform elements hidden from the consumers. This 'skin-deep' diversity fooled consumers to some extent; Buick's LeSabre model, for example, received higher consumer satisfaction ratings than the Oldsmobile 88 and Pontiac Bonneville, which were nearly identical mechanically.

In an attempt to rationalize its assembly plant system, GM created three new groups in 1984, based on its traditional marketing divisions.

The Buick-Oldsmobile-Cadillac (B-O-C) group was designed to engineer, manufacture, and market the company's three higher-priced nameplates. The Chevrolet-Pontiac-GM of Canada (C-P-C) group was given responsibility for the lower-priced nameplates, as well as Canadian production. Truck production and sales were consolidated under the Truck and Bus Group. The GMAD and Fisher Body divisions were eliminated, and management of each assembly and body plant was turned over to one of the three groups.

The 'home' plants in Detroit, Flint, Lansing, and Pontiac, which had remained outside the control of GMAD, were also transferred to the B-O-C or C-P-C. Automobiles other than Buick and Cadillac were produced at Flint and Detroit, respectively, for the first time in 1984, and vehicles other than Oldsmobile and Pontiac were produced at Lansing and Pontiac, respectively, for the first time in 1985.

The logical conclusion initially drawn from GM's reorganization was that C-P-C would concentrate on small cars and B-O-C on large cars, consistent with the traditional position of the nameplates within the two groups. However, GM resisted this concept because it conflicted with the policy of corporate twins. If Pontiac and Oldsmobile, for example, both sold models built from the same platform, who should be responsible for designing and building it, C-P-C or B-O-C?

At first, GM answered the question by assigning responsibility for a range of different-sized platforms to both groups. C-P-C got one each of the company's full-sized, intermediate, compact, and subcompact platforms, plus most of the sporty platforms, while B-O-C took over three full-sized, two compact, and two sporty and luxury platforms. This arrangement proved awkward, because allocating similar-sized platforms to C-P-C and B-O-C hindered reduction in the total number and contributed to underutilization of plants. Consequently, in 1989 GM gave B-O-C responsibility for all three compact platforms, in preparation for building all of them on one platform during the 1990s. To rationalize its production operations futher, B-O-C in 1988 established three divisions, known as Flint, Lansing, and Cadillac. The Flint Division had responsibility for the full-sized platforms, Lansing for the compacts, and Cadillac for the full-sized and sporty platforms sold principally with the Cadillac nameplate.

B-O-C's assembly plants were highly clustered in the US interior (Figure 6.6). In 1991, B-O-C assembled compacts at three plants, including Lordstown, Ohio, Wilmington, Delaware, and two lines at Lansing, Michigan. Full-sized models were assembled at Wentzville, Missouri, plus three Michigan plants, including Flint, Hamtramck, and Orion. Only the Wilmington plant was located in a coastal state. The group faced the possibility of shutting one compact line, depending on the extent to which GM's Saturn model detracted from sales of its other compacts. Annual sales of GM compacts (excluding Saturn) could fall well below 800,000, the level needed to operate four assembly lines at capacity.

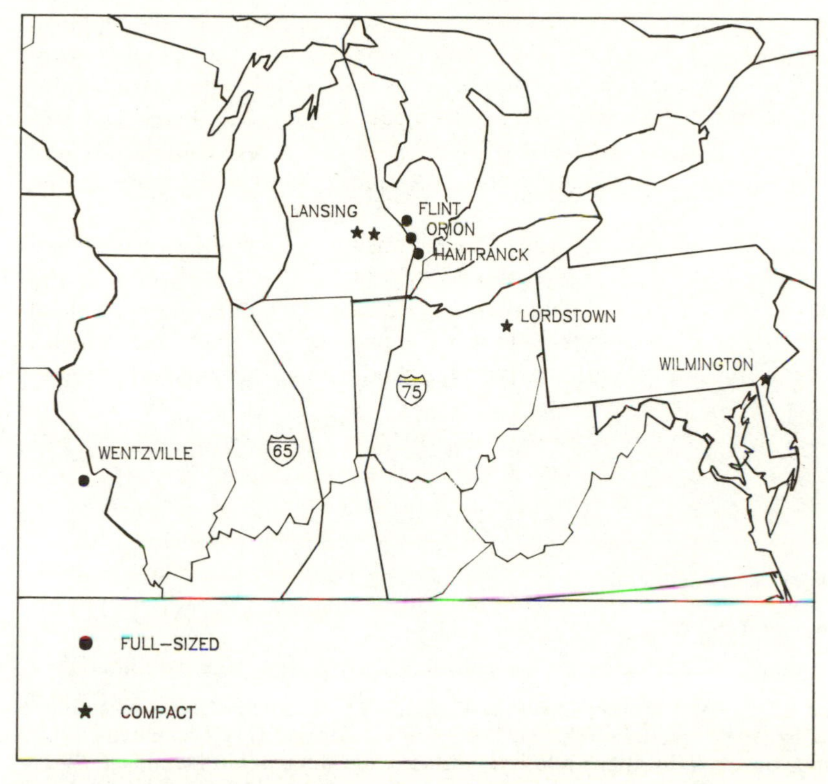

Figure 6.6 General Motors Buick-Oldsmobile-Cadillac Group assembly plants by type of model produced, 1991. B-O-C assembly plants were highly clustered in southeastern Michigan

C-P-C's operations were less clustered in part because of responsibility for the Canadian assembly plants, including Oshawa, Ontario, where intermediates were built on two lines, and Ste Therese, Quebec, which began to build the Camaro and Firebird sports models in 1993. Within the United States, C-P-C operated three centrally located assembly plants, including Willow Run, Michigan, which made full-sized models during the 1990s, and Fairfax, Kansas, and Oklahoma City, which made intermediates. However, the group was also saddled with several peripheral plants that faced an uncertain future, especially given projections of declining sales for the group's models (Figure 6.7).

JAPANESE FIRMS BUILD ASSEMBLY PLANTS IN THE INTERIOR

The number of models assembled in the United States and Canada increased during the 1980s primarily as a result of the opening of Japanese-operated assembly plants. Like their US competitors, Japanese firms

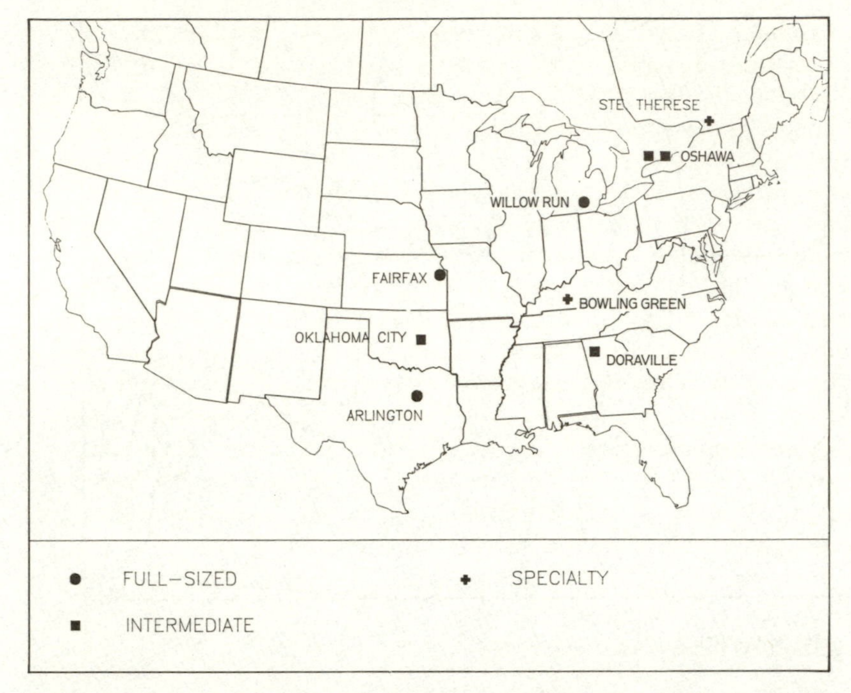

Figure 6.7 General Motors Chevrolet-Pontiac-GM of Canada Group assembly plants by type of model produced, 1991. C-P-C operated GM's remaining assembly plants in locations relatively distant from the automotive industry's core region. However, some of these plants were expected to close during the 1990s

concluded that the optimal location for assembly operations was in the interior of the country.

Import quotas

Japanese auto makers constructed transplants in part because of fear that US sales would be restricted through the imposition of import quotas. Under intense pressure from the newly elected Reagan administration, the Japanese government adopted voluntary restrictions in 1981 on the number of automobiles which could be imported to the United States. The limits were originally set at 1.68 million vehicles, or 22 per cent of the US market, and raised to 1.85 million vehicles in 1983. Following record profits by domestic car makers in 1984, the United States government in 1985 decided not to request voluntary restrictions any more, but the Japanese government maintained a limit of 2.3 million as a sign of willingness to reduce the trade deficit between the two countries.

The Japanese Ministry of International Trade and Industry told each auto maker how many vehicles it could import to the United States during that year. Originally, the government allocated import limits to each company based on its share of the US market immediately prior to the initiation of quotas. This allocation system favored Toyota, Nissan, and Honda, the first three Japanese entrants into the US market. When the import limits were raised after 1983, most of the additional units went to the other companies, such as Mazda and Mitsubishi.

The purpose of the quotas was to allow US car makers time to retool to produce cars that could compete in price, quality, and fuel efficiency with the Japanese imports. However, the main impact of the quotas was to increase profits for Japanese and – in the short term – American producers, as well. Because demand for Japanese cars in the United States exceeded supply, Japanese firms started to import their most expensive models and sell them at relatively high mark-ups. Japanese carmakers further swelled profits by importing primarily vehicles loaded with additional-cost options, and American consumers had difficulty finding 'stripped-down' versions. Meanwhile, US firms also swelled profits by charging higher prices and postponed some of the modernization and product development programs which could have made them more competitive in the 1990s.

Allocation of models to transplants

Japanese producers for the most part emulated the American sourcing strategy of allocating one vehicle platform per line. In the early 1990s, Honda built its compact Accord model at Marysville, Ohio, and its four-door subcompact Civic model at nearby East Liberty. Toyota built the compact Camry at Georgetown, Kentucky, while Nissan built the subcompact Sentra at Smyrna, Tennessee. To minimize the cost of distributing the products throughout North America, all of these transplants were placed in the I-65 and I-75 corridors.

Feasibility studies helped some of the Japanese firms recognize the advantages of selecting interior locations. For example, Mazda in 1983 considered three principal factors in comparing its three finalists – Greenville, South Carolina; Omaha, Nebraska; and Flat Rock, Michigan. First was the handling and distribution costs associated with obtaining components at the three locations. Second, the company looked at differences in labor relations and employment conditions. Third, Mazda compared grants and incentives which the various local governments were prepared to offer.

Given Mazda's criteria, the lower costs associated with obtaining components from nearby suppliers was clearly the most critical locational factor in the selection of Flat Rock. The Flat Rock site was next to the Michigan Casting Center engine plant vacated a few years earlier by Ford. However, Mazda officials denied that the choice had been influenced by Ford, which owned one-fourth of the company (DeLorenzo 1984: 2; Johnson 1985a: 2; 'Toyo Kogyo said to eye assembly in US', *Automotive News* 1983: 1).

Sales of the Honda Accord and Civic, Toyota Camry, and Nissan Sentra were all sufficiently brisk to utilize fully their North American assembly plants. However, given their reluctance to mix more than one platform on the same line, the other Japanese companies had to devise strategies to extract enough sales from a single platform to justify building a US assembly plant. How could an assembly plant be justified in the United States for a model that was likely to sell fewer than 100,000 per year, less than half of the annual capacity of a typical North American assembly plant? Mazda – the first transplant to face the issue – set a precedent by reaching an agreement with Ford to share the output from Flat Rock. In 1987, the plant began production of sports cars for both Ford and Mazda, with the same platform and engine but different bodies. Three years later, Mazda added assembly of a four-door compact based on the same platform.

Like Mazda, Mitsubishi was having trouble projecting the one-quarter million annual sales needed in North America to justify construction of an assembly plant. Consequently, Chrysler, which then owned one-fourth of Mitsubishi, was approached about establishing a joint venture plant. Chrysler had filed a lawsuit challenging the GM-Toyota joint venture, but dropped it in April 1985, when Chrysler President Lee Iacocca flew to Japan to announce the agreement with Mitsubishi. The joint venture was named Diamond-Star, after the symbols of the two parent companies, Mitsubishi's three diamonds and Chrysler's pentastar. The first president and chief executive officer were from Mitsubishi, and the day-to-day operations were managed by the Japanese, but Chrysler nominated the first chairman, an executive vice-president, and the chief financial officer. The joint-venture agreement lasts for ten years, although it can be extended (Hartley 1985: 1; Johnson 1985b: 2; Sorge 1985b: 1; 'Midwest for site?' *Automotive News* 1985: 1).

Officials of the two companies quickly agreed on a site somewhere in the interior, but Mitsubishi ruled out states that already had a Japanese transplant, thus eliminating Ohio, Tennessee, and Michigan. Chrysler's Belvidere, Illinois, plant was rumored to be the choice, but Mitsubishi preferred to build a new plant in the center of the state, near the town of Normal. Production began in 1989 for the 1990 model year cars (Johnson 1985c: 2; McCosh 1985: 2; Sorge 1985b: 1; 'Mitsubishi plant choice narrowed to 4 states', *Automotive News* 1985: 263).

Like Mazda, Diamond-Star has produced a variety of cars that utilize the same platform. Initially, the Normal plant assembled three sports cars, two sold by Chrysler and one by Mitsubishi. Chrysler designed two models to provide new products for both the Eagle division recently acquired from American Motors and the older Plymouth nameplate. A four-door subcompact model based on the same platform was added for Mitsubishi and Eagle in 1991 in order to utilize fully the assembly line.

As the North American motor vehicle market continues to become more fragmented, firms are less likely to produce models whose sales are high enough to require production at more than one assembly plant. With the output of most platforms limited to one assembly plant, production was expected to cluster increasingly in the I-65 and I-75 corridors in order to minimize distribution costs.

7 Just-in-time delivery

GM's 1986 Buick LeSabre was one of the poorest quality cars sold in the United States, according to both independent evaluators such as J.D. Power and Associates market research firm and *Consumer Reports* magazine, as well as GM's internal audits. The LeSabre's windows were especially troubling: they rattled, failed to roll up, and even fell out of tracks on occasion.

The 1986 LeSabre was the first car to be assembled at GM's Flint assembly plant following a several-hundred-million-dollar renovation of the 75–year-old facility. Several Flint-area operations, including a former Fisher Body stamping plant, were consolidated at Buick City, the new name given to the complex. Several outside suppliers located at Buick City, as well, to facilitate timely delivery of parts to the final assembly line.

Buick City was modelled after Toyota City, built up in Japan during the 1950s. And where did Toyota get the idea of consolidating assembly, component, and supplier operations in one complex? According to a 1983 statement by Ford chairman Phillip Caldwell:

> Eiji Toyoda told me himself in Tokyo, last year, there was no mystery to the development of Toyota in Japan. He merely came to see the Ford Rouge Plant in 1950 – and then went back to Japan and built the same thing.
> ('Did Ford Rouge set pattern for Toyota?', *Automotive News* 1983: 42)

Buick City represented GM's most ambitious attempt to date to operate an assembly plant according to just-in-time delivery, known as 'kanban' in Japanese. Under just-in-time delivery, suppliers ship components to the assembly plant as needed rather than to a storage area. By storing parts for only a few hours instead of five days, Buick City expected to lower the value of its inventory to $30 million, instead of $100 million, the level then typical of assembly plants. Reduced inventory also meant that Buick City could be smaller in area than other recently built assembly plants.

At first, Buick City's just-in-time delivery system did not work smoothly. To ensure that parts were available when needed, some suppliers shipped well in advance rather than just-in-time. Because they were not delivering according to the established pinpoint timetable, suppliers' trucks sometimes jammed the loading docks. Crates overflowed the designated storage areas and choked the aisles. Other components failed to arrive when needed and had to be flown in by helicopter (Krebs 1989: 24–25). However, within three years, the just-in-time delivery system was operating smoothly, and Buick City regularly ranked near the top in GM's internal audits of assembly plants. J.D. Power's initial quality and customer satisfaction surveys rated the Buick LeSabre near the top, and LeSabre's sales increased rapidly, in contrast to most other US-made passenger cars.

CHARACTERISTICS OF JUST-IN-TIME DELIVERY

Under just-in-time, suppliers receive schedules of needed components perhaps ten days in advance and then more precise schedules showing the hour when components will be used on the assembly line during the next five days. To meet tight timetables, suppliers may need to locate some production facilities near final assembly plants, typically within 150 km. Because new assembly plants have been located in the I-65 and I-75 corridors, suppliers have been attracted to the area as well.

William J. Hamel, then Cadillac's general manager for manufacturing, pointed to the locational implications of inventory reduction:

> You're going to get to the point of recentralizing not so much because of distance and the time element, but the cost. You can't afford to have your money tied up in pieces of metal or any kind of material and not working, with the cost of capital what it is today.
>
> (Rowand 1982c: 8)

However, just-in-time delivery involves other changes in relationships between final assemblers and components suppliers in addition to tight delivery schedules. As they acquire more responsibility, some suppliers have decided to relocate production outside the I-65 and I-75 corridors, especially to take advantage of lower cost labor. Deregulation and enhanced competition within the truck and rail industries have made more feasible a variety of locational alternatives.

Outsourcing

Traditionally, the large US producers, especially General Motors and Ford, manufactured most of their own components. GM bought independent companies such as Packard Electric and Harrison Radiator in order to assure a supply of reasonably-priced components, while Ford concentrated production of components at the Rouge complex in Dearborn. GM and Ford produced their own engines, drivetrains, bodies, and many of the accessories. The availability of parts from subsidiaries had been considered a source of strength and a principal factor in the long-term market dominance enjoyed by the two corporations. The production of large batches of identical parts enabled Ford and GM to achieve lower production costs through greater economies of scale compared to Chrysler and the smaller independent firms.

Determining the percentage of in-house components is difficult. The percentage of all components which arrive at final assembly plants from in-house suppliers is approximately three-fourths at General Motors, one-half at Ford, and 40 per cent at Chrysler. However, the manufacture of components requires obtaining some inputs, such as steel, from independent firms. If the value of inputs purchased from outside suppliers is taken into account, the percentage of in-house production would decline to approximately one-half at GM, 40 per cent at Ford, and one-third at Chrysler. Some estimates have placed the percentage of in-house components at Chrysler as low as 15 per cent (Vartan 1986: D-6).

To a considerable extent, just-in-time simply transfers the burden of carrying inventory from the assembler to the supplier. Suppliers are forced to carry larger inventories to protect themselves against unexpected surges in demand. Dennis Kessler, vice-president of Fel-Pro, which produces gaskets at a plant in Skokie, Illinois, pointed out how just-in-time has shifted the burden to suppliers: 'The idea is to reduce inventories. But the car makers haven't learned how to forecast their needs enough to give the commitments that the Japanese give to their suppliers' (McCormick 1983: 40).

Increased outsourcing has transferred responsibility for making calculations concerning the optimal spatial division of labor from the handful of producers to a larger number of companies. To minimize production costs, independent suppliers have transferred some production to plants in areas where prevailing wages are relatively low. However, savings in production costs must be weighed against increased freight charges. With increased reliance on just-in-time delivery, independent suppliers must locate other production facilities within 150 km of final assembly plants. This need for proximity has encouraged not only retention of older plants in the southern Great Lakes region which might otherwise have been closed, but also construction of new facilities along the I-65 and I-75 corridors south of Michigan, where most of the new assembly plants have clustered.

The locational strategy adopted by some of the independent suppliers is to construct plants in the I-65 and I-75 corridors where bulky modules, such as seats and instrument panels, are put together from a number of parts. These facilities permit just-in-time delivery to the final assembly plants in the corridor. However, plants responsible for the parts which comprise the modules may be located elsewhere in the country, perhaps in Michigan for skilled tasks and the south for unskilled tasks.

Quality

Traditionally, producers, concerned more about the expense than the quality of parts, awarded contracts to suppliers annually. A firm which held a contract one year could lose it the next if a competitor submitted a lower bid. Product quality suffered, in part because some suppliers cut corners in the production process to reduce costs. More importantly, quality suffered when parts for a particular model were made by more than one supplier from one year to the next. One supplier's batch for a particular model may not match precisely the parts produced by another supplier, as a result of differences in tools, raw materials, or production methods.

General Motors officials, in particular, firmly retained faith well into the 1980s that the annual competitive bid system prevented individual suppliers from gaining too much power and reduced costs (Thomas 1987: 20). To dilute the strength of suppliers further, GM would place an order for a part with more than one company in a given year as a matter of policy (Harney 1989: E30). GM finally abandoned the multiple annual sourcing policy in the late 1980s, primarily because it caused higher prices, not merely lower quality. Suppliers had little incentive to submit low bids to produce particular parts, because they knew that with GM's multiple sourcing policy they would obtain some of the business anyway.

Under the just-in-time system, suppliers are evaluated according to quality. Auto makers achieve lower prices for components not by attacking suppliers' overhead margins but by working with suppliers to reduce production costs. Ford, for example, designates its highest quality suppliers as 'Q-1' and is committed to terminating contracts with firms unable to achieve the 'Q-1' rating within a few years. The term 'world-class suppliers' is generally applied to the leading firms.

David C. Collier, then GM vice president in charge of operating staffs, revealed the importance of improved quality in just-in-time operations.

> The effect on inventory is not the only thing driving just in time at GM, although that's a very handsome fallout. The effect on capital is very significant but most important is the effect on quality. Quality has to be 100 per cent if you're going to run a just-in-time system. The system highlights any problems that are emerging and gets you on top of them. The fallout on the quality side of the game is probably much more important than the working capital savings.
>
> (Rowand 1982c: 8)

The increased emphasis on quality has forced many suppliers to restructure their operations, from engineering, finance, and marketing, through purchasing and manufacturing, and absorb additional costs. Suppliers must offer guarantees against defects and longer-term warranties and submit to random audits. On the other hand, to encourage the development of high-quality parts, producers now give long-term contracts to suppliers, extending perhaps over the life of the automobile or truck model for which the parts are designed. Armed with the stability from long-term contracts, suppliers can invest more in design and engineering to improve the quality of parts and reduce production costs. They also have incentive to invest in new facilities; given the need to deliver parts rapidly to the assembly plants, these new supplier plants are likely to be located in the I-65 and I-75 corridors, where virtually all of the new assembly plants have been built.

As part of the annual contract system, producers designed components and then invited suppliers to manufacture them in conformance with precise specifications; most design and engineering information remained confidential. Under just-in-time, producers share previously proprietary information about vehicle design and engineering, and get the suppliers involved in the planning process several years earlier than in the past. Johnson Controls, which builds seats for Toyota's Georgetown, Kentucky, plant, even receives assistance from a competitor, Arakawa Auto Body, which supplies seats to Toyota's Japanese assembly plants. The companies are willing to share information because they both benefit if Toyotas are well received by consumers (Thomas 1987: 20). In exchange, suppliers bear more of the financial burdens and risks in the design process.

Cooperation extends to the production stage as well. When a part doesn't fit properly, a supplier may send a few hourly workers to the assembly line to diagnose the problem. Rather than a defect in the manufacturing of the part, the problem may be in handling and installing it on the assembly line. To facilitate cooperation, large suppliers need to locate some operations near the executive, research, development, and testing facilities of the large producers, which are clustered along the I-65 and I-75 corridors.

Hierarchy of suppliers

Suppliers to US assembly plants have been organized into a hierarchy or tiers, emulating a long-time Japanese practice. Ford reduced its total number of components suppliers from 3,700 in 1979 to approximately 2,000 in 1990, with a further reduction of several hundred expected during the 1990s (Harney 1989: E30; Versical 1989: E12). GM anticipated reducing its number of direct suppliers from 20,000 to 10,000 (Sorge 1983a: 2). The number of direct suppliers of components to a typical US assembly plant has been reduced by roughly one-fourth, from an average in excess of 1,000 in the early 1980s.

First-tier suppliers in turn obtain parts from firms now known as second-tier suppliers, many of whom previously sold products directly to the assembly plants. Other firms that once supplied assembly plants directly may be dropped even as indirect suppliers if the quality of their products is considered inferior. In Japan, suppliers are organized into as many as ten tiers, but not that many levels have been formally identified thus far in the United States.

First-tier suppliers provide producers with modules like seats, instrument panels, and suspension systems, rather than individual parts. Seats, for example, encompass several hundred parts, such as metal frames, foam, cloth, and motors. Instead of putting these parts together at the final assembly plant, seats are now made elsewhere by first-tier suppliers and shipped as modules ready to install in the vehicles. Suspension systems are shipped as modules rather than as separate parts, such as shock absorbers, struts, and bearings. The tiering system favors larger firms as surviving first-tier suppliers. Winning long-term contracts is an expensive process, involving considerable expenditure of up-front funds, without the assurance of success. Only large companies are in a position to supply the increasing variety of a particular component required by the proliferation of distinctive models.

Because these suppliers must deliver relatively bulky items to producers, transportation costs of products have increased relative to those of inputs. First-tier suppliers therefore opt for locations near their customers – the assembly plants – to minimize aggregate transportation costs. However, second- and third-tier suppliers may not be operating on just-in-time delivery, as is the case for most first-tier suppliers. The lower tiered suppliers may therefore locate in the south, Mexico, or some other low-wage area, rather than in the I-65 and I-75 corridors.

Smaller suppliers, who can't afford to build new facilities next to every assembly plant, are turning to warehousing operations. Components are stored near assembly plants in warehouses run by independent firms who take responsibility for auditing inventory levels and delivering parts as needed to the point of final assembly (Walsh 1983: 16).

IMPACT OF JUST-TIME-ON LOCATION OF SUPPLIERS

The distribution of Japanese-owned suppliers within the United States clearly shows the importance of just-in-time delivery on locational decisions. Eighty per cent of the Japanese suppliers have located in the I-65 and I-75 corridors, for the most part in the late 1980s, to be near the new assembly plants. A few pioneering Japanese suppliers first considered US production in the late 1970s, and by the mid-1980s nearly 50 had trickled in. But the number quadrupled to more than 200 during the second half of the 1980s.

The largest percentage of Japanese suppliers already know their principal customers before they come; typically, they are asked to build plants in order to supply the Japanese-owned assembly plants in North America. In fact, 'asked' may be too polite a term for the process; Japanese producers compel some of their suppliers to build plants in North America as a condition for continuing to buy components from them back in Japan. These 'captive' suppliers can't afford to alienate a producer, who may be purchasing their entire output.

The first four Japanese-owned US assembly plants – Honda, Nissan, Mazda, and Toyota – have each been responsible for bringing over two dozen of these 'captive' suppliers from Japan; altogether, the longer established transplants each purchase supplies – including both components and other materials – from roughly 200 North American plants. Initially, domestic content in transplant models runs between 50 and 60 per cent, but once powertrain plants are built in North America a 75 per cent level can be reached.

Some of the larger Japanese firms, such as Mitsubishi Electronics, Nihon Radiator, and Nippondenso, opened plants in North America in order to sell to the Big Three US-owned producers, all of whom want to expand their outsourcing. These large independent firms are competing to some extent with long-established US-owned suppliers, but most of the contracts are being taken away from components divisions of the Big Three, especially General Motors.

Japanese-owned suppliers may be difficult to identify, because their American operations are usually known by English-language names, such as Stanley Electric and Trim Masters, or by initials, such as NSK, ATR, and KYB. The best sources of information come from state economic development officials, who are not bashful about publicizing their new Japanese plants. Some of the Japanese producers, including Honda, Mazda, and Nissan, publicize the location of their American suppliers to illustrate their commitment to achieving a high percentage of domestic content in their US-made cars.

Distribution of Japanese-operated suppliers

The spatial distribution of Japanese suppliers within the I-65 and I-75 corridors differs from the historic pattern adopted by American firms. The largest number of US suppliers had clustered in southeastern Michigan, with most of the remainder in northern Ohio and northern Indiana, as well as in several large metropolitan areas elsewhere in the corridors, such as Chicago, Cincinnati, Dayton, and Indianapolis.

Japanese firms are more dispersed within the corridors. Ohio, rather than Michigan, has received the largest number of Japanese parts suppliers, roughly 20 per cent, while 15 per cent each have located in Indiana, Kentucky, and Michigan, 10 per cent in Tennessee, and 5 per cent in Illinois. Of the remaining 20 per cent located outside the six states, three-fourths have selected southern or western states and one-fourth northeastern states or Canada (Figure 7.1).

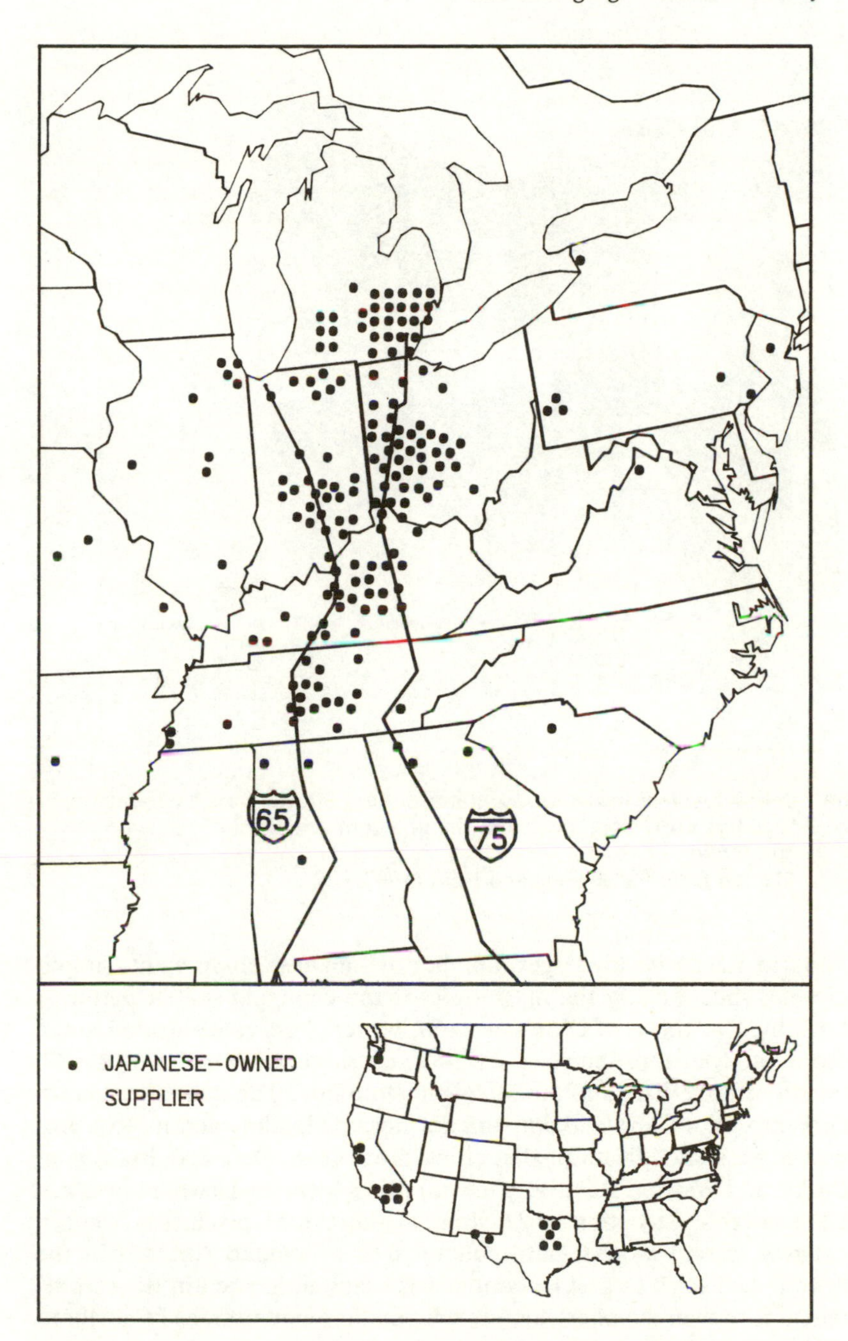

Figure 7.1 Japanese-owned parts suppliers in the US, 1990. Joint ventures with US companies are included but not existing plants in the US acquired by Japanese firms
Source: Rubenstein (1988c)

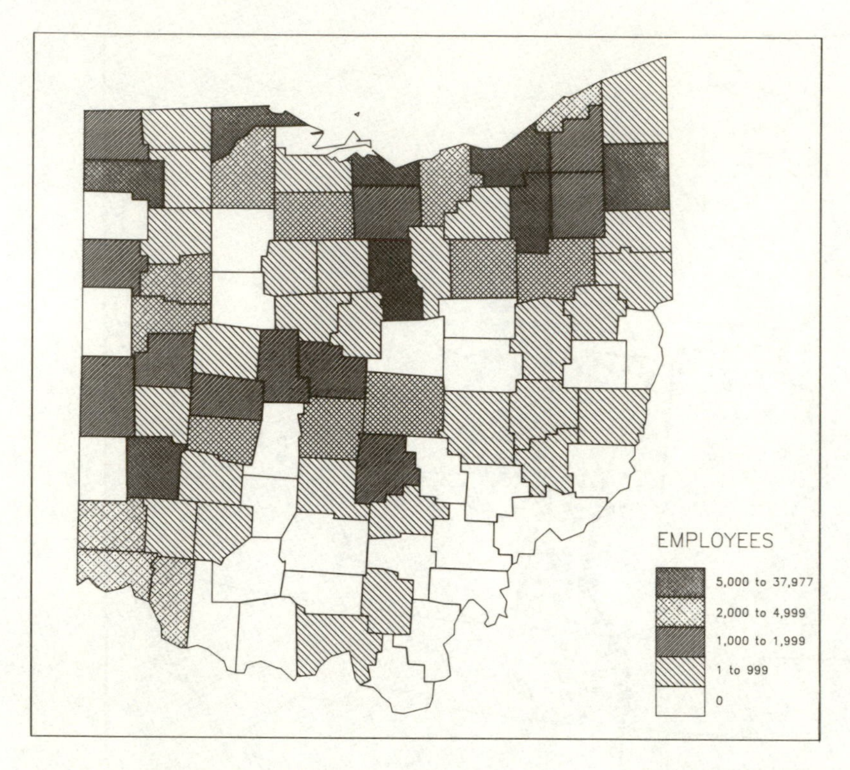

Figure 7.2 Employment in the production of motor vehicle parts by county in Ohio, 1986. Most firms were located in the northern one-third of the state or near Dayton, Ohio
Source: adapted from Rubenstein and Reid (1987)

Ohio, the state with the largest number of Japanese components suppliers, reveals most clearly the magnitude of the changing spatial patterns. Historically, two-thirds of Ohio's automotive suppliers concentrated in the northern one-fourth of the state, a 75-km corridor north of US Route 30 and south of Lake Erie and the Michigan state line. The two largest cities in northern Ohio, Cleveland and Toledo, became leading automotive production centers, along with smaller cities like Akron, Defiance, Mansfield, Sandusky, and Warren. Relatively few suppliers located elsewhere in Ohio, with the notable exception of Dayton, a major GM production center. The state's second largest metropolitan area Cincinnati (located in the southwest) and third largest Columbus (the capital, located in the center) contained far fewer suppliers than much smaller communities in northern Ohio (Figure 7.2).

In contrast, Japanese firms have shied away from northern Ohio, as only eight of the state's first forty Japanese suppliers located north of US Route 30. Furthermore, only one of the plants is located east of the Columbus

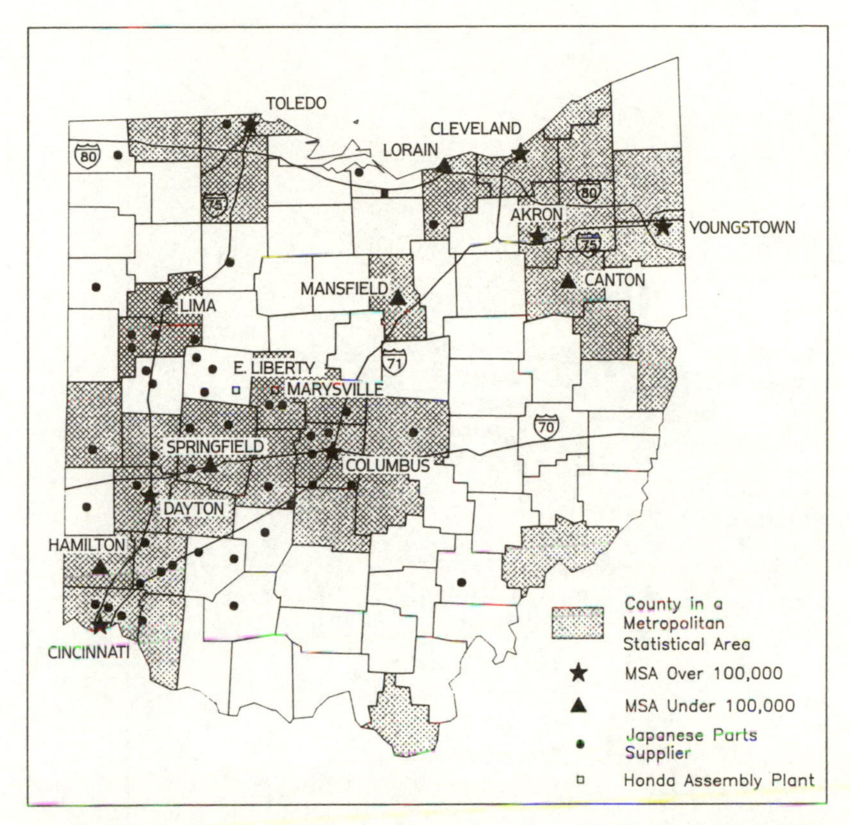

Figure 7.3 **Japanese-owned parts suppliers in Ohio, 1990. Compare with Figure 7.2; Japanese firms shun the northern part of Ohio, where US-owned firms have traditionally concentrated**

metropolitan area. Nor have many Japanese firms selected the Dayton metropolitan area, the state's leading automotive production center outside of northern Ohio. Instead, three-fourths have located in either the Columbus or Cincinnati metropolitan areas or else in nonmetropolitan or smaller metropolitan counties.

The map of Ohio shows how the Japanese firms have spread out among the small towns and counties in the west-central and southwestern areas of the state, one or in a few cases two per community (Figure 7.3). Median population of these Ohio communities with Japanese automotive plants is almost precisely 10,000 inhabitants. Honda's two assembly plants at Marysville and East Liberty, northwest of Columbus, certainly account for some of the bias towards the west-central part of the state, but the map shows that Honda sits at the edge rather than the center of the area preferred by Japanese firms. In the absence of compelling physical features such as mountains and major bodies of water, Japanese suppliers might be

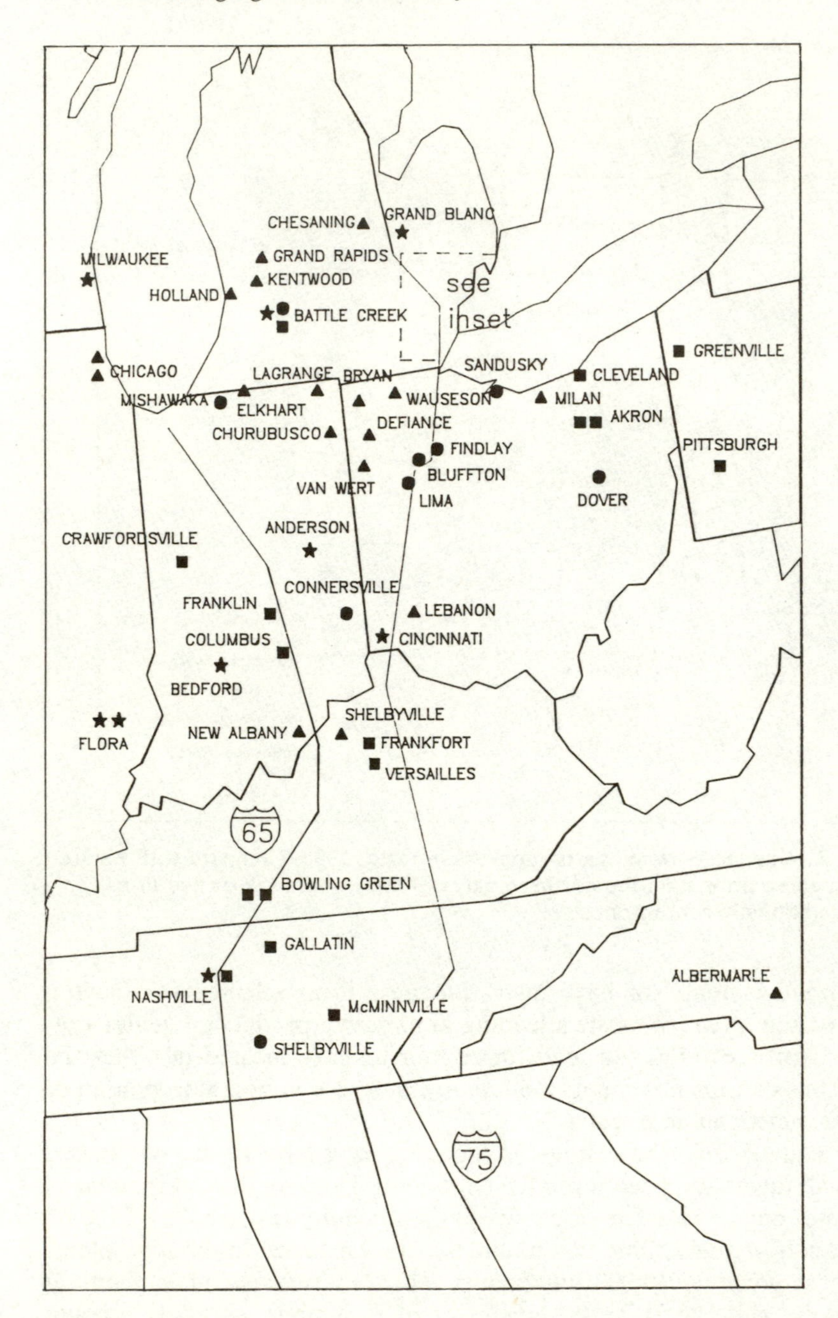
CHESANING
GRAND BLANC
GRAND RAPIDS
MILWAUKEE
KENTWOOD
see inset
HOLLAND
BATTLE CREEK
GREENVILLE
SANDUSKY
CHICAGO
LAGRANGE
BRYAN
CLEVELAND
MISHAWAKA
ELKHART
WAUSESON
MILAN
DEFIANCE
AKRON
CHURUBUSCO
FINDLAY
PITTSBURGH
BLUFFTON
VAN WERT
LIMA
DOVER
ANDERSON
CRAWFORDSVILLE
CONNERSVILLE
FRANKLIN
LEBANON
COLUMBUS
CINCINNATI
BEDFORD
SHELBYVILLE
FLORA
NEW ALBANY
FRANKFORT
VERSAILLES
65
BOWLING GREEN
GALLATIN
ALBERMARLE
NASHVILLE
McMINNVILLE
SHELBYVILLE
75

Fig 7.4

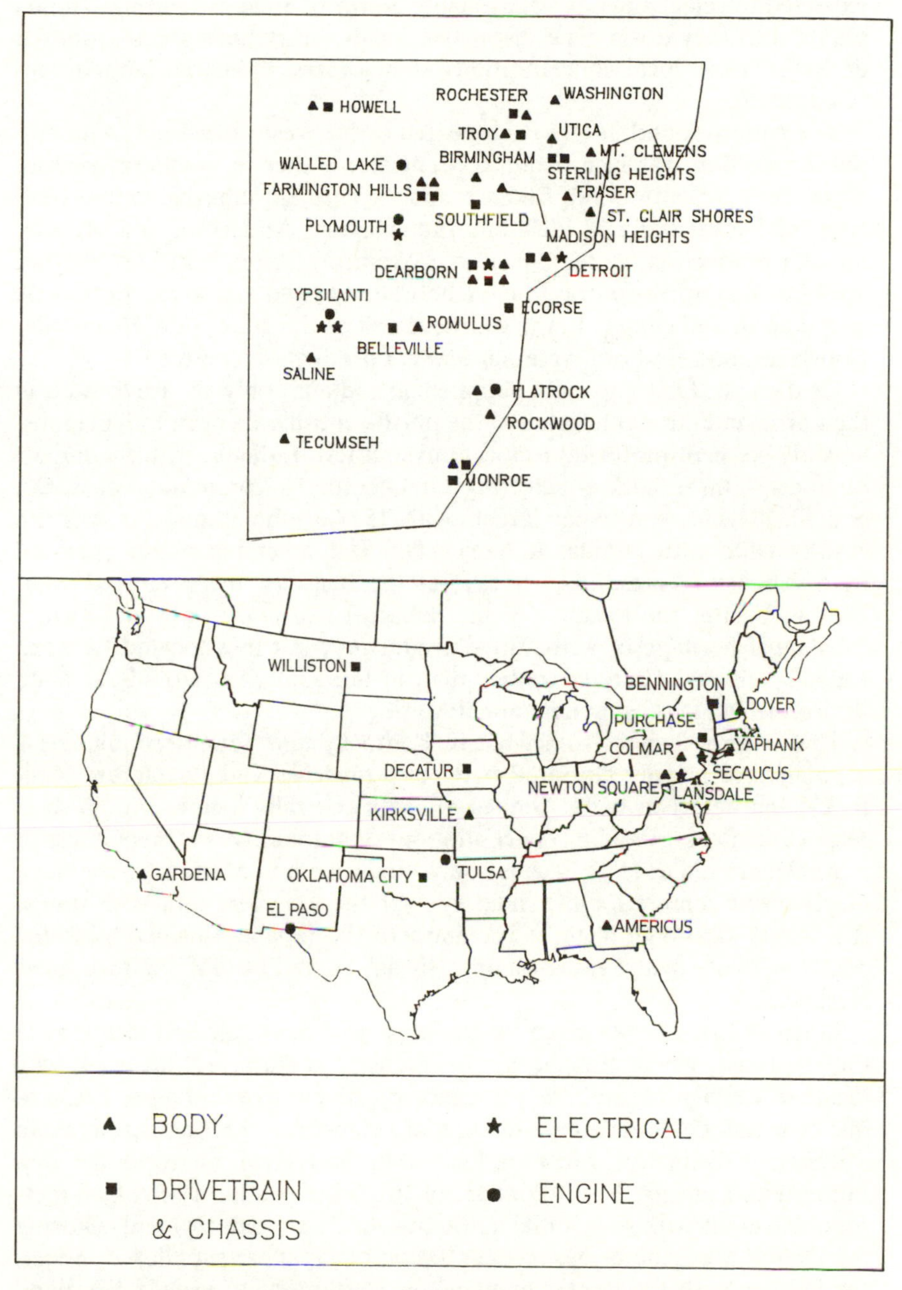

Figure 7.4 Location of US suppliers to Mazda's Flat Rock, Michigan, assembly plant, 1989. Mazda'a suppliers are highly clustered, primarily because the assembly plant is the only Japanese-owned one located in Michigan, the motor vehicle industry's traditional home state
Source: 'Mazda $1 billion to suppliers', *Automotive News* 1989: E-29

expected to select sites in a reasonably uniform circle around the Honda plants. But they don't. The distinctive locational pattern stems from the desire to avoid local concentrations of unionized industrial labor forces (Chapter 9).

The pattern found in Ohio is repeated to the west in Indiana. Automotive production was historically concentrated either in northern communities, such as South Bend, Elkhart, and the Chicago suburbs, or in several cities of central Indiana, including Indianapolis, Anderson, and Muncie. Smaller production centers, such as Kokomo, Marion, and Fort Wayne, are located in north-central Indiana between the two major concentrations in the north and center. In the southern half of the state, only Evansville, Columbus, and Bedford were automotive production centers.

Of the first thirty Japanese suppliers in Indiana, only six are located in the north, and three others are in the north-central area or in Indianapolis. Seventy per cent preferred to locate in southern Indiana, with the largest numbers – three each – selecting Greencastle, a community of 8,000, and Shelbyville, somewhat larger, with 15,000 inhabitants. As was the case in Ohio with Honda, Indiana's Japanese assembly plant, operated by Subaru and Isuzu, does not lie at the heart of the preferred area. East Lafayette, the location of the Subaru-Isuzu plant, may not have a substantial heritage of automotive production, but it is located between Indianapolis and Chicago, rather than in the southern part of the state where the Japanese suppliers are clustering.

The pattern changes somewhat in Kentucky and Tennessee. Japanese suppliers are heavily clustered between Louisville and Lexington, Kentucky, and near Nashville, Tennessee, both centrally located within their respective states. Avoiding concentrations of automotive workers is less of a problem in either state because only Louisville and Memphis have been employment centers for the industry over the long term. In both states, the critical locational factor is proximity to the Japanese-owned assembly plants – Toyota near Lexington and Nissan, as well as GM's Saturn, near Nashville.

Japanese firms in Michigan for the most part have selected locations in the southeast, where the automotive industry is clustered; after all, why locate in Michigan at all if the principal objective is to avoid other automobile companies? Nonetheless, the spatial behavior of the Japanese firms in Michigan is distinctive. Most are located in the Detroit metropolitan area rather than Lansing, Flint, Pontiac, or the other southern Michigan metropolitan areas with substantial automotive industry employment. Outside the Detroit area, the biggest concentration of Japanese suppliers – in fact, the US city with the largest number – is Battle Creek, usually identified with Kellogg's breakfast cereals rather than with automobile production.

Within the Detroit area, Japanese firms – like American ones – have moved to the suburbs, especially in the north and west, such as Farmington and Rochester, rather than within the cities of Detroit or Dearborn. The outer counties, including Oakland to the north and Washtenaw to the west, have received more Japanese parts suppliers than Wayne County, where Detroit is located. Similarly, the Japanese parts suppliers in Illinois have clustered in Chicago's outer suburbs, especially to the northwest.

Joint ventures between US and Japanese firms account for perhaps 20 per cent of the new supplier firms. Taka Fisher, a joint venture between Takata Corporation and General Safety, supplies Honda and Nissan with seat belts from a plant in St Clair Shores, Michigan. Nishikawa Kasei and Standard Products supply sponge rubber products to Mazda from a plant in Topeka, Indiana. AP Technoglass, a joint venture between Asahi Glass, Takahashi Glass Industries, and PPG Industries, provides window glass to Honda and Diamond-Star from a plant in Bellefontaine, Ohio. CKR Industries, a joint venture betwen Kinugawa Rubber Industrial Company and Chardon Rubber Company, supplies Nissan with weather stripping from a plant in Winchester, Tennessee. Sheller-Globe and Ryobi operate a joint venture plant in Shelvyville, Indiana, which supplies aluminum die castings to Ford and GM.

Surveys reveal that only a limited number of US firms not already engaged in a joint venture are interested in starting one. The main incentive for setting up a joint venture is to obtain access to superior technology rather than increase access to new markets. Smaller firms in particular appear content to continue to operate at their current level of production, supplying one of the US car makers. Joint ventures are feared as opportunities for competitors to capture higher market shares (Chappell 1989a: 30).

Locational differences among Japanese companies

Mazda has been the Japanese producer most able to benefit from just-in-time delivery in the United States, because its assembly plant is located the closest to the southern Great Lakes, the traditional core region for production of components. Over half of Mazda's North American suppliers are located within 150 km of the Flat Rock assembly plant. The median distance of Mazda's suppliers is actually only approximately 110 km, as just over half of the suppliers are located in southeastern Michigan or northern Ohio. Another fourth are located in southern Ohio, Indiana, Kentucky, or Tennessee. Half of the remaining fourth are located in Ontario, Pennsylvania, or Illinois, while the others are scattered elsewhere, no more than two per state or province (Figure 7.4).

Nissan, on the other hand, has only one-third of its American suppliers within 150 km of its Smyrna, Tennessee, plant. Tennessee is the state with the highest number of suppliers to Nissan, but only one-fourth of them are there, whereas Ohio and Michigan – states long-associated with parts production – supply much higher percentages to Honda and Mazda. Only 40 per cent of Nissan's US suppliers are located in states in the I-65

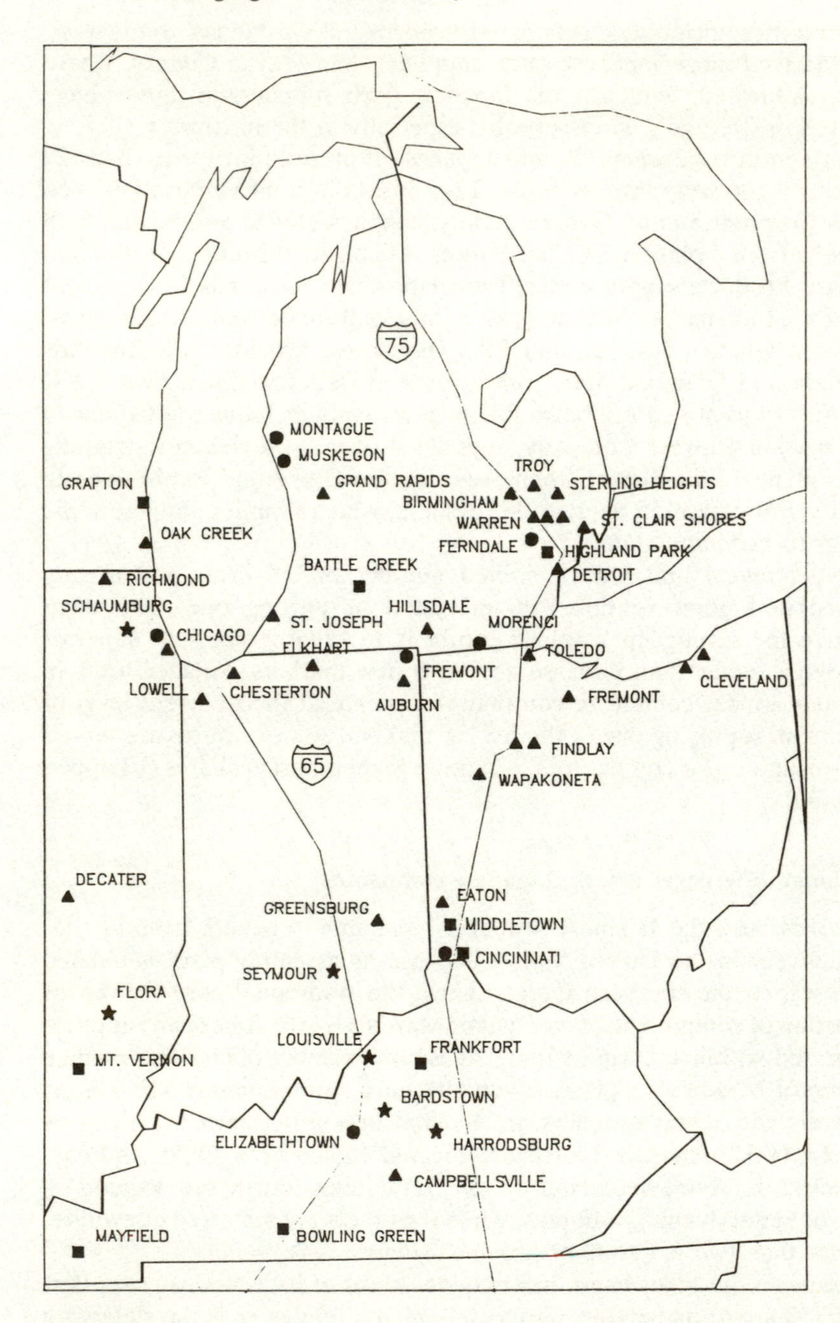
75
MONTAGUE
MUSKEGON
GRAND RAPIDS
GRAFTON
OAK CREEK
BATTLE CREEK
RICHMOND
SCHAUMBURG
CHICAGO
ST. JOSEPH
ELKHART
LOWELL
CHESTERTON
TROY
STERLING HEIGHTS
BIRMINGHAM
WARREN
ST. CLAIR SHORES
FERNDALE
HIGHLAND PARK
DETROIT
HILLSDALE
MORENCI
TOLEDO
FREMONT
AUBURN
CLEVELAND
FREMONT
FINDLAY
WAPAKONETA
65
DECATER
EATON
GREENSBURG
MIDDLETOWN
CINCINNATI
SEYMOUR
FLORA
LOUISVILLE
FRANKFORT
MT. VERNON
BARDSTOWN
ELIZABETHTOWN
HARRODSBURG
CAMPBELLSVILLE
MAYFIELD
BOWLING GREEN

Fig 7.5

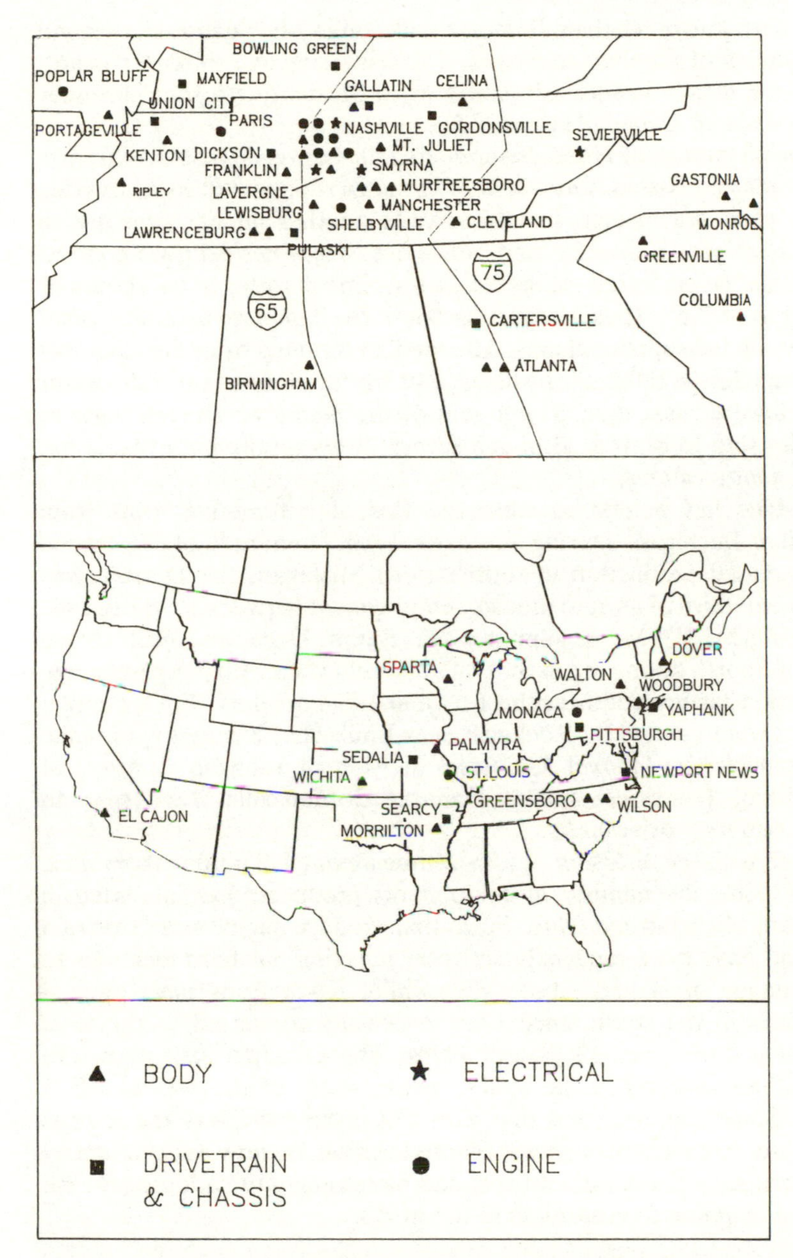

Figure 7.5 Location of US suppliers to Nissan's Smyrna, Tennessee, assembly plant, 1984. Nissan's suppliers are relatively dispersed, as a result of the location of the assembly plant relatively distant from the industry's southern Great Lakes core area
Sources: 'Nissan production-approved US suppliers', *Automotive News* 1984: E-46; 'Nissan Motors US production material suppliers', *Automotive News* 1984: E-26

and I-75 corridors other than Tennessee, although Michigan ranks second in total number of suppliers to Nissan. The remaining one-third of Nissan's suppliers are mostly in southern states or in Mexico, with some in other northern states or Canada (Figure 7.5).

The spatial pattern of Honda's suppliers falls between the two extremes set by Mazda and Nissan. One-third of Honda's suppliers to its Marysville assembly plant are located in Ohio, while another 40 per cent are in Indiana, Michigan, Kentucky, and Tennessee. Forty per cent are concentrated within the preferred radius for just-in-time delivery of two hours or 150 km (Figure 7.6). Thus, the farther south the Japanese assembly plant the longer are the supplier chains. The median distance from the assembly plant to suppliers is 110 km for Mazda, 240 km for Honda, and 320 km for Nissan. However, less than 10 per cent of the plants which ship parts to Nissan also ship to Mazda, although several firms supply both plants but from different locations.

Just-in-time has helped to stem the loss of automotive firms from southeastern Michigan. During the early 1980s, hundreds of US-owned suppliers ceased production in southeastern Michigan; the Detroit area alone lost one-third of its auto industry employment between 1978 and 1981 (Fleming 1982: 174). Major suppliers like Eaton, Dura, and ACF closed plants in the north during the early 1980s and relocated a higher percentage of production farther south in the I-65 and I-75 corridors. For example, new plants were opened by Rockwell near Louisville, Kentucky, to make plastic products; by Hoover Universal division of Johnson Controls at Murfreesboro, Tennessee; and by Dana at Gordonsville, Tennessee, to supply Nissan with driveshafts.

However, with the diffusion of just-in-time among US auto makers since the early 1980s, the number of components producers has increased in southeastern Michigan and Ohio. Firms that produce engine and drivetrain components have been especially active in selecting northern locations to take advantage of skilled labor. Meanwhile, fewer firms have opened new facilities in the south since 1980, especially compared to the level of expansion during the 1960s and 1970s. The principal exceptions are firms that have located in the states farther south in the I-65 and I-75 corridors. Some suppliers find that with just-in-time delivery the savings in labor costs achieved through a southern location are now outweighed by freight penalties – the additional cost and increasingly the additional time involved in shipping to customers in the north.

CHANGES IN SHIPPING VEHICLES AND PARTS

The diffusion of just-in-time delivery and the clustering of assembly plants and components producers in the I-65 and I-75 corridors have altered shipping patterns. In the past, most parts were shipped by rail and assembled automobiles by truck. With just-in-time delivery, most parts

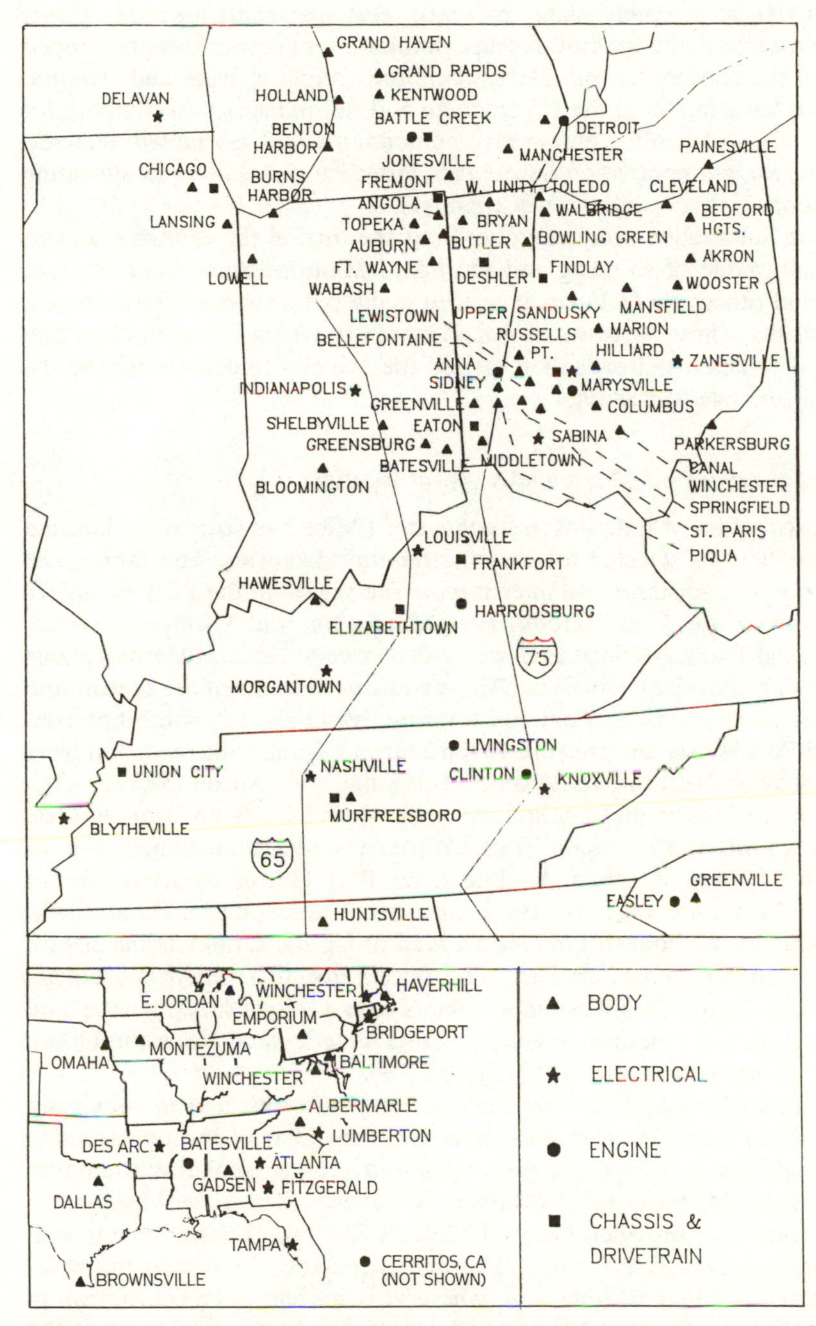

Figure 7.6 Location of US suppliers to Honda's Marysville, Ohio, assembly plant, 1988. As the first Japanese-owned US assembly plant, Honda has more suppliers in the United States than the other Japanese producers

now arrive at assembly plants by truck. But with most assembly plants now clustered in the interior, finished automobiles are increasingly shipped around the country by rail. Deregulation of freight charges and required services has stimulated the US trucking and rail industries to compete for the business of hauling automotive components and assembled vehicles. As a result, automotive producers face less clear-cut choices in allocating shipments between rail and truck haulers.

When automotive production began at the turn of the century, rail was the main mode of shipping and receiving. Manufacturers received iron, coal, and other raw materials by rail to make parts such as axles, bodies, and wheels. These parts were shipped by rail to plants for assembling into finished vehicles; railroads then carried the vehicles to dealers around the country for sale to customers.

Impact of rail on initial automotive plant locations

The distribution of railroads in southeastern Michigan strongly influenced the specific sites selected for early automobile factories. The three most important long-distance rail lines serving the region in the early twentieth century were the Grand Trunk, Pere Marquette, and Michigan Central. The Grand Trunk provided east–west rail service to General Motor's plants in Lansing, Pontiac, and Flint. The main line between Port Huron and Chicago passed through Flint and Lansing, while a north–south line connected Port Huron and Detroit. A third Grand Trunk route dropped from the main east–west line route south to Pontiac and Detroit (Figure 7.7).

Inside the Detroit metropolitan area, the Grand Trunk line from Pontiac – now known as the Grand Trunk Western – runs immediately east of Woodward Avenue, while the line from Port Huron parallels Gratiot Avenue 3 km to the west. The two lines join at a junction near Hamtramck. Assembly plants built in the Detroit area along the Grand Trunk before 1910 included Cadillac, Packard, and Ford's first two. Chrysler's Dodge Main and Lynch Road assembly plants and GM's Clark Avenue and recently opened Poletown assembly plants were located along the Grand Trunk (or successor) after 1910 (Figure 7.8).

The Pere Marquette Railway had one main east–west line and one main north–south line. The east–west line ran from Detroit through Lansing to Grand Haven, where steamers could carry unassembled automobiles across Lake Michigan to Milwaukee, Wisconsin. The Pere Marquette's north–south line provided General Motor's Flint and Lansing plants with their main access to the south. The line, bypassing Detroit to the west, terminated at Monroe, Michigan, where vehicles had to be transferred to other railroads. General Motors had especially close connections to the Pere Marquette: Billy Durant's uncle owned the railroad.

Within the Detroit area, the Pere Marquette ran near the present-day Jeffries Freeway (Interstate 96). Just inside the city limits of Detroit, the

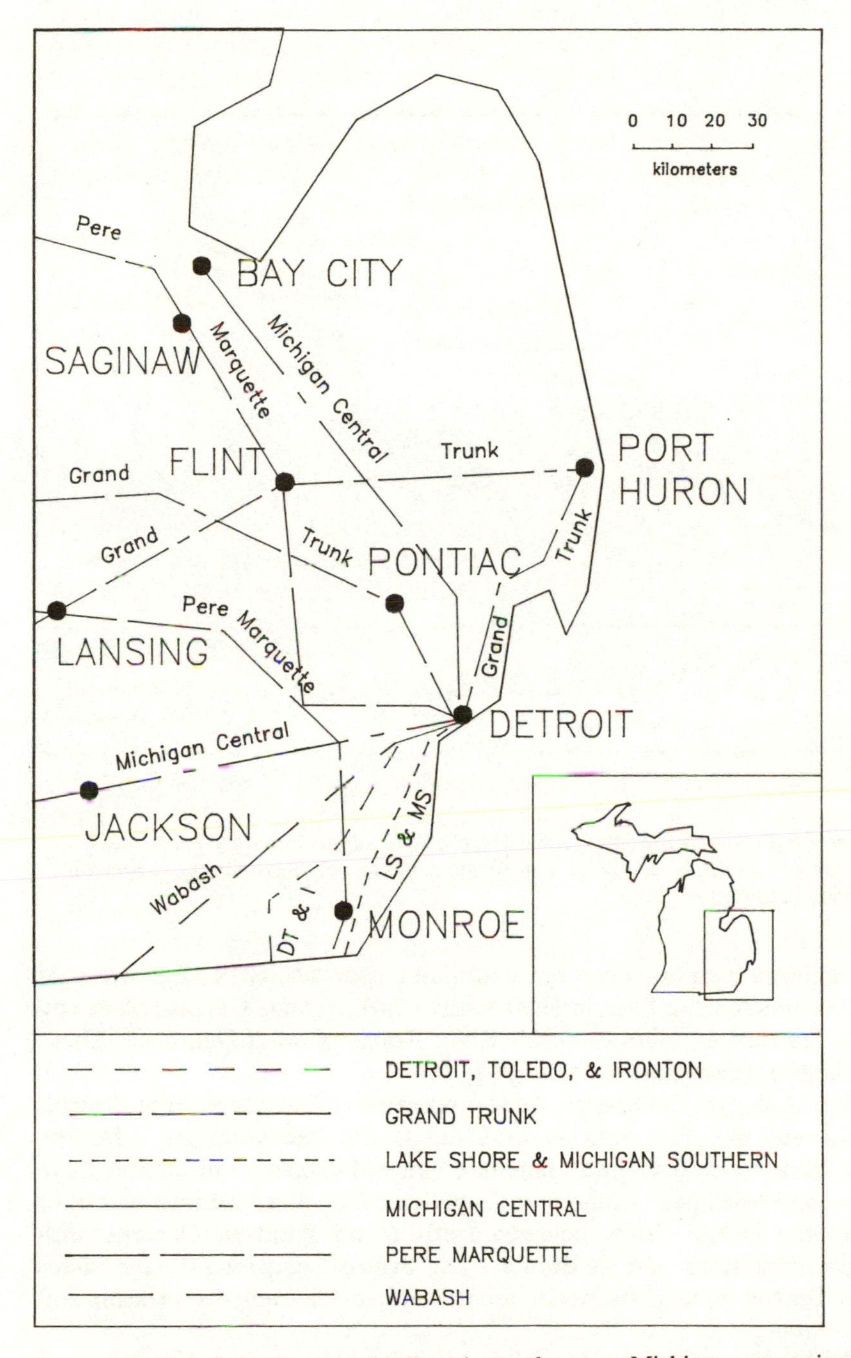

Figure 7.7 Main long-distance rail lines in southeastern Michigan, approximately 1912. These were the principal rail lines which served the state's automobile plants

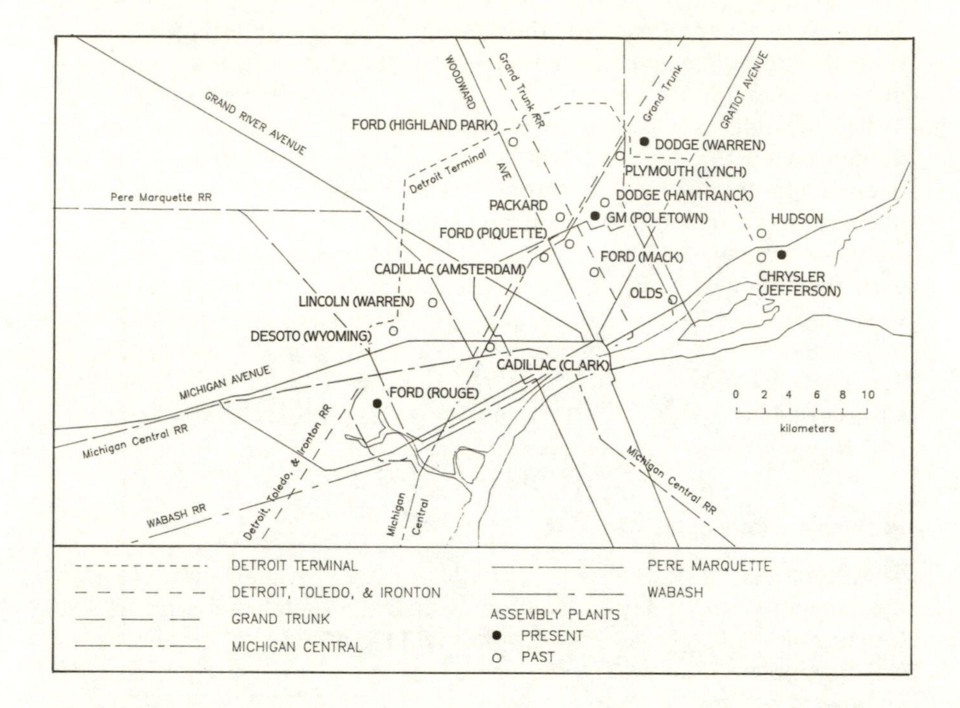

Figure 7.8 Principal rail lines in the Detroit area, approximately 1912. The location of rail lines heavily influenced the selection of sites within the Detroit area for automotive plants

line split into two branches, one continuing near the Jeffries Freeway past the Warren Avenue Lincoln plant to the Clark Avenue Cadillac plant and the other turning south to Ford's River Rouge plant at Dearborn. These two lines are now part of the CSX system.

The Michigan Central provided important connections from Detroit to the east coast as well as to Chicago and the west. In 1915, the New York Central Railway purchased the Michigan Central, the Lake Shore and Michigan Southern and smaller companies; a tunnel was built under the Detroit River between Detroit and Windsor, Ontario, with service continuing east to Buffalo. The railroad had long-distance lines from Detroit to the south via Toledo and to Chicago via Jackson and Ann Arbor.

The Michigan Central's main line from the west reached Detroit along tracks immediately south of Michigan Avenue, crossing over to Canada south of downtown Detroit. The line from the south ran between Jefferson

Avenue and the present-day Fisher Freeway (Interstate 75) and paralleled the Grand Trunk line through Hamtramck, north of the present-day Edsel Ford Freeway (Interstate 94). Most of the plants served by the Grand Trunk were also accessible to the Michigan Central. In addition, a short branch between the two main lines passed through the Rouge. Today, these lines are operated by Conrail.

The location of automotive plants within the Detroit area was also influenced by the route of the Detroit Terminal, a short-haul rail line opened approximately 1920. Barely 30 km in length, the Detroit Terminal made a semicircle around central Detroit. The line began near East Jefferson Avenue and the Detroit River, continued north to a junction with the Grand Trunk and Michigan Central near Hamtramck, turned west through Highland Park, and paralleled Oakman Boulevard south to Dearborn. A number of assembly plants were served by the Detroit Terminal, including Hudson and Chrysler on East Jefferson Avenue, Chrysler's Lynch Road and Wyoming Avenue, and Ford's Highland Park and Rouge. Today, Conrail operates the line.

Problems with shipping by rail

The rapid growth of automotive production early in the century strained the capacity of the rail lines. Railroad companies didn't share the sense of urgency about timely delivery of automotive components: an automobile was incomplete until every component was in place. Delay in delivery of any part could spell disaster for financially-strapped producers in the early years of the automotive industry. Alfred Sloan, as head of Hyatt Roller Bearing Company, maintained agents on the company's payroll to ride the caboose and 'cajole, bribe, or fight, as the occasion demanded,' to assure timely delivery of the bearings (Weisberger 1979: 102).

In the early days, producers distributed automobiles by piling as many as could fit into a freight car and shipping them to customers. The normal practice was to jam three automobiles into a 36-foot (11-m) boxcar or four into a 40-foot (12-m) boxcar, but their shape left a large amount of unusable space inside the boxcar. In 1910, the invention of double-deck freight cars permitted the loading of six automobiles into a 36-foot freight car and seven or eight in a 40-foot one, with the bodies separated from the chassis (*Dodge et al. v. Commissioner of Internal Revenue* 1927a: 62; Goodenough 1925: 183).

As automobile production increased in the first decade of the century from a few thousand a year to more than a million, demand for boxcars soared. Special boxcars were built with wide side doors or end doors, because automobiles could not be easily loaded through the narrow doors of normal boxcars. However, a 1912 trade journal reported a severe shortage of these special boxcars. 'There are twenty times as many boxcars into which automobiles cannot be loaded because they have doors too small' ('The freight car situation', *Automobile Trade Journal* 1912: 79).

Manufacturers complained that once vehicles were shipped from Michigan to the south and west, the scarce wide-door boxcars remained in those distant regions carrying freight that could be accommodated in the more common narrow-door boxcars. Rail lines resisted sending wide-door boxcars back to Michigan empty. 'The situation is going to become critical unless those automobile cars that are being held in the South and West are located and sent back to the automobile manufacturing territory' ('The freight car situation', *Automobile Trade Journal* 1912: 79). Ford's branch assembly plant network was devised in part to get around the shortage of rail cars. Branch assembly plants made possible shipments of motors, gas tanks, tires, wheels, radiators, and other parts in ordinary narrow-door boxcars. By shipping components rather than assembled automobiles, Ford could fully load boxcars with no wasted space.

Pricing rail shipments

More critical to producers than the scarcity of wide-door boxcars was the increasing cost of shipping by rail. From the nineteenth century until deregulation of the rail industry in 1980, freight charges were established through two factors – the classification and the tariff. First, every good was assigned a classification, with goods of similar physical composition or value given the same rating. Second, a tariff was established to cover all goods in a particular classification, normally expressed in terms of cents per mile per hundred pounds.

At first, each railroad created its own classification, with no attempt at national or even regional uniformity. One early line, for instance, had these five classes: heavy goods, light goods, case goods, logs, and whiskey (Widell 1940: 1). At one time in the nineteenth century, 138 different types of classification systems were in use on railroads just in the eastern part of the country (Shannon 1931: 283). Some simplification of rating practices was needed, because if every US railroad set individual ratings for each article moved to every possible destination, approximately 200 trillion classifications would have to be published (Ribe 1944: 1).

The Interstate Commerce Commission (ICC), established in 1887, was instrumental in imposing a simplified classification system on the nation's railroads. Shortly after its creation, the ICC divided the US into three territories, known as Official, Southern, and Western. The western boundary of the Official Classification Territory followed the Mississippi River north from Cairo, Illinois, to the Wisconsin state line and turned east along the Illinois–Wisconsin state line to Lake Michigan. A small portion of southeastern Wisconsin, near Milwaukee, was also included in the Official territory. The southern boundary followed the Ohio River from Cairo to

Cincinnati, a straight line from Cincinnati to Kenova, West Virginia (near the border of Kentucky and Ohio), West Virginia's southern boundary with Kentucky and Virginia from Kenova to near Bluefield, and an irregular line across Virginia from near Bluefield to Norfolk. The Southern Classification Territory lay south of the Official and east of the Mississippi River. The Western Classification Territory comprised the land west of the Mississippi River.

A committee was responsible for allocating goods to one of a small number of classes in each of the three territories. However, each territory had a different number of classes – seven in Official, ten in Western, and twelve in Southern. Thus, a good could be assigned to as many as three different classes nationally. Under the prodding of the ICC, the railroads frequently simplified the rating system during the late nineteenth and early twentieth centuries. The Consolidated Freight Classification, first published by the ICC on 30 December 1919, displayed in one volume the classifications for all three regions, including more than 20,000 items, from abrasives to zinc. Over the next two decades, the ICC worked to increase uniformity in the classes for the three territories, especially during the 1930s Depression, when railroads consolidated or at least were forced to cooperate with other lines to survive. In 1939, the ICC published the first Uniform Freight Classification, although the ratings of some goods still differed among territories.

A good's assignment depended in part on its bulk, weight, ease of loading, vulnerability to damage, and other costs associated with handling, compared to other goods. Goods with higher handling costs were allocated to a lower class, such as Class 1 or 2. Classifications were also based on 'what the traffic will bear,' such as the ability of a good's producer to find alternative transport and competition within a particular industry. The 1939 Uniform Freight Classification contained thirty classes, including Class 100 – the former Class 1 – plus seven classes higher than 100 and twenty-two lower. Other classes were subsequently added before the system was scrapped in the 1980s.

The second factor in determining freight charges was the tariff for shipping between any two points in the country. The tariff was related to distance: that is, rising as the distance between two points increased. The cost per mile generally decreased with greater distances because the charge in part incorporated fixed handling charges at each terminal. However, tariffs for the same distance varied among the territories, as well as among subterritories within each territory. Tariffs were usually lower per mile in the northeast than elsewhere in the country. Tariffs for shipping between two points varied according to the class of the good. Classes other than first class were expressed as a percentage of first class. For example, in the Official Classification Territory, second-class rates were 85 per cent of first class, while sixth-class rates were 27.5 per cent of first class (Lively 1949: 5).

Assembled automobiles were assigned a rating above first class, reflecting difficulties involved in maneuvering them into and out of freight cars, as well as the possibility of damage in transit. In contrast, parts could be shipped at rates as low as sixth class. As a further penalty, railroads imposed additional charges, known as 'less than carload,' for shipping fully assembled automobiles. Railroads justified the 'less than carload' surcharge because, in order to assure a profit in shipping and handling, a freight rate for a particular good was computed on the basis of filling a boxcar to a specified minimum weight. For automobiles, the specified minimum weight was 4,500 kg for an 11-m boxcar (10,000 lb for a 36-ft boxcar) and 5,100 kg for a 12-m boxcar (11,200 lb for a 40-ft boxcar).

Automotive producers could not meet the minimum weight, because early automobiles weighed perhaps 550 kg (1,200 lb), and only three or four could fit in a boxcar. Consequently, charges for shipping automobiles were computed on the basis of the much higher 'less than carload' rate. In essence, automotive manufacturers were charged for shipping the equivalent of 4,535 kg (10,000 lb) of machinery when in fact they were only shipping 1,632 kg (3,600 lb). By shipping parts instead of fully assembled automobiles, Ford benefited in two ways: parts were classified at a lower-cost class than fully assembled automobiles, and the company could load more than 4,535 kg (10,000 lb) of parts, thus avoiding the higher 'less than carload' rate.

Alternatives to rail

Henry Ford, who was especially infuriated with poor service and high rates, searched for alternatives to reduce his dependency on major rail lines. When the Detroit, Toledo, and Ironton (DT&I) Railroad was unable to raise $350,000 in order to replace a bridge across the River Rouge, Ford bought the railroad in July 1920 for $5 million. Applying his automotive marketing strategy, Ford lowered rates on the DT&I by 20 per cent in 1921, amid protests by other railroads, yet by 1923 the line showed a profit. The DT&I, which ran south through rural Ohio, brought Appalachian coal to the Rouge complex at Dearborn, but Ford grew weary of complying with ICC and other federal regulations. He sold the DT&I to the Pennsylvania Railroad for $36 million in 1928 (Nevins 1954: 220–227).

To break the rail monopoly, Ford utilized barges to transport raw materials to the Rouge and components to east-coast assembly plants. Other producers employed water transport to some extent, although they lacked Ford's extreme devotion to it. As early as 1907, Buick transported automobiles across Lake Michigan to Wisconsin, where they were loaded on trains for Minneapolis and the northwest. During the 1930s, General

Motors shipped automobiles from California to the northwest by steamship. Chrysler floated automobiles down the Ohio and Mississippi Rivers from its Evansville, Indiana, assembly plant in the 1930s.

The truck rather than barge emerged as the most effective alternative to rail transport. Automobile producers benefited from the new transport mode in two ways, for not only did they gain the means to break the rail monopoly, but they could manufacture the trucks themselves. Trucks were especially efficient in hauling assembled automobiles to the large percentage of dealers which were located within a couple of hundred miles of one of the branch assembly plants. Truck hauling first made inroads into rail traffic during the 1930s, but the major impact came after World War II. By the early 1950s, trucks carried three-fourths of assembled automobiles, compared to only 10 per cent for rails (US Senate 1956: 777, 1023).

Even after they lost most of their hauling business, though, railroads continued to exert influence over locational decisions in the automotive industry, in part because most components still left by rail from Michigan to the branch assembly plants. More importantly, the rail industry set prices for other modes of transport, as well as its own. To be competitive, truckers charged just under rail rates, and when rail rates increased or decreased, the trucking industry followed suit. Furthermore, lower rail rates from GM and Ford branch assembly plants meant low truck hauling rates from these plants, while high rail rates from Detroit meant high truck rates as well for producers who had to supply the entire country from Detroit (US Senate 1956: 408).

Relationship of freight rates to plant locations

Automotive producers tended to expand the number of branch assembly plants during periods of increasing freight rates. Until 1913, freight rates remained stable, as increased costs of labor and materials during the period were offset by the introduction of more efficient locomotives and larger freight cars. Between 1913 and 1920, freight rates more than doubled in the United States, as the cost of labor and materials increased so rapidly during the period, especially after the outbreak of World War I, that rail lines could no longer keep abreast without frequently increasing rates. This period of rapid freight rate increases coincided with the establishment of the first generation of branch assembly plants by Ford and General Motors for their low-priced high-volume models.

The ICC ordered a 10 per cent rate reduction in 1921, and rates stabilized for the next fifteen years. During this period, the number of branch assembly plants did not increase as rapidly as in the previous period. General Motors added several for Chevrolet production, but Ford for the most part replaced older plants rather than add to the overall number. From the late 1930s through the 1950s, rates increased substantially again, except when frozen by the government during World War II. The cost of shipping an automobile from Detroit to Los Angeles doubled in the immediate post-World World II period, from $144 in 1949 to $281 in 1956 (US Senate 1956: 414). This period of rate increases coincided with the expansion of the branch assembly plant concept by Ford and General Motors to encompass their intermediate-priced automobiles.

Rate increases were somewhat offset by changes in the classification of automobiles. The ICC ruled in 1945 that the shipping costs were too high from Detroit and ordered rail carriers to reclassify assembled automobiles from Class 1 to Class 2, or under the new uniform code, from Class 100 to Class 85. This change reduced the cost of shipping automobiles by 15 per cent beginning in 1946 (US Senate 1956: 409). Smaller automotive producers pressed for still lower classifications on automobiles shipped from the Midwest, in order to reduce the benefits of the branch assembly plants enjoyed by the two largest firms, Ford and General Motors. In 1951, the ICC reduced automobiles produced in the Midwest from Class 85 to Class 75 and increased the ratings on automobiles produced in branch plants to Class 50. To help preserve the smaller producers, the Michigan Central reduced its rates on shipping automobiles from the Midwest to as low as Class 63 (US Senate 1956: 409–410).

Charging consumers for freight

Rather than incorporating freight costs in the advertised price – the approach employed by most industries, such as steel and household appliances – early automotive producers added a separate destination charge to the base price of the automobile. When each producer had only one plant, calculating the destination charge was simple. A dealer paid a producer the actual cost of transporting an automobile from the factory to the showroom and then passed it through to the customer.

Once they built branch assembly plants, the large automotive producers found that the destination charge could no longer be set as the actual cost of shipping. For example, when faced with a sudden surge in demand, dealers in San Diego, who normally received their automobiles from an assembly plant in nearby Los Angeles, might need to be supplied with additional vehicles from a more distant assembly plant, perhaps in Kansas City. Customers would surely complain if identical automobiles on a dealer's lot did not carry the same destination charges because they were shipped from different branch assembly plants.

Instead, automotive producers imposed a destination charge based on the cost of shipping from southern Michigan to the dealer regardless of whether the automobile was made there. For example, a 1954 Chevrolet carried a destination charge of $279 in Seattle, $145 in Houston, $74 in New York, and $11 in Detroit. Outside of Michigan, the destination charge in 1954 was approximately 12 cents per rail mile from Flint, Chevrolet's home plant city. Auto makers justified differential destination charges, arguing that in the absence of branch plants, all production would be concentrated in Michigan, relatively far from coastal customers (US Senate 1956: 899).

Customers outside the Midwest were outraged. Why should Californians pay General Motors to ship from Michigan automobiles which were actually assembled in Los Angeles or Oakland and transported only a few miles? Automotive producers pointed out that destination charges encompassed two sets of transport costs – parts to the branch assembly plants and assembled automobiles to dealers. A higher destination charge for automobiles sold in the south and west was justified because of the increased cost of shipping parts from the Midwest to outlying branch assembly plants. Critics called the method of setting destination charges 'phantom freight.'

The large variation in the final price to consumers in different regions of the country as a result of destination charges led to widespread 'bootlegging' during the 1950s. Entrepreneurs in the south and west would send six youths to Michigan in an old automobile to buy six new vehicles from authorized dealers in a region where virtually no destination charges were paid. The youths got free overnight lodging and a chance to drive new cars home. Used car dealers and other firms not franchised by automotive producers in the south and west could make a profit by selling 'bootlegged' automobiles for a couple of hundred dollars less than the price charged by authorized dealers, who were handling vehicles with high destination charges shipped from the nearest branch assembly plant. One-fifth of the automobiles sold in California during the early 1950s were 'bootlegged' from the Midwest (Cray 1980: 400). Dealers in the south and west claimed that 'bootlegging' unfairly swelled average dealer sales in Michigan at their expense and produced unhappy customers when 'bootlegged' cars needed service.

Reducing regional differences in destination charges

Anxious to prove to the Congress that destination charges did not inflate profits, automotive executives testified during the 1950s about the economic benefits of the branch assembly plants. Frederic G. Donner, then Vice President of General Motors in Charge of Financial Staff and later President, provided evidence at a 1956 Senate hearing that branch assembly plants saved $38 for each Chevrolet. The figure was determined by comparing two sets of freight costs – shipping parts to assembly plants and shipping assembled automobiles to dealers. For the nine Chevrolet branch assembly plants outside of Michigan, the average freight cost per automobile was $40 for inbound parts and $38 for outbound assembled automobiles. Total shipping costs for a typical branch assembly plant thus amounted to $78 per vehicle.

Donner compared the freight costs for the nine branch assembly plants with Chevrolet's home plant in Flint, Michigan. Because of proximity to

Table 7.1 Freight charges saved by GM branch assembly plants, 1955

	Location of assembly	
	At branch plants	*Only at 'home' plant in Michigan**
	($ freight charge per automobile)	
Chevrolet branch plants		
Components from Michigan to assembly plant	40	13
Assembled automobiles from plant to dealers	38	103
Total freight charge per automobile	78	116
Buick-Oldsmobile-Pontiac		
Components from Michigan to assembly plant	87	22
Assembled automobiles from plant to dealers	34	110e**
Total freight charge per automobile	121	132e

Source: adapted from US Senate 1956: 905–906

Notes: * 'Home' plant is Flint for Chevrolet and Buick, Pontiac for Pontiac, and Lansing for Oldsmobile

** Figure is inferred from testimony by F. Donner

e estimated

suppliers, the Flint plant had much lower shipping costs for incoming parts compared to branch assembly plants, $13 per car. Donner estimated that the average cost of supplying every dealer in the country with Chevrolets assembled in Flint would be $103 per car. Thus, if Flint were the only plant in the country producing Chevrolets, the combined costs of receiving parts and shipping assembled automobiles in 1956 would have been $116, or $38 more than the actual freight costs from ten assembly plants (Table 7.1).

That General Motors saved $38 per car by building Chevrolets at branch assembly plants was verified by price changes made in the 1950s in response to complaints about destination charges. In the fall of 1954, GM established a maximum destination charge for all points more than 1,931.1 km (1,200 miles) from the home plant, encompassing the southern portions of Florida, Louisiana, and Texas, and the states west of the 104th meridian. In the case of Chevrolet, approximately 18 per cent of the buyers had been paying more than the new maximum charge of $140. In order to offset the reduction in revenues, the list price of each car was increased by $20, of which GM received $15 and the dealer $5. For Chevrolet, the effect was to reduce the cost of cars by as much as $119 for west-coast buyers, while increasing the price by as much as $20 for customers in the rest of the country. For example, in Albuquerque, New Mexico, the destination charge was reduced from $185 to $140, while the list price increased by $20, for a net reduction to the customer of $25. This change eliminated the net savings in assembly plant costs realized on the delivery of cars in the Pacific coast region.

On 27 January 1956, GM reduced destination charges again for most customers. A new maximum destination charge of $120 was set for Chevrolets delivered beyond 2,575 km (1,600 miles) from Flint – essentially the west coast – and charges were also reduced within the market areas of the other branch assembly plants. Chevrolet's list price was increased $30, of which GM received $23 to offset the reduction in destination charges and the dealer received $7. The increase of $38 per car from the two price changes was represented as the added cost for a manufacturer located solely in Michigan to duplicate Chevrolet's sales pattern from ten branch assembly plants.

Because of lower sales volumes, branch plants for GM's three medium-priced cars produced only marginal economic benefits. The average cost of shipping parts to the seven Buick-Oldsmobile-Pontiac branch plants was $87, and the average cost of shipping automobiles from the branch plants to dealers was $34, for a total average shipping cost per automobile of $121. The actual cost of shipping parts to the three Michigan home plants – Buick in Flint, Oldsmobile in Lansing, and Pontiac in Pontiac – was $22 per automobile. Testimony was not given concerning what the cost per car would be if the entire country were supplied from the three Michigan plants, although the level was said to be somewhat higher than the $103 figure for Chevrolets because the medium-priced vehicles were heavier.

Freight savings were somewhat offset by additional costs of operating several branch plants, such as duplication of management, utilities, maintenance of equipment, inventories, and other overhead items. Donner testified that the added costs of maintaining branch plants amounted to at least $10 to $15 per car, on average, and offset transportation savings to that extent (US Senate 1956: 898). That would reduce the benefit of the Chevrolet branch plant network to approximately $25 per car and the medium-priced cars to virtually nil.

Shipping costs varied widely among branch assembly plants. The cost of shipping parts was calculated by totalling all freight bills into a branch assembly plant and dividing the outlay by the total number of cars produced (US Senate 1956: 789). As most parts were produced in the Midwest, west-coast plants had by far the highest inbound freight costs. Differences in the cost of shipping assembled automobiles to dealers were much less, because the branch assembly plants served roughly comparable market areas (Table 7.2).

Because of freight savings, smaller automotive producers were attracted to the branch assembly plant concept in the 1940s and 1950s. The lack of branch assembly plants placed them at a competitive disadvantage compared to General Motors and Ford. James J. Nance, President of Studebaker-Packard, testified concerning the economic benefits of opening a branch assembly plant in Los Angeles. In 1956, the cost of shipping a Studebaker from the company's home plant in South Bend, Indiana, to Los Angeles in 1956 was $271. In comparison, the cost of shipping parts from South Bend to Los Angeles was $126, and the cost of shipping automobiles from the assembly plant in Los Angeles to dealers in the

Table 7.2 Shipping costs per vehicle for GM assembly plants, 1955

	Components from Michigan to assembly plant $	*Automobile from assembly plant to dealer* $	*Total freight costs* $
Chevrolet			
Atlanta Lakewood	54	37	91
Baltimore	28	32	60
Flint	13	23	36
Janesville	25	37	62
Kansas City Leeds	42	60	100
Los Angeles Van Nuys	105	26	131
Cincinnati Norwood	23	27	50
Oakland	105	27	132
St Louis	28	64	92
Tarrytown	31	22	53
Total	36	37	73
Total (excluding Flint)	40	38	78
Buick-Oldsmobile-Pontiac			
Atlanta Doraville	82	44	126
Arlington	110	32	142
Framingham	65	15	80
Kansas City Fairfax	69	63	132
Linden	65	17	82
Los Angeles South Gate	159	36	195
Michigan 'home' plants*	22	52	74
Wilmington	54	22	76
Total branch plants	87	34	121
Total all plants	66	40	106

Source: adapted from US Senate 1956: 786–787

Note: * Michigan 'home' plants are Flint for Buick, Pontiac for Pontiac, and Lansing for Oldsmobile

region averaged $38. Thus, the Los Angeles plant saved the company $107 per vehicle.

General Motors did not adopt a uniform freight charge throughout the continental United States until the early 1980s. The destination charge for a Chevrolet Caprice was $308 if shipped to Detroit and $435 if shipped to Los Angeles in 1981, compared to $460 for both destinations in 1982 ('GM freight charges to be same for all US', *Automotive News* 1981: 3).

Recent changes in shipping by rail

Freight rates declined during the 1960s but increased again in the 1970s and 1980s (Interstate Commerce Commission: 1913–1980). Higher transport costs encouraged the establishment of branch assembly plants during earlier periods, especially during the 1910s and the decade after the end of World War II, but in recent years produced the opposite impulse: closure of branch assembly plants near coastal population concentrations and construction of new plants in the interior of the country. To some extent, innovations in the rail industry are responsible for the reversal of the traditional relationship between rail rates and distribution of automotive plants.

The rail and automotive industries have laid aside their traditional rivalry and cooperated to construct more efficient hauling services. The two most important changes included introduction of new hauling methods and new ways to set rates. As a result, automotive producers have been able to alter their plant locations without suffering increased freight charges.

The most important innovation in hauling freight by rail has been the introduction of intermodal systems. Instead of competing against truckers, rail lines introduced piggyback cars in the late 1950s, in which trucks load and unload the automobiles. Trucks drive their trailers onto flat rail cars, and the cabs are unhitched and driven off. When the train reaches its destination, cabs can be driven back onto the flatcars for unloading. The growth of piggyback cars was slowed during the 1960s and 1970s by federal regulations, which limited advertising the service, and the lingering reluctance of rail lines to cooperate with trucks. Piggyback service increased rapidly after 1979, though, when the ICC freed it from federal regulations (Mahoney 1985). Trilevel rack cars, introduced in 1960, enabled even more automobiles to be transported in the same space.

Two other acts promoted further deregulation of the rail industry. The 1976 Railroad Revitalization and Regulatory Reform Act, known as the 4-R Act, freed railroads from requiring ICC approval to either lower rates by up to 7 per cent or raise them in areas where they did not have market dominance. The 4-R Act also created Conrail from the debris of six bankrupt northeastern and Midwestern lines and provided $2.12 billion in loans to help the new line undertake a ten-year $6.2 billion rehabilitation program (Association of American Railroads 1977). The 1980 Staggers Act eliminated regulatory review from nearly all rail rate changes (Association of American Railroads 1981).

The second important piece of legislation, the Staggers Act, permitted rail lines to set rates through signing long-term contracts with shippers, distinct from their common carrier service. Contracts typically contain the terms of the agreement, including what is to be carried, how much is to be carried, where it is to be carried, and the rate which is to be paid. Rail lines have filed tens of thousands of contracts with the ICC. As contract rates are lower than common carrier rates, firms such as motor vehicle producers which make heavy use of them have experienced lower rates (Harper and Johnson 1987).

Impact of just-in-time on shipping

At first glance, just-in-time delivery would be expected to favor increased reliance on truck delivery at the expense of rail. Traditionally, trucks were considered more efficient for short, frequent deliveries, whereas the main advantage of rail is shipping large quantities long distances. However, with the deregulation of the transport industry in the 1980s, rail lines have been free to compete for short-haul, just-in-time delivery. Since 1982, Conrail has offered just-in-time delivery to move components among GM plants located several kilometers from each other in the Lansing area ('GM, railroads use just-in-time at Lansing plants', *Automotive News* 1985: 24).

Railroads increased their share of automotive hauling from 50 per cent in 1979 to 55 per cent in 1983, while truck deliveries declined from 50 to 45 per cent. Rail companies began to sign long-term shipping contracts, set rates and levels of service for the length of the contracts, and guarantee specific volumes of deliveries. Multilevel rack cars which deliver parts to GM's Van Nuys plant, used to return empty to the Midwest, but through better scheduling the same railcars can backhaul assembled automobiles to the interior of the country, as well as parts from overseas to the Japanese assembly plants in the I-65/75 corridor ('Railroads hike share of auto hauling', *Automotive News* 1985: 36).

Just-in-time delivery has made producers more dependent on transport systems again. On the first day of a nationwide walkout by rail engineers in September 1982, GM was forced to close its St Louis assembly plant because of lack of parts. By the third day, Janesville, Leeds, Lordstown, and Wilmington were operating on half days, and Chrysler president Iacocca wired the White House to urge intervention. On the fourth day, the US Congress ended the strike by passing legislation forcing the engineers to accept contract terms recommended by an emergency fact-finding board and agreed to earlier by other rail unions (Rowand 1982b: 1).

When workers at Delco Electronics plant in Kokomo, Indiana, went out on strike in November 1986, GM assembly plants around the country were closed. Because radios from the plant were being shipped on a just-in-time basis, assembly plants had little inventory on hand with which to work once the strike began. The strike was called to protest Delco's sourcing of some work to a plant in Mexico. GM felt compelled to return the work to Kokomo in order to get a supply of radios flowing into the assembly plants again ('GM's chief: Strike may shut all plants' *Dayton Daily News* 1986: 1).

While minimizing transport costs has become more critical for auto makers, the proliferation of just-in-time delivery and the development of more diverse products have caused a reversal in the long-time split between rail and truck traffic. With the development of branch assembly plant networks, auto makers depended on rail for long-distance shipment of parts from the interior to coastal assembly plants, while trucks hauled away the automobiles to the nearby dealers. Now, trucks are more likely to bring in parts on a just-in-time basis to assembly plants in the interior, while trains carry the assembled vehicles to distant customers.

Part III

Reasons for recent locational changes: community scale

Part II demonstrated that decisions to construct new auto plants in the I-65 and I-75 corridors were heavily influenced by least-cost transport principles. However, neoclassical location theory does not offer much help in explaining why specific communities along the I-65 and I-75 corridors have been selected for new plants. Further, while least-cost transport principles played a role in decisions to close coastal plants, other factors have become more important, especially now that most coastal plants have already been closed.

Chapter 8 reveals that the location of new plants may be influenced to some extent by governmental subsidies, especially for infrastructure, training, and reduced tax burdens. As communities place greater importance on promoting economic development, local officials increasingly believe that financial incentives are necessary to attract and retain jobs. However, officials who are competing to attract new auto plants are aware of the prevailing levels of subsidies generally provided for similar facilities in other communities. Consequently, governmental subsidies have not generally proved critical in decisions to close or retain older automotive plants or to build new ones. A few cases have arisen when older plants have closed because of lack of local support and new plants have been attracted by exceptionally high levels of subsidies.

At the national scale, some North American automotive production has shifted from the United States to Canada as a result of governmental policies. Especially important to auto makers is the much lower level of contributions for employee health care insurance required in Canada as a result of that country's national health insurance program. The savings derived from Canada's lower-cost health care system amount to several hundred dollars per vehicle.

In general, governmental policies are not as critical as local labor climate in the selection of specific sites for new plants. As Chapter 9 shows, for some automotive firms, especially suppliers of components, the most critical factor of production is minimizing labor costs. US auto makers have transferred some production from in-house divisions to independent suppliers who can produce parts more cheaply because they have lower labor costs. Given that auto-industry labor rates do not vary widely among regions within the United States, manufacturers seeking significant wage savings have relocated production to so-called 'maquiladora' plants in Mexico's border towns.

However, for most US auto firms, minimizing labor costs is less critical than avoiding concentrations of militant workers. Chapter 9 recounts that the US automotive industry has a long history of labor–management confrontation. In recent years, Japanese-owned companies have been especially eager to identify locations that avoid concentrations of militant workers. To find suitable sites, Japanese firms engage in careful study of community attitudes towards work stoppages and foreign investors. Japanese firms carefully screen potential employees, frequently with the cooperation and financial support of state governments, to assure that they will adapt to the anticipated working environment inside the plant.

Japanese plants have been especially attracted to small towns on the periphery of metropolitan areas in the I-65 and I-75 corridors that have relatively low concentrations of automotive workers and union members. These peripheral locations also are characterized by smaller percentages of ethnic minorities, who are regarded by some Japanese officials as harder to train in their production methods.

The local labor climate is also a critical factor in decisions to close or retain older plants. Chapter 10 shows that auto makers are basing retention decisions in part on measures of the relative quality of the vehicles or components that each plant produces. Plant quality, in turn, is increasingly perceived as a function of the willingness of workers to adopt more flexible work rules, such as the team concept. The prevailing attitude among union officials and members is that adoption of more flexible work rules can help US firms compete more successfully against Japanese firms. However, some union dissidents have opposed concessions, especially after American companies closed plants where flexible work rules had been adopted.

8 Governmental impact on locational decisions

Rumors spread quickly through a small Midwestern community: the Japanese are coming. Older residents who remember World War II may regret the invasion: 'We won the war and lost the peace.' Some veterans protested when the Subaru-Isuzu plant in Lafayette, Indiana, was situated on the Bataan Memorial Highway. Younger residents welcome the job opportunities: 'We don't care if the plant is run by green, three-eyed Martians, as long as it means jobs for us.' Facts are scarce; the only concrete evidence may be when a group of foreign-looking individuals get out of a limousine and walk along Main Street holding video cameras.

Local officials may have heard rumors from the state government or regional electric company that the community is on a short list of sites being considered by a Japanese firm. But no direct communications occur until representatives of the firm telephone for an appointment with the responsible local leader, such as the city manager or mayor. When the appointment is made, the official is told that a Japanese firm – which prefers to remain anonymous – is considering locating in the community.

When the Japanese businessmen – not women – arrive, one of the first rituals is exchange of business cards, but these do not clarify who among the Japanese are in charge, because the English translation of the titles all include the word 'manager.' Gifts are exchanged, even though the practice violates the ethics under which US local government officials operate. Negotiations are slow, because one man speaks in English on behalf of the Japanese group and then translates everything said by the Americans into Japanese. The Japanese describe the anticipated size of the plant, including number of employees and floor area, but do not reveal the firm's name or principal products. However, at the first meeting, the priority is learning about the atmosphere in the community, such as attitudes towards Japanese and development and recreational opportunities. The Japanese also want to know about the personal lives of the local officials – their families, houses, and leisure activities.

If the community is not the firm's first choice, then local officials hear nothing further, no letter of regret. If the community emerges as the leading choice, then subsequent meetings are held to discuss details, such as financial incentives and timetables. Only after the deal is concluded do the other Japanese participants reveal that in fact they understood English; on-the-spot translations were designed more to set a leisurely pace to the negotiations than to provide information. On hindsight, the Japanese spokesman may not be the highest ranking representative; the man who sat in the corner saying nothing may have been the critical decision-maker. The name of the company is certainly revealed by this time, but even after the commitment is made, the precise products to be manufactured at the plant may still not be described. Local officials tolerate this process – in fact, welcome it – because in many small Midwestern communities promoting economic development is an increasingly central expectation of their jobs.

THE GOVERNMENT BAILS OUT CHRYSLER

The bailout of Chrysler in the early 1980s set a precedent for subsidizing automotive firms. Chrysler had come through the Depression of the 1930s and the World War II era second only to General Motors in total sales but failed to move aggressively to capture a large share of the booming post-war automotive market. Sales slipped behind Ford in the early 1950s, and the company never seriously challenged again for the second place position. The company was inefficiently organized and built poor quality cars not attuned to changing consumer tastes. One of the most visible signs of Chrysler's problems was the stockpiling of tens of thousands of unsold cars at the Michigan fairgrounds and the Windsor, Canada, Raceway, after rolling off the assembly lines. Inefficient management procedures had permitted the production of too many cars, and their poor design and quality made them unsellable.

Faced with the likelihood of bankruptcy, Chrysler appealed to the federal government for financial assistance. The Chrysler Loan Guarantee Act, passed in late 1979, called for the creation of a loan guarantee board which was authorized to issue up to $1.5 billion in loan guarantees to Chrysler over a two-year period. The loans were secured by Chrysler's assets, which were estimated by the government to be worth $2.5 billion if liquidated. The board issued the first $550 million to Chrysler six months after passage of the act, in June 1980, and ultimately issued $1.2 billion in guarantees.

Chrysler's plant closures during the early 1980s

In exchange for receiving the loan guarantees, Chrysler agreed to close several plants. Chrysler's principal criterion for targeting closures was age of the plant rather than location. Most vulnerable were four Michigan assembly plants, all of which dated from before World War II, whereas the four elsewhere in the United States had opened during the 1950s and 1960s.

First targeted for closure in the late 1970s was the Jefferson Avenue plant, the company's oldest, built by Chalmers in 1907. However, Chrysler aborted the closure notice and spent $17 million to convert production from full-sized cars to light trucks ('Chrysler plans to close old Hamtramck plant', *Automotive News* 1979: 3; Irvin 1979a: 1). A decade later, the Jefferson Avenue plant was closed, but it was replaced by a new facility next door.

The Hamtramck plant, known as Dodge Main, was closed instead in January 1980. Nearly as old as the Jefferson plant, Dodge Main was originally opened in 1910 for the Dodge Brothers to manufacture parts under contract to Ford, and automobile assembly started six years later. Hamtramck was Chrysler's largest plant, with two separate assembly lines, designed to turn out more cars than the company was likely to sell in the future, but the overhead on the large plant was too high (Sorge 1981b: 42).

The selection of Hamtramck may have also been influenced by the militant reputation of the plant's labor force. After the 1967 civil disorders in Detroit, the Hamtramck plant became the home base for the Dodge Revolutionary Union Movement (DRUM), a group of militant blacks who sometimes referred to Dodge Main as the plantation and talked of burning the plantation. Some of the most outspoken DRUM leaders later moved to leadership positions in the United Auto Workers (UAW) union ('Chrysler plans to close old Hamtramck plant', *Automotive News* 1979: 3).

Chrysler closed two other assembly plants during the early 1980s, including Lynch Road in Detroit and Fenton No. 2 in Missouri. Lynch Road had opened in 1929 for production of the company's newly introduced low-priced Plymouth model. The plant was sold in December 1982 to the City of Detroit, along with two closed components plants, for a total of $10.

In late 1982, Chrysler tried unsuccessfully to sell its 14-year-old Fenton No. 2 assembly plant to Mitsubishi, which was looking for a North American site at the time (Bernstein 1983: 8). When demand for its full-sized car increased in 1983, Chrysler reopened the Fenton plant. One month later, President Reagan stopped by to declare that the imminent reopening of the plant was a sign of economic recovery ('Reagan sees recovery sign at St Louis Chrysler plant', *Automotive News* 1983: 6). The UAW, having demanded in its summer 1982 contract negotiations that the plant be reopened, cooperated by agreeing to new work rules (Sorge and Bernstein 1983: 2; 'UAW to demand St Louis reopening', *Automotive News* 1982: 1). Four years later, Chrysler retooled the plant for production of compact vans.

The bailout of Chrysler met strong opposition among politicians and other business leaders, who argued that Chrysler should file for bankruptcy and reorganize as a more efficient corporation under the protection of Chapter 11 of the bankruptcy act. However, bailout advocates argued that Chrysler was unlikely to survive bankruptcy. The company would face years of administrative problems under the leadership of a court-appointed bankruptcy judge. For example, no capital expenditures for future products could be made without the judge's approval. Meanwhile, consumers would be unlikely to continue buying Chrysler products, and prospects for regaining consumer confidence would not be good.

Studies by several economic forecasting organizations, including Data Resources for the Congressional Budget Office, the Transportation Systems Center for the US Department of Transportation, and Chase Econometrics for Chrysler, agreed that the impact of a Chrysler bankruptcy would be severe for the nation's economy. The US gross national product would decline by approximately 0.5 per cent and unemployment would increase by between 0.5 and 1.09 per cent. The nation's trade balance would decline by $1.5 billion, while the annual loss of tax revenue would reach $500 million. Public assistance programs to unemployed Chrysler workers would cost the government an additional $1.5 billion, plus the Pension Benefit Guarantee Corporation would face a liability of $800 million. The loss of business for Chrysler's suppliers was not estimated (Moritz and Seaman 1981: 279; US Department of Transportation 1981).

Chrysler returned to profitability in the mid-1980s, spurred by brisk sales of its new 'K' car model, plus restraints on Japanese imports. On 15 June 1983, the first day it could legally do so, Chrysler repaid the first $400 million of its government-guaranteed loans. After closing three of its ten assembly plants in the early 1980s, the company expanded capacity in 1983, first by reopening the Fenton plant and then by launching a new plant at Sterling Heights, Michigan (Figure 8.1) (Sorge 1983b: 1).

FROM POLETOWN TO SATURN

Michigan, where most of the automotive plant closures were clustered in the early 1980s, moved aggressively to stem the losses. Sensitive to the charge that it was abandoning Detroit, GM chairman Thomas A. Murphy approached Mayor Coleman Young in early 1981 for help in identifying a suitable site in the community to consolidate Cadillac's Clark Avenue assembly plant and Fleetwood body plant, which both dated from the 1920s. An attractive site was offered: the recently abandoned Dodge Main plant, within the City of Hamtramck, a 5-km^2 enclave surrounded by Detroit. The Chrysler Corporation made its contribution to the area's revitalization in November 1980 by donating the plant, which was promptly demolished.

Problems at Poletown

GM officials pointed out one fundamental flaw with the 53-hectare Dodge Main site: it was too small for a 300,000-m^2 one-storey assembly plant. Anxious not to lose the Cadillac plant, the cities of Detroit and Hamtramck together offered to assemble 188 hectares by expanding the site to the south

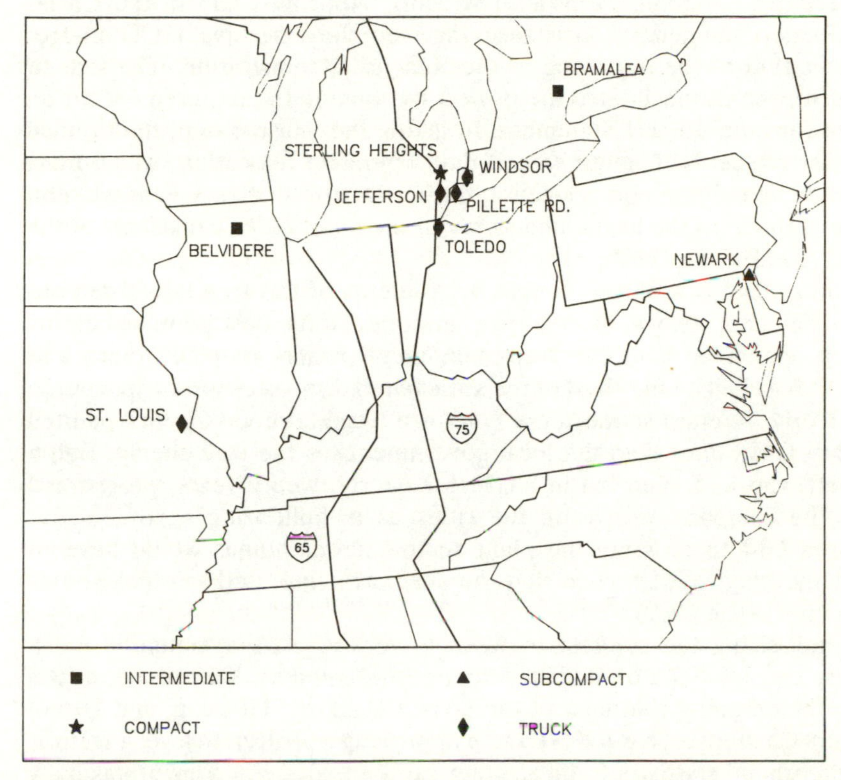

Figure 8.1 **Chrysler Corporation assembly plants by type of models produced, 1991. Like other firms, Chrysler clustered its assembly plants in the US interior. However, Chrysler also maintained a relatively high percentage of production in Canada**

into Detroit. GM would get 144 hectares, and the rest would be used for rail yards and an industrial park (Sorge 1980b: 2). GM accepted the proposal. The two municipalities then faced the awkward problem that the plant site was filled with people and businesses. In order to present GM with a cleared parcel of land, complete with utilities, roads, and rail sidings, the two cities had to acquire 1,675 structures and relocate 150 businesses, 1,500 households, several churches, and a hospital (Sorge 1980a: 42). While General Motors could not compel people to sell their property, the local governments could, under eminent domain, as long as just compensation were paid and the compulsory purchases were shown to be for a legitimate public purpose.

Unable to absorb all of the site preparation costs, the cities received more than $90 million from the US Department of Housing and Urban Development (HUD) for property acquisition, relocation of residents, and demolition of buildings, including a $30 million Urban Development Action Grant and a $60.5 million loan guarantee ('GM plant funds OK'd for Detroit', *Automotive News* 1981: 40). Approved late in the Carter presidency, the federal assistance survived the extensive HUD budget cuts in 1981 at the beginning of the Reagan administration. The federal government also facilitated the project by waiving the requirement for an Environmental Impact Statement. In taking the unusual step, the Council on Environmental Quality argued that 'complete relocation, which must precede demolition and relocation during Detroit's often severe winters, poses a threat to the health and safety of the many elderly residents of the area' (Kelderman 1980b: 3).

The project was bitterly fought by residents of the area facing demolition. The area, known as Poletown, in honor of its most populous ethnic group, contained a mix of well-maintained owner-occupied homes and gutted-out shells. One-third of the anticipated displacees were expected to be elderly. To enlist support, the Poletown Neighborhood Council painted mighty GM rather than the local governments as the true enemy. Ralph Nader, who had been battling GM for nearly twenty years, proclaimed that the company was using the cities as a 'bulldozing agent.' Nader pressed GM to redesign the plant so that fewer homes would have to be demolished. GM replied that the site was being used as efficiently as possible (Wylie 1989).

Emphasizing the contrast between a working-class community and a luxury car, US Senator William Proxmire (Democrat, Wisconsin), at the time the outgoing chairman of the Senate Banking, Housing, and Urban Affairs Committee, blasted HUD for approving a project 'to level a Detroit neighborhood so that a Cadillac plant can be built . . . . This brings back memories of the worst excesses of the old urban renewal program.' According to Proxmire, HUD's own experts warned that the city had severely underestimated the project's costs and was incapable of handling a workload of 1,675 property acquisitions adequately (Kelderman 1980b: 3).

The Poletown Neighborhood Council initiated court proceedings to stop the plant, but the Circuit Court in late 1980 permitted the cities to proceed with property condemnation. The project met the legal test of serving a public purpose, the court ruled, because the new plant would improve the community, retain jobs, and enhance the tax base. The Michigan Supreme Court agreed to review the case in early 1981, bypassing the state Court of Appeals, but it too found a legitimate purlic purpose in the exercise of eminent domain (Sorge 1981a: 6). With legal remedies to block the project exhausted, construction began in 1981.

Poletown's image of an oppressed working-class neighborhood valiantly fighting against mighty General Motors was sharply compromised by racial issues. The City of Hamtramck was widely known as a hostile environment for blacks and had been convicted of discriminatory housing practices. With blacks comprising over one-third of Detroit's automotive workers, civil rights groups placed jobs ahead of preserving a community widely viewed as racist. The Detroit Coalition of Black Trade Unionists sent Nader a letter which called the plant 'a giant step forward,' vowing not to see the plant 'hobbled by an emotional dispute that threatens to hide (1) the critical need for more industrial jobs as soon as possible . . . and (2) the spawning of other businesses and services that will certainly spring up around an industrial workforce that will be growing – not fading' ('Black unionists slap Nader effort to halt GM plant', *Automotive News* 1981: 30). UAW President Fraser gave GM credit for remaining in Detroit at a time when other new plants were being built in the south (Sorge 1981d: 8).

GM for its part tried to lie low and deflect criticism to the cities. Chairman Thomas A. Murphy offered Poletown's residents 'compassionate understanding,' but pointed out that the cities had selected the site and demolished the properties (Sorge 1981c: 26; 'Murphy: Cadillac will renew Detroit', *Automotive News* 1980: 56). As late as September 1983, GM let people attend a prayer service on the site of the demolished Immaculate Conception Church. The company also preserved and renovated the Beth-Olem Cemetery, the oldest Jewish cemetery in Detroit, in deference to Jewish law, which forbids distinterment (Sorge 1983c: 16).

Changing attitudes towards subsidizing auto plants

A decade later, Poletown looked like a bargain in comparison to Detroit's outlay to retain Chrysler, when the company decided to close its 75-year-old Jefferson Avenue assembly plant. The plant had received a cosmetic face-lift and retooling in the early 1980s but remained inefficient. Bodies were welded and painted in a building on the north side of Jefferson Avenue and shipped across the street by way of an overpass to the hard-trim shop on the second floor of the main plant. The bodies then had to be transported down to the first floor, where the engines, chassis, and transmissions were raised into the bodies.

The City of Detroit convinced Chrysler to build a replacement facility next door for assembly of new Jeep models. By 1989, the city had spent $117 million to acquire the land, $13 million to relocate the businesses from the site, $87 million to demolish the buildings, and $22 million for planning, engineering, legal fees, and administration. In addition, the city spent $34 million to remove toxic wastes and prepare the site in compliance with environmental regulations ('Chrysler cleanup costs climbing', *The Detroit News* 1989: 1A).

When automotive firms began to reinvest in the I-65/75 corridor in the early 1980s, locational decisions were not usually announced formally, and public subsidies were kept discreet. Local government initiatives to attract automotive plants, such as Poletown, generated widespread opposition. But within a few years, politicians were tripping over each other to attract automobile plants to their communities, and auto companies were enjoying the attention (and incentives) bestowed on them.

The turning point in the behavior of governmental officials was the competition to lure General Motors' Saturn plant in 1985. Local government efforts to attract or retain automotive plants certainly began prior to the Saturn competition, but awareness of the need for industrial development policies was less widespread and recruitment of Japanese investment highly dependent on the behavior of local politicians.

Instead of discreetly planning for the new product, GM chairman Roger Smith chose to shower lavish publicity on Saturn at a preliminary stage in its development. These moves may have been designed to forestall charges that General Motors was not moving rapidly to meet the Japanese challenge. By announcing that Saturn would be built by a new division of General Motors at a new plant, Smith unleashed a fierce battle among localities to attract the facility. Smith may have deliberately encouraged the competition in order to secure higher concessions, but it was probably unwittingly, because GM did not go where the financial package had been most attractive.

During the first half of 1985, states and localities organized campaigns to attract Saturn. Youngstown, Ohio, recruited three local sports personalities – the boxers Boom Boom Mancini and Harry Aroyo and the owner of the San Francisco 49ers football team (and shopping center developer) Edward J. DeBartolo – to send messages to General Motors. Chicago paid for a billboard along a Detroit freeway to support its application. Cleveland residents sent GM 'We Want Saturn' coupons found in their newspapers and signed petitions circulated by a local television station. All 1,700 schoolchildren in New Hampton, Iowa, wrote letters to GM executives.

Governors of the competing states made sure that their voices were being heard in Detroit. They travelled to Michigan to confer with GM officials; they offered outlandish packages of tax incentives. Michigan's governor announced that his state would match any other state's financial package; Illinois' governor in turn announced that his state would top any offer made by Michigan. The governors of Illinois, Iowa, Minnesota, Missouri, Ohio, Pennsylvania, and Texas joined GM chairman Roger Smith on Phil Donohue's popular day-time television show to plead their case. One telephone caller to the program said that the competition to land Saturn 'is turning into a circus.'

STATES TAKE THE LEAD

Local government officials learned from the Saturn experience that aggressive marketing practices to attract or retain industries were not only acceptable behavior, they had become expected. Overseas trips by politicians, once considered thinly-veiled holiday junkets, became a necessary undertaking to attracting investors. Not only states and provinces, but individual cities and counties as well, began to send trade delegations abroad in search of new investment. Secret deals were replaced by highly publicized signing ceremonies involving local officials, corporate executives, and union representatives.

Honda comes to Ohio

One of the first state officials to campaign unashamedly for new industries was James Rhodes, who served as governor of Ohio from 1962 until 1970 and again from 1974 until 1982. A native of a rural Appalachian community, Rhodes attracted voters as a bumbling, rumpled alternative to slick, packaged politicians geared to television. The conservative Republican's pro-business, low-tax policies also appealed to voters at a time when Great Society programs flowed from Washington.

Had he offered the vice-presidency to Rhodes in 1976, Ronald Reagan could have wrested the Republican presidential nomination away from the incumbent Gerald Ford instead of waiting until 1980. As governor of the most populous Republican-controlled state, Rhodes appeared to control enough delegates to swing the nomination from Ford to Reagan, but according to Jules Witcover's account of the 1976 election, at the moment of truth even Reagan's fiercest partisans considered Rhodes 'too old, a terrible public speaker, and slightly unsavory.' Witcover quoted David Keene, one of Reagan's top advisers, as saying 'Who wanted to run in the general election with Jim Rhodes?' And Reagan's campaign chief John Sears defended the rejection of Rhodes even though it may have cost his candidate the nomination, telling Witcover, 'You've got to have some responsibility in this business' (Witcover 1977: 460).

Politicians outside of Ohio may have had difficulty comprehending Rhodes' appeal, but his folksy style appealed to Japanese auto makers, at a time when other politicians were still reciting their World War II combat experience. Rhodes went out of his way to flatter Japanese executives by frequently flying to Tokyo and issuing hearty invitations to visit Ohio while many Americans still found recruiting Japanese industries distasteful.

Rhodes flew to Tokyo with Ohio's director of development within twenty-four hours of first hearing that a Japanese auto maker was interested in building a plant in the United States. Not knowing which Japanese firm was considering a US plant, Rhodes visited each in turn until he learned that Honda was the one. He lured Honda to Ohio in 1977 with a promise of $22 million in new highways, site improvements, and tax abatements, a commitment considered by many at the time as excessive, but judged a remarkable bargain in the 1990s compared to outlays by other states. Honda's total direct investment in Ohio, including two automobile assembly plants, an engine factory, and a distribution center, were estimated to be worth $1.7 billion in 1991.

Honda was a logical company to build the first Japanese-owned automobile assembly plant in the United States. The company had already tested the US waters by opening a plant in Ohio in April 1978 to assemble its popular motorcycles. Honda was known for innovative engineering, but its market share in Japan was low, because it did not begin producing automobiles until 1962. As a result, Honda from its inception was highly dependent on overseas sales. At the same time, Honda's Japanese plants were operating at full capacity in the late 1970s, while Toyota and Nissan both could expand production (Lienert 1978: 1).

In its role as the first Japanese auto maker to plunge into the uncharted waters of assembly in the United States, Honda never formally announced its intention. At the April 1978 groundbreaking ceremonies for a 24,000-m^2 motorcycle plant in Marysville, Ohio, a Honda official revealed that a US automotive plant might be in the company's future (Lienert 1978: 1). Confirmation of Honda's intention came in October 1978, from a notice in the *Federal Register*, in which the Greater Cincinnati Foreign Trade Zone Incorporated sought a foreign trade subzone in Columbus. The application stated that the tract for which the subzone would apply would initially house a motorcycle assembly plant and 'the second phase of development would include an auto assembly plant' ('Honda's Ohio car plant confirmed', *Automotive News* 1978: 1). The application clearly referred to Honda; who else was building a motorcycle plant near Columbus, Ohio? The motorcycle plant showed that Honda could count on high quality labor ('Honda paves way for US auto plant', *Automotive News* 1980: 6). However, because the motorcycle plant did not generate widespread publicity when it was built, Honda did not feel compelled to explain publicly why Ohio was selected.

In early 1980, Kiyoshi Kawashima, president of Honda, announced that a study to determine the feasibility of assembling automobiles in the United States had entered its final stage ('Honda moves towards US car assembly', *Automotive News* 1980: 1). The first public acknowledgement that an automobile plant would definitely be built did not come until July 1980, when company representatives showed up at Governor Rhodes' office to receive a formal transfer of the title to some of the land. Groundbreaking for the automobile plant took place in late 1980, and the first cars rolled off the line two years later. By the late 1980s, Japanese producers and American local government officials held highly visible joint press conferences to reveal investment decisios, but back in the late 1970s and early 1980s, Japanese companies had not yet adjusted to American-style press conferences and were still not sure that building plants in the United States was a good idea.

Nissan nearly selects Ohio

The second Japanese motor vehicle announcement of a US assembly plant came in April 1980. Like Honda, Nissan moved carefully, and did not begin with production of automobiles. Rather than motorcycles, Nissan chose to start American production with light trucks, which were simpler to build and did not require annual model changes. In contrast to the situation with Honda, Nissan did not immediately announce the state where the factory would be located. Given Honda's positive experience, Ohio was on Nissan's short list, as were other southern Great Lakes states. But Nissan officials hinted that the company was looking at sites in southern states, as well, with Georgia the leading candidate (Irvin 1980: 1; Kelderman 1980a: 1; 'Nissan to build in US?', *Automotive News 1980: 3).*

Governor Rhodes nearly brought Nissan, the second transplant, to Ohio in 1980, an act which would have altered the emerging landscape of Japanese automotive investment in the United States. He flew to Tokyo on two days' notice in 1980 to create a favorable impression with Nissan officials and invite them to visit Ohio. When Rhodes met Mitsuya Goto, deputy general manager for Nissan's international division, at the Columbus, Ohio, airport, the first thing he asked was whether he liked popcorn. 'I didn't know what to say, but I told him I did. A few minutes later, when we went to the governor's car, which was very ordinary, he asked me to sit in front and there on the seat was a bag of fresh popcorn.' Rhodes then took Goto to lunch at a private club, where the governor ordered his 'standard lunch, bean soup' (Irvin 1980: 46).

Goto feared that Rhodes would urge Nissan to consider a former Army tank plant in Brook Park, too close to heavily industrialized and unionized Cleveland to suit the Japanese. Volkswagen had recently rejected the Brook Park site in favor of Sterling Heights, Michigan, for its second US assembly plant, which was ultimately abandoned. In discussions after lunch, though, Rhodes showed that he knew Ohio's locational assets which really appealed to Japanese. According to Goto:

> [Ohio officials] talked about Honda's experience and how they thought we ought to use the cornfield approach – go into an area with no other large industries and be a community leader. They told us we could get a lot of young workers and if these workers got a Christmas bonus (traditional in Japan) they would be very happy and would never want to join a union.
>
> (Irvin 1980: 46).

At 4 p.m., the governor and a few aides drove Goto to the Columbus airport, where they boarded a private plane for a tour of the state.

> On the way to the airport, the governor stopped at a mom-and-pop grocery store and came out 10 minutes later with a bag of groceries. When we got in the plane, the governor started making ham sandwiches and passed them to me and his men. Then he offered us beer or Coke. Then he offered me some hard candies for dessert. I thought to myself that maybe this is how Honda was sweetened up. But I really like him.
>
> (Irvin 1980: 46).

Goto contrasted Rhodes with California Governor Jerry Brown, who also visited Japan. Brown objected to his Tokyo hotel, the Okura, located across from the US embassy, on the grounds that it was too plush and expensive. He insisted that he be moved to an inexpensive Japanese inn, but Goto claimed that nothing suitable was available in Tokyo to foreigners. Consistent with this preference, Brown had refused to occupy the California governor's mansion, preferring to live in a small apartment in Sacramento. 'The whole while I saw him he never smiled. He always had a very serious look and it would have been much better if he had smiled – like Gov. Rhodes of Ohio' (Irvin 1980: 46).

In the end, Nissan announced its choice of Smyrna, Tennessee, outside Nashville, in November, 1980. Production of trucks began in 1982, and automobiles followed in 1985. Nissan was attracted to the south by the weakness of labor unions; the company gained a reputation as the transplant with the most rabidly anti-union attitude and successfully withstood union organizing efforts during the 1980s through vigorous opposition. Had it lured both the Honda and Nissan plants, Ohio would have comprised an overwhelmingly dominant core of Japanese automotive investment, but Nissan selected Tennessee in late 1980, setting the pattern for a more dispersed arrangement within the I-65 and I-75 corridors.

States compete for auto plants

By the mid-1980s, states had become more aggressive in soliciting new auto plants. The Saturn site selection process demonstrated that leadership in attracting and retaining large projects, such as automobile plants, must reside at the level of the governments of the fifty states. Overwhelmed by requests from hundreds of communities, General Motors turned to the states to bring order to the selection process. States were asked to coordinate the flow of material from individual communities to GM. More significantly, state officials were asked by GM to rank the suitability of each proposed site and to indicate which sites appeared to the state to be most attractive.

States have also been forced into a leadership role because they are manageable units for foreign investors to comprehend. Japanese banks may have maps hanging on the walls showing the fifty states in different colors, while the tens of thousands of cities, counties, townships, parishes, school districts, and other local government units remain a bewildering muddle. Japanese banks or producers may direct smaller parts suppliers to locate in a particular state. Ohio, for example, is widely perceived to be 'Honda's state,' and smaller firms who wish to supply Honda may be directed by their banks to contact Ohio development officials. State economic development officials then help firms create a short list of specific communities.

Within a state, electric companies also play a key role in identifying finalists for the plant. The most accurate and up-to-date information on individual communities may be in the hands of the electric company, and they may have personnel especially trained to work with potential new Japanese customers. Electric companies see themselves as the most neutral advisers capable of giving the best advice: because they are providing power throughout the region, they don't care which individual community gets the plant.

The lack of strong national policies to guide industrial location decisions has also forced states into assuming leadership in promoting economic development. Even relatively modest past programs aimed at encouraging industrial development in multi-state regions, such as the Tennessee Valley in the 1930s and Appalachia in the 1960s, had long since faded as national government priorities. Efforts to promote cooperation among more than one state in a region, such as the Midwest, have been swamped by the recognition that neighboring states were competing for the same firms.

However, the most compelling reason for states to take on a stronger role in attracting and retaining industries is financial. The level of subsidies offered to firms may exceed the fund-raising capabilities of localities. Tennessee offered Saturn $30 million for highway improvements and $20 million for job training. In comparison, Nissan had received $20 million a few years earlier, as had Honda from the state of Ohio. However, the going rate for a Japanese assembly plant soon escalated far beyond Tennessee's outlay for Saturn. The first major Japanese-managed assembly plant after Saturn – the Chrysler-Mitsubishi Diamond-Star joint venture – received $86 million to locate in Normal, Illinois, including $40 million for training, $10 million for site acquisition and grading, $13 million for water and sewer construction, and $20 million for highway improvements (Pastor 1986: A-1). Indiana matched the Illinois figure for the joint Subaru-Isuzu plant in 1986 (Holusha 1986: A-1).

Kentucky raised the stakes even higher to attract Toyota. The state promised $157 million in direct investment, including $68 million for job training, $40 million for new roads and sewers, and $49 million to buy the 400 hectare site. Further, Kentucky agreed to subsidize the interest payments on the debt incurred by Toyota to build the plant, estimated at $168 million (Fiordalisi 1989: 18). Kentucky's subsidy may work out to as much as a quarter-million dollars for each job created at the Toyota plant.

FINANCIAL INCENTIVES

Three principal types of financial incentives are being offered by state governments to attract and retain automobile plants. These include making improvements to the site, subsidizing programs to train workers, and reducing tax obligations.

Infrastructure

The most expensive form of financial assistance is likely to be for site improvements, such as transport, electricity, water supply, and waste disposal. As the public sector has an obligation to maintain and improve infrastructure, little controversy is likely to surround the commitment. The major impact may be to reshuffle the state's priorities, causing delays in the completion of other projects.

Roads must normally be built to connect the plants to existing highways. For plants located immediately adjacent to interstate highways, such as Mazda and Toyota, both visible from I-75, and Saturn, immediately off I-65, the state's principal obligation is to construct new interchanges and perhaps a short service road between the interchange and the plant entrance. At the other extreme, Ohio took on the obligation of building a 30-km four-lane highway to connect the Honda plant with the nearest interstate highway, the I-270 ring road around Columbus.

To retain GM's Tarrytown assembly plant, New York State agreed to reconstruct a 185-km rail line from the plant by raising the height of twenty-three bridges over the tracks by 2 feet and enlarging seven rail tunnels. Improvement of this line permitted General Motors to move triple-stacked rail carriers, which are 6-m (19 feet 6 inches) high, to reach Albany, where they could be switched to already improved lines for distribution around the country. When Tarrytown served as a Chevrolet branch assembly plant, trucks were used for most deliveries, because few vehicles needed to be shipped more than 300 km. But when the plant was converted to production of minivans, which had a national market, distribution became much cheaper by rail than truck. The cost of the project was initially estimated at $22 million, but actually took over $40 million. The biggest single problem was modification of the train station at Ossining, which straddled the tracks. The entire station had to be raised 1 m (3.5 feet) on hydraulic jacks, at a cost of $12 million. (Feron 1987: B-2; Henry 1989: 41).

Training

The second principal method by which local governments provide financial assistance to automotive plants is support for worker training. Training programs operate in retooled or modernized plants, as well as newly constructed ones. State government officials consider training grants to be an attractive form of subsidy because the principal beneficiaries can be shown to be local citizens rather than large – sometimes foreign – corporations.

Henry Ford was proud of the fact that he could pluck people off the streets and put them to work on the moving assembly line with no training. By the 1980s, with the reskilling of automotive production, auto makers faced an inadequately trained workforce. Japanese producers hesitated to invest in US plants without assurance that the workers could be adequately trained, while US firms concluded that training programs were essential to make their plants more competitive with the Japanese.

The bill for training programs in the Japanese-operated US assembly plants at the time of initial operations ran to the order of $40 million, or $15,000 to $20,000 per worker. Michigan provided $19 million of the cost at Mazda, while Illinois and Kentucky picked up higher percentages to lure Diamond-Star and Toyota, respectively. Training grants have been more modest for existing plants, to the order of several million dollars.

Training programs frequently include classroom-style instruction away from the shop floor, perhaps at a local community college. Much of the teaching is done by outside instructors drawn from the local universities, in order to help convince workers that the course material is independent of company propaganda, although plant mangers and labor representatives may give some of the lectures. Classroom – or so-called 'soft' – training introduces workers to the harsh realities of contemporary automotive production – global competition, excessive plant capacity, higher quality standards – and the role of their plant in the national and global production picture. Workers are taught the basic principles of the company's philosophy and the relationship of these principles to specific work tasks in the plant.

Public subsidies frequently pay for the salaries of the instructors – either directly or through release time from other obligations – as well as instructional materials, such as manuals and videos. The local university may provide classrooms, laboratories, and other facilities. Workers in Japanese-operated plants may be sent to Japan at public expense to observe first-hand distinctive management practices in operation. To assure availability of training facilities, producers have often cited proximity to a university as a major factor in selecting plant locations. General Motors officials claimed Tennessee's educational system influenced the company's selection of Spring Hill for the Saturn plant. However, given the proliferation of community colleges, branch campuses of major state universities, and vocational schools, most plant sites in the Midwest are within commuting distance of adequate facilities.

Public subsidies also permit on-the-job training of workers in both new and existing plants to operate unfamiliar machinery, such as robots; which now dominate tasks such as painting and body welding once done by hand. For many jobs physical exertion and brute strength have been replaced as the most critical skills by judgments concerning when to push buttons and how to adjust electronic controls.

Publicly subsidized retraining programs contributed to GM's decision to retool its Baltimore assembly plant for minivan production. The plant closed for modernization on 30 March 1984, a Friday, and the following Monday morning all 3,200 employees began attending a forty-hour course at nearby Essex Community College. The curriculum was developed by a four-member team which included both union and management representatives; it dealt largely with non technical subjects, such as why change was needed at the plant, the nature of the new product, and ways to improve communications and quality. Technical training was also offered in areas such as electronics and robotics. The course was taught by thirty-two employees, including sixteen hourly and sixteen salaried workers, who had been trained by instructors at the community college.

Training programs at Japanese plants do not live up to the expectations of some workers. In principle, Mazda workers started with three weeks of 'soft' classroom training on the basics of teamwork and Mazda's philosophy, followed by seven more weeks of classroom and hands-on training in a particular area, such as the body shop. Workers were then assigned to a place on the line by the unit leaders but given three or four more weeks of job-specific off-line training before actually working on the line. Mazda viewed the training program as another application of just-in-time, in contrast to the situation in American plants, such as nearby Poletown, where workers were trained so far in advance of need that they forgot much of what they had been taught by the time production began.

Workers, however, claimed that Mazda's training program delivered far less than it promised in both quality and quantity. The training program was considered disorganized and slipshod, a series of unrelated classes which dealt inadequately with sophisticated subjects, such as operation of plant robots. The company allegedly placed inadequately trained workers on the line but reported to the state that training took place during periods when in reality production had temporarily halted for other reasons. A worker who was supposed to be training others stated that:

> It was protocol to record more training hours than actual training. Even if we cut classes short, we'd put the same total training hours on the sheets to turn into the state. The important thing was to get the line time going.
>
> (Kertesz 1989a: 52)

In 1988, Evan Bayh, the Democrats' candidate, was elected governor of Indiana, a state which normally votes for Republicans, and in the same year George Bush carried it by a two-to-one margin. Bayh was elected in part because of his opposition to state subsidies for the Subaru-Isuzu plants, although he was also the son of a popular politician. After becoming governor, Bayh blocked the appropriation of training funds for the plant.

When state officials had lured the Subaru-Isuzu plant to Indiana back in 1986, they signed a 'memorandum of understanding' which committed the state to providing $34 million in funds for training programs at the time the plant expanded from its start-up level of 1,700 employees and 120,000 vehicles per year. The companies originally anticipated expanding to 3,200 employees and annual production of 240,000 vehicles in 1995 or

1996. Instead, the companies decided to expand to an intermediate level of 2,250 employees and production of 170,000 in 1991. The state was asked for $8.8 million in 1991 to assist with training, but Governor Bayh claimed that by expanding to a lower level the plant was not living up to the spirit of the memorandum (Chappell 1990a: 11).

Despite the supposedly massive reskilling of the work place, Ronda Hauben, an injured Mazda worker, was told in her job interview that 'no job here takes more than a half-hour to learn' (Kertesz 1989a: 52). Has Henry Ford's vision of Taylorism come full circle?

Lower tariffs

The most controversial financial incentives are tax breaks. Reductions are made primarily for two types of obligations – tariffs on imported parts and taxes on property. Tariff levels are controlled by the national government, while property tax rates are set by local governments.

Tariffs on imported parts are reduced when the US Department of Commerce designates the land surrounding a factory as a foreign trade subzone. Foreign trade zones were established by the Congress in 1934 to encourage export of US goods. The program was originally intended to benefit processors of raw materials and warehousing operations rather than manufacturers. So that firms could minimize the transport costs associated with importing raw materials and exporting finished goods, most foreign trade zones were located near ports or airports.

Congress amended the law in 1952 to permit designation of a manufacturing plant as a foreign trade subzone within a foreign trade zone, even if the plant was actually located outside the zone. Volkswagen became the first automotive producer to request foreign trade subzone status in 1977 for its New Stanton assembly plant then under construction, and Honda followed with a similar request for its Marysville facility two years later. The other Japanese-owned and joint venture assembly plants subsequently built in the United States routinely asked for foreign trade subzone designation during construction.

US-owned producers realized that foreign trade subzone status could benefit them as well. The first designated US-owned plant was American Motors' Kenosha facility, in 1981, followed by Chrysler's Jefferson Avenue plant in Detroit the following year. A flood of applications hit from the Big Three producers between 1983 and 1985. By the beginning of 1986, foreign trade subzone status had been granted to eleven of GM's assembly plants, nine of Ford's, and four of Chrysler's, as well as to all of the foreign-owned assembly plants. A number of components plants had received designation as well. In the 1990s, automotive plants accounted for nearly half of all manufacturing facilities designated foreign trade subzones and five-sixths of the value of goods flowing through subzones (Chappell 1989b: 25).

Foreign trade zone legislation was originally intended to encourage exports, but ironically in the automotive industry has encouraged increased imports instead. The reason stems from the fact that duties are lower on assembled automobiles than on individual parts. According to foreign trade subzone legislation, a firm defers payment of duty owed on foreign parts brought to a US assembly plant until the finished vehicle is then exported, perhaps to Taiwan or Europe. In reality, of course, the overwhelming majority of parts imported to the United States are attached to vehicles sold inside the country rather than abroad. A vehicle assembled and sold in the United States with foreign parts is subject to a tariff. The tariff in such a case is the same as that levied on an imported vehicle, although duty is applied on the US-built vehicle only on the percentage of the value of the finished product judged to derive from foreign content.

Taxing the foreign content after assembly rather than the individual parts benefits the motor vehicle industry, because import tariffs are lower on finished vehicles than on components. According to Dennis Puccinelli, Foreign Trade Zone Board analyst at the US Department of Commerce, the imported content of a finished vehicle is taxed at the rate of 2.5 per cent, compared to 3.2 per cent for imported engines, 8 per cent for radios, and as high as 11 per cent for other parts. Thus, assuming that tariffs on individual parts are designed to offset the actual cost savings of foreign production, then the more a US automotive assembly plant imports the greater the cost savings. US-owned assembly plants, which typically import less than 10 per cent of their parts, realize modest savings of perhaps $5 per vehicle from foreign trade subzone designation. In contrast, Toyota, which planned to import roughly half of the parts to its Georgetown, Kentucky, assembly plant, revealed in its foreign trade subzone application that designation would save the company $41 per car. The foreign trade zone at Lafayette, Indiana, would save Subaru and Isuzu $23 per vehicle (Chappell 1989b: 53).

A Government Accounting Office study prepared for the House Ways and Means Committee concluded that the impact of foreign trade zones was mixed. The report claimed that:

> [The automotive industry] would still import parts if zones did not exist but added costs would render them less competitive . . . Benefits of continued operation of final assembly plants and the continued purchase of many US-made components offset the negative impact of reduced employment related to a small per cent of components.
>
> (Sundstrom 1984: 51)

The Foreign Trade Zone Board, which oversees the program, claims that manufacturers are not required to export their finished products, even if the original intent of the legislation was to encourage exports. Consequently, the board cannot deny an application for a new subzone solely on the basis of where the product is likely to be sold. Trade groups representing US parts suppliers have mounted campaigns to change the law to apply only to the vehicles produced in subzones which are actually exported.

The national government has also influenced the distribution of components production, as well as final assembly through its Corporate Average Fuel Efficiency (CAFE) standards. Each motor vehicle produced in the United States or Canada is assigned a figure computed by combining the average miles per gallon (mpg) achieved in city and in highway driving. An overall CAFE level is computed for each company, based on the performance of each model weighted by sales of that model. Companies failing to meet the mandated CAFE level for that year face fines. The minimum CAFE level was set at 20 miles (mpg) in 1980 and was expected to rise to 27.5 mpg in 1985; however, because the government delayed imposing higher standards on several occasions, the CAFE did not reach 27.5 mpg until the 1990 model year.

To count as North American for CAFE purposes, a vehicle has to contain at least 75 per cent of domestically produced components. This requirement has induced US producers to adopt curious policies. The percentage of North American parts in some full-sized models with low fuel-efficiency ratings has been reduced to below 75 per cent so that they would be considered foreign for CAFE purposes. On the other hand, domestic content has been increased to over 75 per cent in some subcompact models with high fuel efficiency ratings previously produced overseas.

GM originally created the Geo name in the late 1980s primarily to embrace a number of models it was importing from Japan. However, to increase its CAFE rating, GM began to build two of the Geo models at its joint venture plant with Suzuki in Ontario, while the domestic content was increased to over 75 per cent for a Geo model built at the joint venture plant with Toyota in California. Meanwhile, Ford's full-sized Ford Crown Victoria-Mercury Marquis model – one of the last vestiges of the large American car – was considered foreign, because domestic content was reduced to below 75 per cent, while the subcompact Ford Escort-Mercury Tracer – a twin to a Mazda model – was assembled in Mexico but counted as US-built because more than 75 per cent of the components were made in the United States (Versical 1990: 61).

Local government incentives

The most important tax incentive which local governments can offer to automotive manufacturers is reduction or abatement of property taxes (rates). The value of property derives from two elements – the value of the land and the value of the improvements, such as structures and equipment located on the site. Raw agricultural land carries a lower value than land containing a factory which has been improved with utilities, roads, and other infrastructure. Typically, a community reduces an industry's property tax obligation by continuing to value the land for a period between twelve and fifteen years at the level before improvements were made, that is the value of the land for agriculture.

A Japanese firm may buy far more land in the United States than it needs to build a factory, in part to keep away rival firms, bars, or other unwanted activities. However, the principal attraction of buying large quantities of land is cost: land prices in the rural Midwest are barely 10 per cent of Japanese levels; in an uncertain world, buying American land appears to be the safest possible investment to many foreigners, not just Japanese.

The cost of land is not a critical factor in the choice of state or locality for Japanese firms, but within the preferred community, the specific parcel finally selected will be influenced by price. For example, once Mitsubishi Electronics identified Mason, Ohio, as the first choice for its plant, four parcels were examined. The asking price was $30,000 per hectare for three of the parcels and $18,000 for the fourth; Mitsubishi preferred the low-cost site. However, once the company took an option on the parcel, it informed public officials that the site was too small and asked for help from the city to 'request' adjacent property owners to sell their land, so that a sufficiently large parcel could be assembled. Of course, the fair price for the additional parcels, Mitsubishi argued, was the same $18,000 per hectare paid for the original one. Why, city officials wondered, did Mitsubishi need 25 hectares to build a 6,000-m^2 factory? To protect the possibility of future expansion was the response.

Mitsubishi perceived another problem with the selected parcel. In the middle of the 25-hectare site was a 1-hectare lake. This 1-hectare area was perceived a waste because it could not be built upon. The city pointed out that the lake could serve as the stormwater-retention center, something the site would need anyway. Further, the city hired a local architect to sketch how the lake could be integrated into an attractive setting for the factory. Mitsubishi officials responded that the sketch was so attractive that they wanted the city to arrange for the architect to complete detailed drawings for the lakeside factory – another unexpected public expense.

While communities commonly will forgo for a period of time the additional taxes which would result from increasing the value of the land from agricultural to urban, policies vary with regard to the improvements. Some tax the building and the factory machinery at full value, while other localities reduce, perhaps by one-half, the liability for some or all of the improvements. Mason officials calculated that the property tax abatement it gave Mitsubishi was worth $4 million over fifteen years. However, the company would still pay $500,000 annually in local taxes based on the value of the raw agricultural land and the equipment inside the plant.

Tax abatements are frequently offered for plant expansions as well as new construction. In Ohio, tax abatements for improvements to existing plants are available within areas designated as state enterprise zones if the project generates additional jobs. Similarly, Michigan provides twelve-year abatements to firms which increase their payrolls or at least demonstrate efforts to maintain existing employment levels (Rowand 1982a: 2).

General Motors was the biggest beneficiary of Ohio's enterprise zones, receiving $1.4 billion of the first $2.5 billion invested or pledged in the program. Abatements to GM in the Dayton area, where most of the company's Ohio plants are clustered, exceeded $300 million in 1987 alone, including up to $247.7 million in Enterprise Zones. Abatements to Dayton area GM plants included $16 million to Detroit Diesel Allison and $7.2 million for five years to Delco Products to cover planned acquisition of new machinery and equipment. Harrison won't pay taxes on as much as $41.4 million of a total planned investment of $84 million in new equipment. The Moraine truck assembly plant won't pay taxes for ten years on up to $190 million of a planned $220 million investment, while Inland received a ten-year abatement worth $52 million at its Dayton plant. In return, GM promised to invest $380 million in Dayton-area plant improvements and save up to 1,166 jobs.

General Motors has also received tax abatements from the county and city governments in the Dayton area. Montgomery County, which includes Dayton, has granted General Motors an eight-year abatement on the buildings valued at $24.3 million, for a tax saving of $2.6 million over the period. Eight-year tax abatements were also given by the city of Kettering on $6.3 million worth of Delco Products property and by the city of Moraine on $18 million worth of buildings for truck assembly, diesel engines, and air-conditioning compressors. Kettering estimated that over an eight-year period, its abatement would save GM $735,452 in taxes, more than two-thirds of the company's liability.

Tax abatements to stimulate industrial growth have proved controversial in some communities. The mayor of Flat Rock, Ted Anders, described Mazda's demand for property-tax abatements, worth $2.5 million annually for fourteen years, as 'industrial blackmail.' He made the deal anyway and was promptly voted out of office. His successor Richard Jones secured $150,000 in aid from Michigan by arguing that the Mazda plant benefited the entire state. For its part, Mazda pledged to contribute $100,000 to Flat Rock, partially to offset a $200,000 budget deficit incurred in the first two years after construction of the plant began. Mazda also donated two police cars and a pickup truck to the city and $30,000 to the local historical society (Bachelor 1990: 97).

Individual communities are usually able to prove that new industries contribute more in local taxes over the long run than they receive in grants and abatements. However, at a regional or state-wide scale, the impact of tax abatements to new industries is less clearly beneficial, because inevitably the flood of tax breaks to new industries has led to a call for lower taxes on already existing plants, which are often located in different communities than the new ones. General Motors has been especially aggressive in pushing for lower property taxes at its plants. During the mid-1980s, the company appealed many of its property tax assessments, arguing that assessed values were illegal because they were based on excessive fair market values. GM charged that many of its properties had been assessed at rates ranging from $2 to $7 per square meter, at a time when the going rate for industrial property was approximately $1 per square meter.

The auto makers have had some success in arguing their case before Michigan's Tax Tribunal. Following a five-year battle, Ford won a $29.3 million property tax refund, plus $3.25 million in interest, from the City of Dearborn in 1985. General Motors sought reduced valuations from $41 million to $12 million on its Lansing engine plant, from $38 million to $16 million on its parts warehouse in Delta Township near Lansing, and from $34.5 million to $10 million on its stamping plant in Comstock Township near Kalamazoo. Thirty per cent reductions in property taxes were sought in Pontiac (Krebs 1985a: 26). As of 1985, Ralph Nader found that GM had filed appeals against fifteen Michigan communities, asking for $100 million in refunds and future reductions of $37 million a year at eleven plants. Nader pointed out that these appeals were filed one year after corporate profits had reached $8.2 billion (Kahn 1985: 39).

Flint, where GM plants accounted for 40 per cent of the city's total property tax revenues, was especially hard hit. The company argued before the Michigan Tax Tribunal that the assessments levied by the city, as well as the surrounding county and local school districts, should be reduced for the period 1983 through 1985 by $32 million on several properties, including Buick City, AC Spark Plug, CPC engine plant, and the Truck and Bus Group's assembly plant. In response, Flint city attorney S. Olof Karlstrom began an audit of GM's equipment and other non-real-estate properties in the city, claiming they were undervalued, and urged state legislation to clarify how business properties should be valued.

In one community, GM even protested an assessment at the same time as it requested a tax abatement. In 1983, the company filed an appeal in the Cuyahoga County Board of Tax Revision protesting the county auditor's as 1982 valuation of its plant in Parma. The company claimed that the plant's market value should be $19.2 million rather than $55.2 million, as set by the auditor, and that the annual tax obligation consequently should drop from $942,662 to $327,331. GM argued that the factory, originally built in 1947, with six additions, was obsolete and not energy efficient, and if placed on the open market would not sell for anything near $55 million because few customers would be interested in such a large building. The local school board, the main recipient of property taxes in most Ohio communities, opposed the request (Rutti 1985: 1–A).

While the tax case made its way through the local courts, GM also had the plant designated an enterprise zone in order to obtain a 50 per cent tax abatement for ten years on $406 million worth of planned improvements. GM claimed that the modernization program would generate an additional $30 million in the local economy, the same amount as the value of the abatements (Gabe 1985: 12–A).

COMPETITION FROM CANADA

Governmental policy differences have influenced the allocation of production between the United States and Canada. For decades Canadians feared that they would lose their automotive industry, but in recent years the trend has reversed, and US officials charge Canada with amassing more than its 'fair share' of North American automotive production.

Prior to the early 1960s, Canada's automotive industry was organized separately from that of the United States, even though US firms owned the Canadian production facilities. Canadian tariffs of 17.5 per cent on vehicles and 25 per cent on parts effectively limited imports from the United States. The Canadian government also required that at least 60 per cent of the content in domestically-built cars be produced in a British Commonwealth country. Despite the barriers, motor vehicle consumption outpaced production in Canada, and production declined from 400,000 in 1955 to 350,000 in 1960.

The government in 1960 appointed a Royal Commission headed by Vincent Bladen to investigate the competitive position of the Canadian automotive industry. The Bladen Commission in 1961 reported that the fundamental problem was the limited size of the domestic market. Canadian assembly plants and suppliers incurred high costs because they had to produce small batches of a wide variety of distinctive models. Unless the market for Canadian products were expanded, the commission concluded, the plants would become unprofitable and would likely be closed in the near future by their US-based owners. The solution was to integrate the Canadian and US automotive industries.

Consistent with the Bladen Commission's recommendation, the Canadian government began to reduce tariffs on vehicles and parts. In October 1962, the Drury Plan, named for the Minister of Industry C.M. Drury, suspended the 25 per cent tariff on transmissions and engines; other tariffs were reduced in July 1963. Finally, President Johnson and Prime Minister Pearson signed the Canada–US Automotive Products Trade Agreement on 16 January 1965. The trade agreement called for elimination of tariffs on most vehicles and parts passing between the two countries (US–Canadian Automotive Agreement Policy Research Project 1985).

While free trade was the most highly visible element of the agreement, more significant for the future spatial distribution of production was a complex series of letters of understanding sent to the Canadian government by the subsidiaries of the US-owned producers guaranteeing a minimum level of production in Canada. In essence, US companies agreed to produce

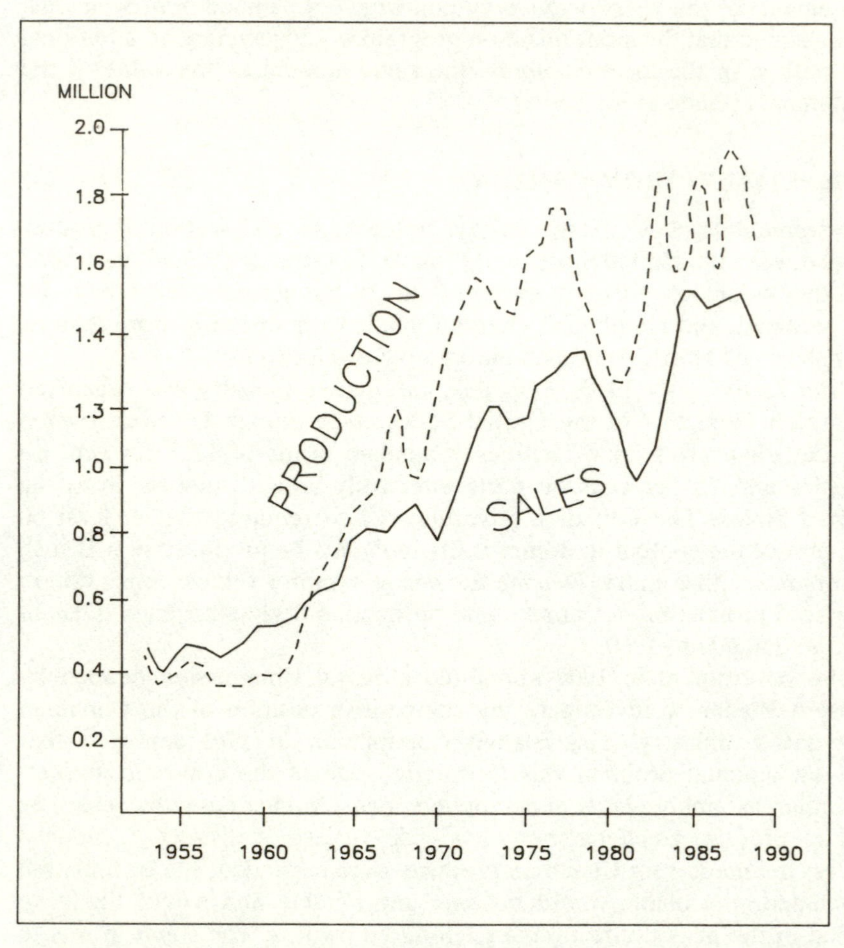

Figure 8.2 Production and sales of automobiles and light trucks in Canada, 1952–1990. Canada exports more vehicles to the United States than it imports

at least as many vehicles in Canada as they sold there. Canadian production climbed rapidly immediately after the agreement, from 400,000 in 1961 to 1.2 million in 1968 and 1.6 million in 1973 (Figure 8.2).

Canadian critics claimed that the temporary production increase from the trade agreement was offset by long-run adverse effects. Production in Canada fell from 1.6 million in 1973 to 1.1 million in 1981 in reaction to declining demand triggered by escalating fuel prices. Canadian analysts argued that a spatial division of labor had emerged between the United States and Canada, with US firms relegating the lower-skilled, lower-paying jobs north of the border. Canada had lost its research and design facilities and was in danger of losing the low-skilled jobs to Latin American countries, where wage levels were much lower (Holmes 1983).

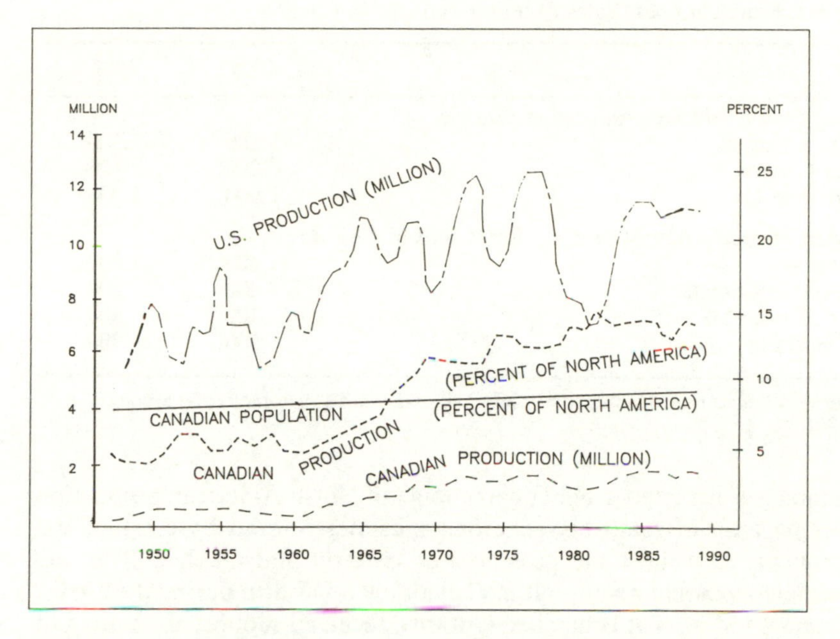

Figure 8.3 Production of automobiles and light trucks in Canada and the United States, 1947–1990. Since the mid-1960s, Canada has produced a higher percentage of vehicles than would be expected from its share of the North American population.

In reality, the restructuring of North American production has benefited production in Canada at the expense of the United States (Figure 8.3). Since the trade agreements of the early 1960s until the early 1980s, Canada's level of imports from the United States was roughly comparable to its level of exports. In 1981, for example, Canada exported approximately 1 million cars and trucks to the United States and imported approximately 800,000. By the late 1980s, Canada was exporting more than twice as many vehicles to the United States as it was importing. In 1989, for example, approximately 700,000 US-made vehicles were sold in Canada, but the same year, approximately 1.6 million Canadian-made vehicles were sold in the United States (Table 8.1).

While the Big Three US firms have reduced productive capacity by more than 10 per cent in recent years in the United States by closing a number of plants, at the same time they have expanded facilities in Canada. Assembly plants in Canada, as in the United States, have been converted from regional branch facilities to specialized plants producing one or two models for distribution throughout the two countries. By the early 1990s, Canadian plants were producing half of Chrysler and GM intermediates, half of Ford's compacts, all of Ford's full-sized models, and a substantial percentage of various truck models.

Table 8.1 Production and sales of motor vehicles in Canada

	1989	*1981*
Sales of motor vehicles produced in Canada		
total production	1,800	1,100
sold in Canada	200	100
exported to US	1,600	1,000
Location of production of motor vehicles sold in Canada		
total sales	1,400	1,200
produced in Canada	200	100
imported from US	700	800
other imports	500	300

Sources: Adapted from Holmes 1983: 263–264; and Ward's Automotive Yearbook 1989: 201

Canada has attracted a higher percentage of North American production in part because of relatively generous subsidies offered by the national government, as well as the provinces of Ontario and Quebec. The one automobile assembly plant built in Canada by a US firm during the 1980s, by American Motors at Bramalea, Ontario, received support amounting to $1,212,000,000, paid equally by the national and provincial governments. [All figures in this and the next three paragraphs are in Canadian dollars.] Technically, the government support was called a loan, because American Motors agreed to pay back 1 per cent of the value of each vehicle produced at Bramalea. If production fell below 158,000 vehicles in a year, the company would pay an additional 1.5 per cent of the value of each vehicle produced (Clark 1987b: 49; Sorge 1984b: 2). When it took over Bramalea in 1987, Chrysler, in addition to honoring AMC's commitment to the Canadian government, also agreed to produce at least 300,000 vehicles per year based on Renault models or pay a substantial penalty to the French company (Connelly 1990b: 43).

Some rough arithmetic concerning future sales and prices reveals the extent of the Canadian subsidy to Bramalea. Anticipating future price rises, Chrysler would have to sell approximately four million intermediate-sized cars to repay $1.2 billion. If Bramalea operated at capacity, roughly 200,000 per year, then the loans would be repaid in twenty years. In essence, the Canadian financial support would amount to the equivalent of a twenty-year virtually interest-free loan of over one billion dollars, a saving of several billion dollars for Chrysler compared to the cost of borrowing the money at commercial interest rates. However, as low sales for the plant's products have kept annual production well under 100,000, a substantial percentage of the 'loan' will never be repaid.

General Motors was also the recipient of generous Canadian support during the 1980s. The national and Quebec governments provided GM with a total of $220 million in the form of thirty-year interest-free loans to modernize the Ste Thérèse assembly plant. The loans contributed to a $450 million plant modernization program, with half of the funds earmarked for replacement of the outdated paint facility. US trade officials criticized the agreement as an unfair subsidy under the trade agreement (Burns 1987: D-13).

GM had threatened to close the Ste Thérèse plant in 1982 unless it received the loans. In exchange, GM agreed to keep the plant open for at least seven years and to provide the other $230 million. Ste Thérèse was then awarded responsibility for building some of the then-new A-platform intermediates. Ste Thérèse was threatened with closure again in the early 1990s, as sales of A-platform cars declined. However, the plant was designated as the production site for the redesigned Camaro/Firebird sporty models, known as the F-platform, previously built at Van Nuys, California. Under Canadian law a company must give sixteen weeks notice if it thinks a plant might be closed.

Production has also moved from the United States to Canada because of differences in government policies towards health care insurance. Big Three per capita expenditures on health care coverage for employees and retirees quadrupled during the 1970s in the United States from $0.6 billion to $2.4 billion and more than doubled again during the 1980s to roughly $5 billion (Sorge 1984a: D-1). The motor vehicle industry's health care bill actually grew more slowly than the national average for all industries, primarily because auto makers were relatively quick to adopt innovations, such as health maintenance organizations and prepaid group coverage. Nevertheless, by 1989, auto makers spent almost 50 per cent more per capita on health care in the United States than in Canada, amounting to a savings of $300 to $500 per vehicle. US health care costs per capita were twice as high as in Japan (Maher 1989: C-3). Not surprisingly, US automotive executives have been in the forefront of calling for a form of national health insurance, a position otherwise associated in the United States only with very liberal Democrats.

ARE COMMUNITIES PAYING TOO MUCH FOR AUTO PLANTS?

Plagued by uncertainties and contradictory evidence concerning the optimal location for a plant, Toyota eventually threw open its site selection process to the highest bidder, and Kentucky won by offering far more subsidies than any other state. Critics charge that Kentucky will never collect enough taxes to recoup the subsidies given to Toyota.

Toyota's site selection process

Toyota agonized between building its own plant and setting up a joint venture with a US firm. To buy time, the company announced in April 1980 that it was spending more than a million dollars to decide whether it should set up a US plant by hiring outside consultants, including Boston-based Arthur D. Little, SRI International in Stanford, and Nomura Research Institute, Japan's biggest think tank. The three consultants submitted reports in March 1981, but Toyota did not make public the results or announce its intentions. The sticking point in the feasibility study was having to pay 50 per cent higher wages in the United States than in Japan (Hartley 1980: 16; 'Toyota orders study on assembly in US', *Automotive News* 1980: 2; 'Toyota still studying reports on possibility of US plant', *Automotive News* 1982: 6).

Toyota opened talks with Ford in September 1980 about building a joint US assembly plant, but rejected the venture in early 1982. Instead, Toyota reportedly decided to built an assembly plant in Kansas, although a formal announcement was never made. Then in March 1982 a new rumor surfaced: Toyota was talking about a joint venture with GM instead of Ford; dismissed at the time, the rumor was confirmed within a month. Toyota president Eiji Toyoda had flown to New York to meet with GM chairman Roger Smith on 1 March, and Toyota Motor Sales Chairman Kato admitted that he had discussed with Smith the possibility of a joint venture the previous September at the US Open tennis tournament (Hartley 1982: 2; Rowand 1981: 2; 'Toyota near decision to build US plant', *Automotive News* 1982: 2).

Details of the joint venture filtered out during the summer of 1982, mostly from General Motors officials, and after eleven months of negotiations, the joint venture was officially announced in February 1983. To facilitate shipment of parts from Japan, the two companies agreed on a west-coast site for the plant. After touring a number of potential sites, the joint venture team narrowed the choice to GM's two recently closed west-coast plants at South Gate and Fremont, before settling on Fremont during the summer. Adjacent to the assembly plant, a new 16,000-m^2 stamping plant was constructed during 1983. Toyota contributed $250 million to New United Motor Manufacturing Inc. (Nummi), while GM added $20 million, plus the Fremont plant. Tatsuro Toyoda, grandson of Toyota's founder, was named president (Hamilton 1982: 2; Pelgue 1983: 1; Sorge 1983d: 62; Sorge and DeLorenzo 1983: 1; 'GM-Toyota decision due in fall, Smith says', *Automotive News* 1982: 42; 'Smith 90 pct. sure of GM–Toyota tie', *Automotive News* 1982: 2).

Despite Nummi, Toyota still needed more production capacity in the United States to compete with Nissan and Honda. Toyota apparently generated a short list of sites, including Kansas City, Kansas, and Lebanon, Tennessee, a northern suburb of Nashville. Kansas City had been in contact with Toyota for nearly a decade, from the time that the company first considered a US plant, prior to setting up Nummi with General Motors, and Toyota had established a regional parts distribution center in Kansas City. At the same time, Toyota looked at several sites in the Nashville area, because parts manufacturers had already built plants there in order to supply Nissan and Saturn. To maintain independence from the other two plants, Toyota looked at sites on the north side of Nashville rather than the south, where Smyrna and Spring Hill were located. Toyota asked the state of Tennessee for preliminary data on traffic figures, rail lines, roadways, utilities and water suppliers (Bernstein 1985: 22; DeLorenzo 1985: 1; Krebs 1985b: 64).

Toyota backed off from Kansas City and Nashville and decided to throw open the choice to the highest bidder. Communities wishing to be selected were required during the summer of 1985 to complete a lengthy questionnaire concerning labor, transport, and the supplier base. Thirty communities responded, and a North American Project Office was established in Tokyo to make the decision. Through the fall of 1985, Toyota sifted through the information it had collected from the surveys. Important factors included proximity to parts suppliers and markets, availability of highways and railroads, population of community, and availability of utilities. In November 1985, Toyota reportedly preferred a site in Tennessee, although in extreme southeastern Monroe County, rather than near Nashville. Kansas City remained in the running, as did sites in Missouri, Indiana, and Kentucky (DeLorenzo 1985: 1; Gawronski 1985a: 1; 1985b: 2; Hamilton 1985: 8; Johnson 1985d:3; Kelderman 1985: 5; 'Autumn decision due on US Toyota plant', *Automotive News* 1985: 3; 'Toyota near decision on US plant site', *Automotive News* 1985: 1; 'Toyota officials in Knoxville on plant search', *Automotive News* 1985: 5; 'Toyota to Kentucky; new subsidiary likely', *Automotive News* 1985: 1).

Toyota's choice of Georgetown, Kentucky, announced in December 1985, initially came as a surprise. But when the magnitude of Kentucky's incentive package was revealed, the logic underlying the choice became clear. According to Kentucky state officials, Toyota's incentives amounted to $157 million, far more than the $86 million recently provided by Illinois and Indiana to other transplants, including $68 million for job training, $40 million for roads and sewers, and $49 million for other capital improvements, principally acquisition of the 580-hectare site. However, the subsidy may climb to $325 million, with the addition of a state obligation to assist in paying interest on debts that the company may incur in conjunction with the plant construction (Glickman and Woodward 1989: 230–32).

A University of Kentucky study concluded that the Georgetown plant would generate $632.6 million in property, sales, and income taxes during its first twenty years, nearly twice the value of the subsidies. However, economist Larry Ledebur placed the expected total tax revenue at only $267.5 million. The large difference resulted from the university's higher estimate of the number of new businesses attracted to the state by the Toyota plant. Ledebur claimed that the university used an excessively high multiplier of eleven, that is estimating for each new job at Toyota eleven new jobs would be created at other industries and services, whereas economists rarely employ a multiplier higher than two (Fiordalisi 1989: 18).

Kentucky clearly did not have to offer as much as it did in the mid-1980s to lure a Japanese-owned assembly plant. Japanese producers had already calculated that new plants should be located in the I-65/75 corridor, and by the time Toyota searched for a site, Kentucky and Indiana were the only states in the region without a Japanese-owned assembly plant. Had Toyota gone to Indiana instead, undoubtedly Subaru/Isuzu may have picked Kentucky. However, Kentucky officials feared that the state was not yet regarded by manufacturers as a credible location for large plants in terms of ability to deliver services and infrastructure, and states a bit further from the Midwestern core automotive region, such as Missouri and Georgia, were actively pursuing Japanese investment, as well.

Framingham loses its GM plant

The city of Framingham, Massachusetts, learned that lack of cooperation with automobile companies can result in plant closures. GM's Framingham assembly plant had long suffered from a reputation of having poor labor relations; reflecting characteristics of the New England labor force in general, workers were more politically liberal and better educated at Framingham than elsewhere in the country. Grievances piled up at the rate of 1,000 per month in the 1970s, many related to alleged health and safety hazards, air pollution from the paint shop, excessive noise, and other environmental issues. Absenteeism was sometimes so high at the plant that GM had to hire college students in the Boston area to maintain operations. Several wildcat strikes hit the plant during the 1970s, sometimes in defiance of the national union, and workers rejected a quality of work life scheme. The plant manager took to calling his workers 'unruly, militant kids,' even though several hundred female parents worked the night shift. GM finally idled the plant in October 1982 (Plotkin 1984: 41).

Attitudes changed after GM reopened the plant in March 1983. A GM spokesperson claimed that the six-month closure 'awakened a lot of people to harsh reality' (Plotkin 1984: 41). With grievances down to ten or fifteen a month and vehicle quality improved, GM made a commitment to retain the plant. In 1985, construction began on a $250 million expansion of the paint facility. Framingham's future appeared even brighter in May 1986, when GM awarded it the company's most glamorous new product, the plastic-body APV minivan. A $60 million plant to make body panels and other plastic parts would be built next to the assembly plant and leased to a supplier, DiversiTech General (Plotkin 1986a: 55).

Did the citizens of Framingham rejoice that their largest employer was expanding? No, they were outraged by GM's high-handed manner. Back in 1985, Massachusetts Governor Michael Dukakis had agreed to sell

14 hectares of state-owned land in Framingham to General Motors in exchange for a commitment from the company to build the paint plant on the site. Framingham officials resented the deal, because the state had previously promised to transfer the land to the city. Now, scarcely one year later, GM wanted to expand again (Plotkin 1986b: 1).

Citizens raised questions at a hastily called town meeting, and local officials found a way to stop the deal between GM and the state. The city refused the rezoning needed to permit parking for the plant on a 6-hectare parcel of land still owned by the state pending completion some day in the future of a time-consuming environmental impact statement covering traffic, air quality, water supply, sewerage, and drainage at the site (Cooperman 1989: D-3).

DiversiTech General President Al Schmeiser expressed willingness to build a facility elsewhere in New England, perhaps in Salem, New Hampshire, where the company already operated a research facility. But a General Motors spokesperson pronounced the deal dead. A few months later, in June 1986, the plastic minivan was awarded to Tarrytown, and within two years the Framingham assembly plant closed. John Judge, Governor Dukakis's aide for Economic Development Affairs, believed that Framingham's actions 'could not have happened anywhere else in the United States.' He blamed the rezoning refusal on the town's then-booming economy (Plotkin 1986b: 1).

9 Avoiding militant workers

US Representative Mary Rose Oakar visited Japan in January 1985. Her mission was to convince officials of Mitsubishi Motors to locate its planned joint assembly plant with Chrysler in her northeastern Ohio district, which included the City of Cleveland. Like the other Japanese auto makers, Mitsubishi was looking at locations in the Midwest in order to minimize freight costs, and sites in Ohio were rumored to be on its short list. In April, Representative Oakar was informed that Mitsubishi was not interested in building a plant in northeastern Ohio.

A letter to Representative Oakar signed by S. Osakatani, assistant general manager in the office of the president of Mitsubishi Motors Corporation, explained why the Cleveland area had been rejected.

> As you pointed out the Greater Cleveland and Northeast Ohio region is a leading area for the automotive manufacturing industry. The rule of thumb we have been using in our site selection process is to avoid going right into the heart of any existing heavily automotive industrial region. Candidly, we see little likelihood of our being able to add the Greater Cleveland Northeast Ohio (region) to our list of prospective sites.
>
> (Jensen 1985: 1–E)

In December, Mitsubishi announced its choice of Normal, Illinois.

LORDSTOWN

Two decades earlier, General Motors had reached the opposite conclusion about northeastern Ohio. GM's only new assembly plant built during the 1960s (excluding a replacement for an older facility in the San Francisco Bay Area) was located in Lordstown, Ohio, 70 km southeast of Cleveland, and only a few kilometers south of the headquarters of the company's Packard Electric division in Warren. The assembly plant opened in April 1966, originally for production of full-sized Chevrolets.

Within a few years, Lordstown became a national symbol of factory workers revolting against numbing, mindless, repetitive tasks. GM Chairman Richard C. Gerstenberg acknowledged the problem in 1972: 'When

Blue Collar Blues are discussed, the auto industry is often singled out as the seat of the problem, the assembly line worker as the chief victim, and the new Vega plant in Lordstown as the case in point' (Risen 1987: D-5). Fear of 'Lordstown Blues' has dominated a generation of locational decisions in the automotive industry since the early 1970s.

Shortly after the plant opened, GM picked Lordstown to produce all of the company's first subcompact, the Chevrolet Vega. Although brand new, Lordstown was retooled with automated equipment to increase efficiency in the plant. Production began in 1970, and *Motor Trend* bestowed its 'Car of the Year' award on the 1971 Vega. But the Vega quickly displayed a critical problem to GM officials. For a company which had always generated profits by selling large, lavishly equipped vehicles, the small, spartan Vega was a money loser. The solution was to transfer management from Chevrolet to the General Motors Assembly Division (GMAD).

GMAD began life as the Buick-Oldsmobile-Pontiac Assembly Division, created after World War II to manage the five new assembly plants built in Arlington, Doraville, Fairfax, Framingham, and Wilmington, as well as the plants dating from the 1930s in Linden and South Gate, which had been originally managed by the Buick and Pontiac divisions, respectively. As the policy of producing corporate twins expanded during the 1960s, the B-O-P Assembly Division was viewed as the logical manager of the former Chevrolet branch plants, as well.

The division's name was changed to GMAD in 1965, when it took over its first Chevrolet plant, a new one in Fremont. Three years later, GMAD gained control of six other Chevrolet assembly plants, plus the companion Fisher Body plants, including Baltimore, Janesville, Lakewood, Leeds, Tarrytown, and Van Nuys. Chevrolet's four remaining assembly plants – Norwood, St Louis, and Willow Run, plus the new Lordstown facility – were transferred to GMAD in 1971. GMAD thus managed every assembly plant except for the four Michigan home plants still in the hands of the Pontiac, Oldsmobile, Buick, and Cadillac sales divisions.

After it took over Lordstown in October 1971, GMAD brought in time-study experts to figure out how the plant could turn a profit producing Vegas. Resulting changes at the plant included dismissal of 700 workers, reassignment of hundreds of others, and elimination of 300–500 line jobs considered too easy. Line speeds were increased from 60 to 101 cars per hour. To meet production goals with fewer workers, shifts often ran eleven or twelve hours, including mandatory overtime and weekends. Supervisors issued thousands of disciplinary warnings and laid off or fired militant workers. Lordstown cranked out nearly 400,000 Vegas in 1972 and again in 1973.

Sixteen thousand grievances were filed at Lordstown, an average of two per worker (compared to perhaps a few hundred in recent years). When the formal grievance procedure proved ineffective, workers staged wildcat strikes and other actions to slow the line. Hooded pickets sometimes blocked the plant gates, threatening supervisors and preventing workers from entering. Finally, a strike in March 1972 completely shut the plant for twenty-two days. *Playboy* magazine dismissed the Lordstown workers as long-haired, spaced-out hippies (Cray 1980: 475). True, the average age of the Lordstown workers was 22 or 23, but Vietnam veterans far outnumbered flower children. After surviving in the jungles of Vietnam, the last thing these workers were willing to do was be pushed around by intimidating foremen. 'Most of them had been in Vietnam, fought and came back with the attitude, who is this punk foreman?' recalled one union official (Risen 1987: D-5).

The Vega itself quickly left GM with a much bigger headache than unruly workers. Within days of the strike's settlement, in April 1972, the US government issued a mandatory recall on the Vega, because the muffler could easily rupture, posing a fire threat. A second recall followed quickly, because the accelerator was subject to jamming in an open position. Then the third recall within three months hit: rear wheels were liable to drop off due to a production error that left the rear axle a fraction of an inch too short. The problems did not surprise GM officials, because the Vega was known to be a poorly designed car from the time that the front end of the first prototype fell off at the test track. The car suffered from an inadequate cooling system, an easily warped engine block, rapidly wearing front disc brakes, and a faulty carburetor (Cray 1980: 473). GM soon killed the Vega, but memories of the 'Lordstown Blues' remained.

THE UNIONIZATION MOVEMENT

Avoiding concentrations of militant workers influenced locational decisions even in the early days of the automotive industry. After the Olds plant in Detroit burned in 1901, a nearby foundry which had escaped the fire was transformed temporarily into a machine shop. However, two months after the fire, thirty-five workers who were members of the Machinists Union joined a national strike for higher wages and a reduction of the working day from ten to nine hours. Several days later, 500 sympathizers of the strike gathered on a vacant lot near the Olds factory and moved towards the plant. Three people were injured in a scuffle, before police arrived to escort home Olds' twenty-five non-union workers. Shortly thereafter, Olds decided to build a new plant in Lansing, considered sufficiently far from the unrest of Detroit.

Detroit is now known as a 'closed shop' city, thanks largely to collective bargaining in the automotive industry, but at the beginning of the century the city had a different image, which attracted early car makers. By 1914, when Ford's moving assembly line transformed automotive production from a skilled to an unskilled occupation, the open shop concept – that membership in a union is not a condition of employment – had become strongly entrenched in Detroit and the automotive industry in particular.

Credit for making Detroit an 'open shop' city went principally to an aggressive campaign waged by the Employers' Association of Detroit (EAD) to eliminate closed shop agreements. The EAD set up a centralized employment agency which screened job applicants for union sympathies and supplied firms hit by strikes with nonunion workers. Thanks to the EAD, union membership in Detroit, which had risen from 8 per cent in 1901 to 11 per cent in 1904, declined to 9 per cent in 1911. Union membership in Detroit, which had increased from 8,000 in 1901 to 14,000 in 1904, reached only 15,000 in 1911, during a period when the city's total labor force rapidly expanded (Babson 1986: 19–21).

UAW recognition

Initial attempts to unionize automotive workers concentrated on the minority involved in skilled trades, such as metal finishers, painters, and tool makers. The country's leading labor organization, the American Federation of Labor (AFL), had little interest in organizing unskilled mass production workers on an industry-wide basis, because it had originated in the nineteenth century as an amalgamation of workers organized around crafts or skilled trades. Finally in 1935, the AFL offered a charter to an industry-wide automotive union, the United Automobile Workers (UAW). However, the AFL provided the UAW with older, cautious leaders and postponed democratic elections of officers by the rank-and-file.

The AFL actions proved a case of too little, too late. Once they gained voting rights, in April 1936, UAW delegates replaced the AFL-appointed president, Francis J. Dillon, with Homer Martin, a former Baptist minister and automotive worker from Kansas. The UAW left the AFL to join the CIO, organized around industries, such as mining and clothing. Headed by John L. Lewis, President of the United Mine Workers and former AFL vice president, the CIO (which originally stood for Committee for Industrial Organization and later Congress of Industrial Organizations) was more militant than the craft-oriented AFL unions.

The UAW was ultimately successful in securing recognition because it understood the linkages among producers and suppliers in the automotive industry. The union determined that GM was most vulnerable to strikes at two Fisher Body plants in Cleveland and Flint. The Cleveland plant supplied all of the bodies to Chevrolet assembly plants throughout the country, whereas Fisher plant number 1 in Flint produced the bulk of body stampings for the company's other divisions (Cray 1980: 295). With supplies from these two plants cut off, GM assembly plants throughout the country would be forced to close.

Simultaneous strikes were planned for the Cleveland and Flint plants in January 1937, but 700 of Cleveland's 7,100 workers became infuriated when the plant manager cancelled a meeting to discuss reducing piece-work rates. They jumped the gun, walking out on 28 December 1936. Two days later, 200 members of the Flint local on the night shift, tipped off that GM intended to move its body-stamping dies out of the plant, seized control of Fisher 1, while another 100 captured nearby Fisher 2. Within a week, General Motors was paralyzed: 106,000 of GM's 150,000 workers were idled by 10 January, 140,000 by 1 February. Cleveland workers subsequently agreed to terminate their strike in exchange for a pledge from GM not to resume production at the plant until a settlement was reached in Flint.

The union's chief weapon was the controversial sit-down strike. Until declared illegal by the Supreme Court in 1939, the sit-down strike proved more effective than a walkout or picket line because workers remained inside the factory, while employers, scabs, police, national guard, and hired thugs – all left outside – could regain control only by force.

GM secured an injunction ordering the workers to vacate the Flint plant, but Michigan's Governor Murphy, the former mayor of Detroit, refused to call out the National Guard to enforce the injunction. Sympathetic to the aims of the strike, Murphy was more interested in avoiding bloodshed than with removing the workers, and without Murphy's support for the use of force, GM had no lever to end the occupation. On 11 February 1937, the 44-day strike ended with a union victory.

In the seventeen plants which had been the focus of the strike, the UAW was granted exclusive bargaining rights for six months, during which time the union was free to sign up as many members as it could. GM agreed to negotiate with the union over the issues related to working conditions raised in the strike and was barred from retaliating against the strike's organizers or participants. GM was permitted to make a verbal rather than a written promise to bargain exclusively with the UAW, but as Governor Murphy was prepared to enforce the agreement, the verbal promise proved binding and gave GM little more than a modest face-saving device. Still believing that workers could be enticed to join a company-sponsored union, GM secured the right to negotiate with other unions in its other plants if it wished, but by October 1937 the UAW had enrolled 40,000 workers, predominantly at GM, and was well on its way to becoming the industry's dominant union.

Unrest continued in the US automotive industry after World War II. Hundreds of strikes disrupted suppliers, and within ten weeks of V-J Day, the UAW began a strike against General Motors which would last 113 days. Shortly after the end of the 1946 strike, the UAW elected as its president Walter Reuther, one of the most creative visionaries of the American labor movement and a long-time prominent leader for civil rights and other progressive causes in the United States. Reuther also presided over an era of unparalleled economic benefits for auto workers. Every three years, as contracts were about to expire, Reuther concentrated negotiations on the one auto maker most likely to meet the union's demands. At the same time, he quietly offered the other producers no-strike guarantees as long as they agreed automatically to match the contract signed by the targeted company (Serrin 1973: 18).

Reuther accurately concluded that if only one company were threatened with a strike, it would be willing to sign a generous contract in order to avoid a strike during which its competitors would thrive. Chrysler (the weakest of the Big Three) was targeted in 1961 and 1964, and Ford (the most conciliatory once Henry Ford II took over) in 1955, 1958, and 1967. The union avoided direct confrontation with GM, which had the financial reserves to withstand a strike and was content to match contracts signed by its smaller competitors (Serrin 1973: 174).

'OPERATION APACHE' AND THE 'SOUTHERN STRATEGY'

Walter Reuther died with his wife in an airplane crash in May 1970, and was replaced as president by Leonard Woodcock. The UAW in the 1970s continued to focus on higher hourly wages and more generous benefits, such as cost-of-living adjustments, company-funded pensions, and unemployment benefits. Also maintained was the policy of targeting a national strike against one company – GM for fifty-nine days in 1970, Chrysler for nine days in 1973, and Ford for twenty-eight days in 1976.

However, the labor climate in the US automotive industry was increasingly dominated in the 1970s by issues raised under the banner of the 'Lordstown Blues,' with significant geographic consequences for both the union and producers. The UAW's geographic response to the 'Lordstown Blues' became known as 'Operation Apache,' while GM's geographic response was the 'Southern Strategy.'

The UAW's 'Operation Apache'

The UAW rediscovered the same weakness at GM as those it had exploited in 1937: a strike at one or two key supplier plants could disrupt the entire company. The union would strike a plant which produced an essential component, such as the only source of a hinge or steering gear. Cutting off the flow of needed parts disrupted production at assembly plants, forcing them to close as well. Shipments got so fouled up that one time boxcars arrived at the Lakewood assembly plant full of toilets instead of car doors.

Critical to the success of the strategy was the ability to strike at short notice. The UAW had the right to strike over grievances related to local production standards, health, and safety, but only those which had reached the third stage in the appeal process, and only after issuing a five-day notice of intent to strike. To increase its strike options, the union made certain that a large number of grievances were always held at the third stage in each plant.

However, once a five-day notice of a strike intent was filed, the union was still obligated to sit down with local management in an attempt to settle the problem without a strike. The union therefore would present its first grievance, and management would agree to resolve it in the union's favor. The union would then pull out a second grievance, and a third one, and as many as it took until management realized that the union could never be satisfied at the bargaining table, because it was intent on calling a strike.

When the strategy was explained at a UAW convention, a delegate stood up and said that it reminded him of the hit-and-run tactics used by the Apaches to attack US cavalry. Thereafter, the strategy became known in the union as 'Operation Apache.' 'Apache' strikes were popular with the workers because they were usually scheduled to give them long weekends. Once the Janesville local called to ask if an 'Apache' could be scheduled for a particular weekend. The reason? Hunting season was opening in Wisconsin. The request was denied.

GM's 'Southern Strategy'

GM also learned valuable geographic lessons from its confrontation with the union in the late 1960s and early 1970s. First, the company concluded that it probably could not speed up lines and reduce employment enough in US plants to build subcompacts profitably. It made one last stab from 1976 through 1987, with the Chevette, which was built at Wilmington and Lakewood rather than Lordstown. But by the mid-1980s, GM turned over production of subcompacts to Japanese partners or to joint-venture transplants in North America.

The second geographic lesson learned by GM as well as other firms – both within and outside of the automotive industry – was that the local labor climate could be more critical than transport costs in selecting new plant sites or deciding which older plants to modernize. Sympathy for the striking Lordstown workers ran high in heavily industrialized and unionized northeast Ohio. Consequently, GM decided to build new plants where union membership was low, and local residents would be less receptive to militant workers. In older plants, the 'Lordstown Blues' induced GM to look for ways to minimize the number of mindless, repetitive tasks, such as installation of robots and adoption of more flexible work rules. Once overcapacity hit the North American industry in the 1980s, GM and other auto makers tied innovative labor relations to increased productivity, and adoption of new work rules became a condition – although not a guarantee – of an older plant remaining open.

GM's spatial response to the 'Lordstown Blues' in the 1970s became known within the UAW as the 'Southern Strategy.' During the 1970s, GM built or planned fourteen plants in the south, primarily in rural areas or small towns. Four were built in Mississippi, three in Louisiana, two each in Alabama and Georgia, and one each in Oklahoma, Texas, and Virginia. Plants were located in Laurel and Meridian, Mississippi, and Albany and Fitzgerald, Georgia, by Delco-Remy; in Brookhaven and Clinton, Mississippi, by Packard; in Athens, Alabama by Saginaw Products; in Tuscaloosa, Alabama by Rochester Products (now Harrison Radiator); in Monroe, Louisiana, by Fisher Guide (now Inland Fisher Guide); in Wichita Falls, Texas, by AC (now AC Rochester); in Fredericksburg, Virginia, by Delco-Moraine (now Delco-Moraine NDH); and in Shreveport, Louisiana by Delco Electronics. In addition, GM planned new assembly plants in Oklahoma City and Shreveport during the 1970s.

GM was claiming that the new southern plants were needed to accommodate a projected increase in the US market from 9.3 million vehicles in 1975 to 15 million in 1985. However, believing that the US sales would grow modestly over the next decade, to perhaps 10 million, the UAW feared that GM intended to close its older unionized plants in the north. (Actual US sales in 1985 fell between the two forecasts – 12.1 million – and GM still utilized most of its northern plants.)

At first, the UAW had difficulty organizing the southern plants. The union won its first election at the Monroe plant, after it showed that Fisher Guide was paying higher wages in the north. Once the Monroe plant was organized, the UAW negotiated the same pay scale in effect at Fisher Guide's largest plant, at Anderson, Indiana. The only other southern components plant organized during the 1970s was the Delco-Electronics facility at Shreveport, where the union won a representation election in February 1979. The plant was subsequently closed (Sorge 1979a: 6).

As an indication of the environment which attracted GM to the south, the anti-union campaign in Monroe was organized by the local Chamber of Commerce. However, the Chamber president showed up at a dance celebrating the first anniversary of the union victory to declare that his opposition had been misguided. As the owner of the town's Pontiac dealership, he had observed that higher wages had stimulated local sales of new cars.

By 1976, the UAW regarded GM's search for union-free locations as a grave threat. As the 15 September contract deadline neared, GM and the UAW had reached an agreement on all issues with one exception. The sticking point was GM's refusal to remain neutral during union organizing drives in the southern plants. The union declared its intention to strike over the issue and prepared walkouts at fifteen or sixteen key plants. George Morris, chief GM negotiator, agreed that the company would remain neutral with regard to production and maintenance workers but not white-collar workers.

GM adopted a new hiring policy in 1978, giving 'preferential treatment' to UAW members seeking jobs in the south, including the company's two largest southern plants, the assembly plants under construction at Oklahoma City and Shreveport. When the Oklahoma City plant opened in 1978, 12 per cent of the workers had been UAW members at other plants. The UAW was recognized as the bargaining agent for workers at

GM's Rochester Products carburetor plant in Tuscaloosa and Delco-Remy generator plant in Albany. The UAW said that new hiring policy 'makes it clear that GM's "Southern strategy" has been abandoned' ('New hiring procedure for GM southern plants', *Automotive News* 1978: 3).

But the union had reason to doubt GM's neutrality pledge. UAW complaints centered on GM's campaign tactics in an election scheduled for Wednesday, 19 July 1979, at the Oklahoma City plant to determine whether the workers wanted the UAW as their bargaining agent. The Friday before the election, GM officials in Oklahoma City placed anti-union literature at all work places. Over the weekend, the company promised to have the literature removed, but on Monday morning it was still there. Worse, plant officials had stacked cartons of anti-union T-shirts in an executive garage and were helping to distribute them.

The union had a strong card to play: Monday morning by coincidence marked the scheduled start of negotiations with GM concerning the new three-year national contract. When they learned immediately before the start of the first bargaining session about the continued violations of the neutrality pledge at Oklahoma City, union officials refused to sit down with company negotiators until they met first with GM Chairman Murphy. Negotiations were impossible, they informed Murphy, at a time when the company's basic integrity was being called into question, even if a strike resulted.

Murphy, respected by union officials as a truthful and honorable man, was upset by the reports from Oklahoma City. UAW President Douglas Fraser reported, 'We presented to them overwhelming evidence that a conspiracy existed between Oklahoma City plant management and anti-union forces there. They said they didn't tolerate this, and I don't believe they are a part of it.' During the meeting with union officials, Murphy ordered the Oklahoma City plant manager to correct the problem immediately or be fired. Murphy also sent officials from Detroit to Oklahoma City to assure compliance. Two days later, the election was held, and 69 per cent of the workers voted for UAW representation (Irvin 1979b: 3; 'UAW win at OKC', *Automotive News* 1979: 46).

Once evidence of compliance was received, the union returned to the bargaining table but refused to discuss any other issues until the issue of union representation in the southern plants was clearly resolved. At the same time, recalling 'Operation Apache,' strikes were authorized at seven plants, in order to shut down production of the company's two best-selling models, the subcompact T-platform (Chevette) and compact X-platform (Citation). Plants included Chevrolet Metal Fabricating in Flint; Flint Engine, which built all of the X-platform four-cylinder engines; Flint Manufacturing, sole supplier for all X-platform gas tanks; Fisher Body in Mansfield, which made X-platform floor pans; Rochester Products in Rochester, which made T-platform and X-platform carburetor and emission controls; Lakewood, which assembled the T-platforms; and Willow

Run, which assembled the X-platforms (Sorge 1979b: 41). A strike was averted, and the union's national agreement with GM was extended to all of the southern plants.

Mack Truck's Southern Strategy

Mack Truck tried to emulate GM's Southern Strategy during the 1980s, with similar results. In January 1986, Mack threatened to move its largest assembly plant out of Allentown, Pennsylvania, unless the UAW agreed to a long list of concessions. When the union refused, Mack made good on its threat, relocating 1,800 jobs to a new plant in Winsboro, South Carolina. The company refused to give the union automatic recognition as a bargaining agent in the new plant, insisting that workers had to vote for union representation. Confident that southern workers would reject the union, an anonymous Mack official crowed that the UAW 'called our bluff and it backfired.'

Stung by the loss of jobs in Allentown, officials of UAW Local 677 negotiated a six-year contract which conceded most of Mack's demands, in exchange for company guarantees to give laid-off Allentown workers priority in hiring at the South Carolina plant and to retain the remaining 2,300 jobs in Allentown during the contract period. Similar concessions were negotiated at Mack's other northern plants in Hagerstown, Maryland, and Somerset, New Jersey, and overwhelmingly ratified by the members. However, officials at UAW headquarters in Detroit refused to approve the contracts, claiming the union's national bargaining position would be eroded by the extensive concessions, including a wage freeze, a no-strike clause, and length of contract. Local 677 president Kim Blake was incensed: 'What good is having the best contract in the entire industry if it is in your back pocket in the unemployment line?' (Stevens 1987: A-24).

The union won the next round, though. To the surprise of Mack officials, workers at the Winsboro plant voted overwhelmingly in 1989 to accept UAW representation. The union taunted the company with the claim, 'you can run from us but you can't hide.'

General Motors and Mack both learned from their failed southern strategies that plants could not be relocated within the United States from the northeast to the south if the primary purpose was to pay lower wages. Unions successfully organized instinctively hostile southerners by demonstrating that unionized workers in the north received higher wages for comparable work.

MAQUILADORAS

With savings in labor costs no longer sufficiently high to offset additional freight charges, the south became less attractive to US auto makers during the 1980s. To find lower-cost labor, US firms moved some production to Mexico. Production transferred from the United States to Mexico to take advantage of laws known as maquiladoras. The term, derived from the Spanish verb 'maquilar,' meaning to take measure or payment for grinding or processing corn, was originally applied to a colonial tax (Biederman 1989: 53). Maquiladoras – commonly shortened to maquilas – depend on two sets of laws, regulation of foreign investment by Mexico and preferential tax treatment by the United States.

Under the maquiladora, or in-bond, laws, Mexico permits foreign firms to import components duty-free, assemble them in Mexico, and export them to the United States. Firms do not have to pay duty on the equipment, raw materials, or subassemblies brought into Mexico. The Mexican government permits foreign investors to own 100 per cent of the in-bond enterprise, whereas other businesses must be at least 51 per cent Mexican owned. The government also streamlines the approval process, to as little as four weeks, compared to a year in the 1970s.

When the components are shipped back to the United States for final packaging and distribution, Sections 806.3 and 807 of the US Tariff Code requires that duty be paid only on the value-added during assembly in Mexico, which is principally the cost of labor. In other words, the duty is the difference between the value of the finished product and the sum of the value of the American-made parts. It doesn't matter whether the Mexican factory is owned by a US, Mexican, or other foreign firm so long as the components imported into Mexico were made in the United States.

Origin of maquiladoras

Antonio J. Bermudez, the one-time head of Pemex, the state petroleum company, is known as the 'godfather' of the maquiladoras in recognition of his role in developing the idea. In the early 1960s, Bermudez was named by President Adolfo Lopez Mateos to head a program aimed at improving the appearance of Mexican border towns, including his hometown of Ciudad Juarez. Concerned with economic growth, not just physical redevelopment, Bermudez and other officials in Ciudad Juarez hired the American consultant Arthur D. Little to conduct a study on the possibility of attracting new industry; the firm's report was published around 1963. The Mexican government initiated the Border Industrialization Program in 1965, although for several years little was accomplished as government interest waned, especially after Bermudez resigned as head of the program.

Bermudez's great-nephew, Sergio Bermudez, became the head of the largest industrial development firm in Ciudad Juarez, employing more than 400. The Bermudez group sells industrial real estate, builds plants, contracts with manufacturers to occupy the buildings, and operates four industrial parks in the city. In a typical week, the firm receives inquiries from fifteen firms interested in establishing a maquiladora in Ciudad Juarez. Sergio Bermudez's father, Bermudez Cuaron, was elected mayor of Ciudad Juarez in 1986.

In 1968, RCA became the first large American company to open an in-bond plant in Ciudad Juarez. It was a decade later before US automotive manufacturers took advantage of the maquiladora system, beginning with General Motors. Its Packard Electric Division established Conductores y Componentes Electricos to make wire harnesses in Ciudad Juarez in 1978. Later that year GM's Inland Division opened Vestiduras Fronterizas in Ciudad Juarez to make seat covers and interior trim. In 1979, Inland told its unions that brake-hose work would be shifted to Mexico. The unions didn't fight the threatened move because they didn't believe it would happen. Later that year, Inland's Compontentes Mecanicos plant in Metamoros began to make brake-hose assemblies, dashboards, and steering wheels.

Maquiladoras boomed after 1982, when the peso collapsed as a result of a drop in oil prices. The peso declined from 26 to the dollar in 1982 to 1,000 in early 1987 and over 3,000 in November 1987. The number of maquiladora plants in Mexico grew from approximately 600 in 1982 to 1,000 in 1986, 1,250 in 1987, 1,550 in 1988, and 1,800 in 1989. Employment increased from 70,000 in 1982 to 360,000 in 1988. US firms, which owned or controlled two-thirds of the plants, invested more than $2 billion in maquiladoras during the 1980s.

Automotive plants accounted for only approximately 129 of the 1,500 plants as of 1988 but nearly 78,000 of the employees (Scheinman 1990:122). Electronic equipment accounted for 34.7 per cent of the maquiladoras, followed by textiles and clothing at 14.6 per cent; transportation equipment accounted for 8.6 per cent (Perez 1989: 60).

Attraction of maquiladoras

The principal attraction of locating in Mexico is the very low wage rates compared to the United States. Most workers at maquiladoras earn Mexico's minimum wage, which in mid-1988 was approximately 1,000 pesos (43 cents) per hour. In comparison, automobile workers in 1987 earned an average of $19.87 in the United States and $13.25 in Japan.

It is impossible to state exactly the level of wages being paid, either in dollars or pesos, because the Mexican government frequently raises the minimum wage in response to high inflation, after consulting with private employers. For example, in 1987, General Motors workers in Mexico earned 381 pesos per hour in 1987, equal to 39 cents. One year later, hourly wages had increased to 1,157 pesos, but because of 121 per cent inflation in Mexico and a 241 per cent devaluation of the peso against the dollar, the new wages were worth 52 cents per hour. Almost all plants pay the minimum wage because earnings above that level are taxed at more than 5 per cent, resulting in lower take-home pay for workers.

Outsourcing to Mexico results in higher freight charges and makes 'just-in-time' delivery difficult. Mexican production is therefore economically justifiable only if labor savings outweight the shipping problems. According to David Collier, at the time GM vice president in charge of operating staffs:

> You can attain, in some cases, an economy of scale that makes up for distance. For example, there's the GM engine plant being built in Mexico for an economy of scale that should give optimum competitive costs. Obviously, it creates inventory problems, but they are not sufficient to offset the economy-of-scale efficiencies.
>
> (Rowand 1982d: 8)

Maquiladora plants specialize in assembly work which can be done by relatively unskilled workers needing little training and instruction. More than half of the automotive plants assemble wire harnesses, including fourteen plants performing this function for GM's Packard Electric Division. Independent companies, including Essex and American Yazaki, also mass-produce wire harnesses for US cars. With the rapid growth in the use of electronics, auto makers have determined that they can no longer afford to pay US workers $20 an hour to build wiring harnesses, which requires little more than clipping and bundling color-coded wires.

Electronics assembly other than wire bundling comprise the second largest group of maquiladora plants. These electronics plants assemble radios, turn signals, dashboard controls, heating and air-conditioning units, and solenoids. The third group of firms produces plastic trim and other parts, including seat covers, belts, and bumpers.

Most of the automotive-related maquiladoras are operated by the Big Three US car makers. As of 1989, General Motors had twenty-five, including twelve opened in 1987, employing more than 25,000 workers. Ford had ten plants with 7,000 workers, and Chrysler four plants with 5,000 workers. GM's Packard Electric Division operated fourteen plants with 15,000 workers in order to assemble wire harnesses (Table 9.1). Two Delco-Remy Divisions made engine controls, turn signals, and solenoids. Two Delco Electronics Division plants, employing 3,000, assembled stereos and stereo parts, air conditioners, engine controls, electronic dash displays, and circuit boards. Two Delco Products Division plants assembled ceramic magnets, ISC motors, and other electronic components, including power anttenas for Cadillacs. Two Inland Division plants produced brake-hose assemblies, plus interior trim such as steering wheels, dashboards, and seat covers. A Fisher Guide plant made plastic bumpers, a Rochester Products Division made valves and nozzles, and a truck plant made wiring connectors (Downer 1988: 20).

Table 9.1 General Motors 'maquiladora' plants in Mexico

Division	*Plants*	*Employees*
Packard Electric	14	14,068*
Delco Electronics	2	3,079
Inland	2	3,113
Delco Remy	2	1,923
Delco Products	2	1,294
Fisher Guide	1	1,127
Rochester Products	1	*
GMC truck	1	*
	25	24,604*

Sources: adapted from Roberts 1988: 4; and Downer 1988: 20

Note: * includes four plants – two Packard Electric, one Rochester Products, and one GMC truck – where employment was estimated

The United Auto Workers union maintains that it is economically unhealthy for auto makers to move their capital to countries where workers cannot afford to buy the products they manufacture. However, Donald Michie, Director of the Institute for Manufacturing and Materials Management at the University of Texas at El Paso, supports maquiladoras. 'There is no exploitation of the Mexican worker in the maquiladora industry. You don't find a better wage/benefit package within the Mexican economy and yet by American standards the wage rate is low. You can't impose American standards on Mexico' (Downer 1988: 22).

Distribution of maquiladoras within Mexico

The overwhelming majority of maquiladoras have been located in Mexico's northern states which border the United States. Of the first 1,132 maquiladoras, approximately 1,040, or 92 per cent, were in the six states bordering the United States, including 461 in Baja California, 103 in Sonoro, 247 in Chihuahua, 79 in Coahuila, 21 in Nuevo Leon, and 129 in Tamaulipas. Within these six states, maquiladoras were highly clustered in a handful of cities, with more than half in three cities, Tijuana, Mexicali, and Ciudad Juarez. Other leading centers include Matamoros, Nogales, and Nuevo Laredo.

The leading center for motor vehicle suppliers is Ciudad Juarez, where more than two dozen, or over half, have located (Figure 9.1). Ciudad Juarez is the most popular location for auto makers because of good transport links to the United States. Three bridges cross the Rio Grande River to El Paso, Texas; from El Paso, limited access interstate highways carry trucks to the Midwestern auto producing region. Matamoros, across the Rio Grande from Brownsville, Texas, is only slightly farther from the Midwest than Ciudad Juarez, but no interstate highway reaches the Mexican border at that point.

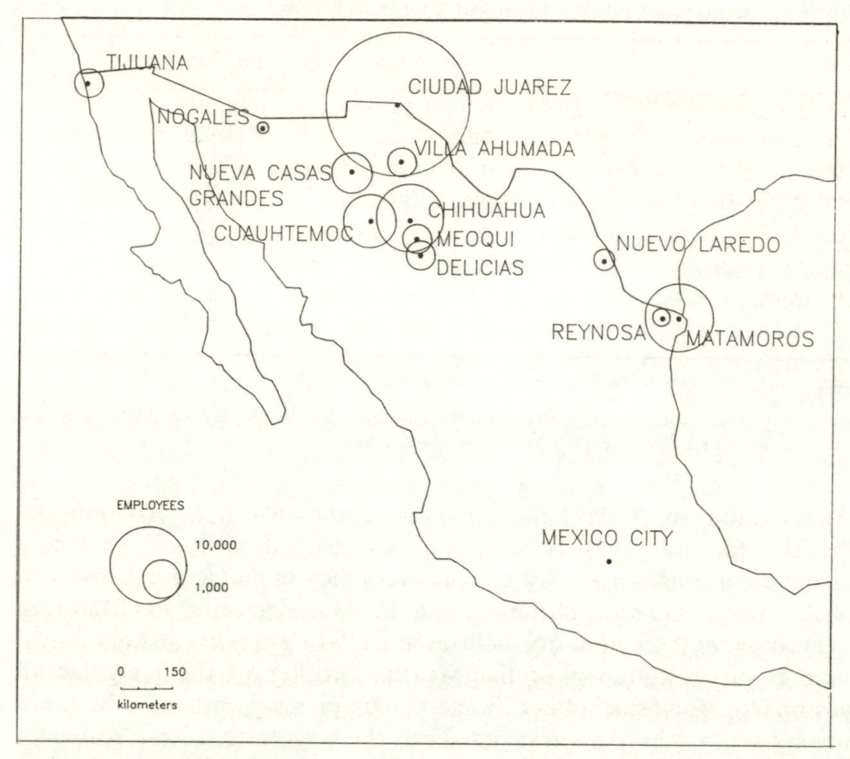

Figure 9.1 Location of automotive industry 'maquiladora' plants in Mexico. The plants are clustered in the far north near the US border to facilitate shipments to and from the United States

Both Ciudad Juarez and Matamoros have main rail line service to the United States. Many components are shipped by rail because of the relatively long distance to the Midwest; further, trains are less likely than trucks to be delayed by Mexican customs officials at the US border.

Mexico's border towns were traditionally isolated from the country's major urban areas, which are concentrated in the interior. Far from major markets, the border towns possessed little of the country's industrial base. However, as Mexico's internal market stagnated during the 1980s, industrial development transferred to the border towns, which provided relative proximity to the growing US market. While Mexico's economy declined 3.7 per cent in 1986, Tijuana's expanded by 7 per cent. Mexicans are flocking to Tijuana at the rate of more than 50,000 a year, even though housing is unavailable, but some maquiladoras still have trouble finding enough workers. The maquiladoras have spurred construction of new offices, factories, and warehouses, as well as luxury houses for the plant managers, straining water and other public services. The old sleazy red-light district along Avenida Revolucion remains, but high-class shopping boutiques have expanded to attract American shoppers taking advantage of the depressed peso.

Quality of maquiladora workers

The maquiladora program has generated two sets of issues and problems – the quality of the work and the impact on the US economy. Firms have difficulty attracting and retaining qualified workers as a result of an acute labor shortage in the border towns where the maquiladoras cluster. Employee turnover rates are high at maquiladoras, as high as 20 per cent per month in some cases. Among all maquiladoras, monthly turnover rates were 13 per cent during 1986 and 9 per cent during 1987. Workers leave either to go to work at another plant or to withdraw from the local labor market altogether.

The labor pool has expanded rapidly in the border towns as a result of the maquiladoras. The population of border cities like Ciudad Juarez, Nogales, and Tijuana doubled during the 1980s, as a result of rapid immigration. At first, migrants to Ciudad Juarez came primarily from rural areas elsewhere in Chihuahua State or from adjacent northern states, such as Coahuila, Zacatecas, and Durango. But as economic conditions deteriorated elsewhere in Mexico by the late 1980s, most immigrants came from central and southern Mexico, such as Chiapas, Oaxaca, and the Mexico City region. The US crackdown on illegal immigration induced more Mexicans to try their luck in the towns on the south side of the border.

Companies encouraged migration to the border towns by recruiting for labor in the interior. However, many of the migrants soon return to their home villages. The reason that many people don't stay long in the border towns is the severe shortage of affordable housing. According to Steven A. Whittlesey, president of an association which represents 55 of the 65 maquiladoras in Nogales, the housing shortage was the principal contributor to the high employee turnover rates. The association has complained that Nogales has not received its fair share of federal funds for subsidized housing.

To retain workers, some maquiladoras pay above the national minimum wage. Salaries at the maquiladoras in Tijuana, where the labor shortage is especially acute, start at $7 a day, twice the official minimum wage. Rather than pay higher wages, which are taxable, most companies try to entice workers through benefits, such as free or heavily subsidized meals, bus passes, athletic activities, day-car centers, monthly raffles, and bonuses for perfect attendance in a week. Fringe benefits bring the average wage to the equivalent of 70 cents per hour. However, by offering higher benefits, one maquiladora may attract workers primarily from other maquiladoras nearby, thereby contributing to the overall high turnover rates.

Maquiladoras have also tried to attract workers by building plants near residential areas, especially because most of the workers are young women, who prefer to commute shorter distances. Females comprise an average of

57 per cent of the workforce at the maquiladoras, and most of the males are teenagers rather than husbands and fathers.

Evidence is mixed concerning the level of quality of the workers at the maquiladora plants. On the one hand, GM claims that the technology gap is narrowing in some of its plants. For example, Delmex uses state-of-the art cutters and sequencers to build tens of thousands of circuit boards a day. Until 1987, Delco products bought circuit boards that automatically level car suspensions from Lectron, an independent supplier in Kokomo, Indiana. After its Delmex plant in Mexico took over the operation, warranty claims were cut in half in one year. As a result, Delco Products may give Delmex responsibility for manufacturing the entire component system, not just the subassembly. GM's warranty records show that some parts produced in Mexico are equal in quality to the United States, and workers at comparable plant follow the same procedures and standards. Firms that came originally only for cheap labor, find workers who are flexible and quick-responding as well.

Other observers find a large gap in skills between Mexican and American workers. To a considerable extent, the gap derives from the fact that the average age of workers is much younger in Mexico. The average age at Delco's Delmex plant is 27, compared to 40s and 50s at its US plants. The Bermudez group in Ciudad Juarez works with technical schools and universities in the United States and Mexico in order to get a supply of trained personnel into Mexico. Inland, for example, moved its brake-hose production back from Mexico to Dayton. The division had been making brake hose for fifty years in Dayton and had developed high standards of quality there, with no major recall or product failure. Inland decided that a relatively skilled workforce was essential to making a good hose.

Ford had difficulty maintaining production at its Mexican plants during 1990 as a result of labor unrest. Some of the workers at Ford's Cuautitlan plant were unhappy with the leadership of the local union's general secretary Hector Uriarte. To forestall militant action by the dissidents, Uriarte hired a gang of thugs to wait at the factory gates and beat up the dissidents when they arrived for work on Monday morning. One of the workers was shot dead and seven others wounded by gunfire.

After the shootings, about 2,400 workers – two-thirds of the total labor force – occupied the Cuautitlan plant in protest, halting production. Ford then issued dismissal notices to the workers involved in the take-over. The strike spread to other Ford plants in Mexico, including the Chihuahua engine plant and the Hermosillo assembly plant.

Considerably complicating a resolution of the dispute was the manner in which Ford, like other US firms in Mexico, hired workers. The local union leader typically recruited the workers, although the company claimed that it could refuse employment to workers on a case-by-case basis. Further, the local was affiliated with the Confederation of Mexican Workers, an arm of Mexico's long-time ruling Institutional Revolutionary Party, and Uriarte was known to be close to the confederation's national leader, 90-year-old Fidel Velazquez.

Ford found itself in a crossfire. Rehiring the dissident workers could offend the national union, with which Ford has to work given its national political power. Allowing the national union to continue to control hirings could provoke further unrest in the plants, given its apparent unpopularity.

While labor quality is the chief obstacle to locating in Mexico, US firms also face problems of inadequate support services, such as telephones, telefax capabilities, and microwaves. It can take a long time to get contracts to provide these services, up to a year to get a telephone, for example (Downer 1988: 24).

Impact on US economy

The second issue with regard to maquiladoras is the extent to which they take away or save jobs in the United States. The United Auto Workers claims that maquiladoras have cost US workers 40,000 jobs. The program may also create a back door by which Japanese companies circumvent US trade barriers, because Item 807 does not require Mexican plants to be owned by US firms. Twenty Japanese-owned factories had located in Mexico as of 1987.

On the other hand, supporters claim that maquiladoras result in a net increase in US jobs rather than a loss. The argument is that maquiladoras protect jobs in the United States by saving entire product lines. By sending its final assembly of wire harnesses to Mexico, Packard cut a few hundred jobs at US plants, but by making their products more competitive, the division saved tens of thousands of jobs in Ohio and Mississippi. Delco Electronics lost business during 1970s, when dealers removed the stereos from their automobiles and replaced them with lower-cost Asian models. After Delco Electronics moved the subassembly of circuit boards and other stereo components to Mexico, Delco reclaimed 96 per cent of GM's radio installations and saved hundreds of jobs in Indiana and Wisconsin.

Delco Products Division hired 1,000 new workers at its Rochester, New York, plant as a result of increased sales of its windshield wiper systems outside General Motors. The division attributes its sales growth to cost-reduction moves, especially moving subassembly work to its Delmex plant in Ciudad Juarez. According to Michie, the maquiladora 'is a strategy that enables American industry to achieve price competitiveness in the US economy . . . . Where labor costs for various production processes are not competitive with the rest of the world, these have been shipped offshore' (Downer 1988: 24).

Several studies have generated figures to support the contention that the maquiladoras are helping the US economy. A US Department of

Commerce study estimates that 500,000 jobs were created in the United States as a result of maquiladoras, principally in initial manufacturing processes and reprocessing. William L. Mitchell, marketing vice president for Grupo Bermudez Industrial Park in Juarez, claims that between 1982 and 1986 the number of US plants supplying parts and raw materials to maquiladoras rose from 6,000 to 14,159, while the number of employees increased from 100,000 to 219,000 during the period. As a result, Mitchell estimated that for each new job created in a maquiladora, 2.5 jobs were created or maintained in the United States (Hayes 1987: 28).

The economic growth in the United States allegedly generated by maquiladoras has not been confined to the border states. The Department of Commerce estimated that 5,714 firms in forty-four states supplied components, materials, and equipment to maquiladoras. Only 38 per cent of these firms were located in border states.

Nonetheless, US border cities have clearly felt much of the economic impact from maquiladoras. Lucinda Vargas, research manager for Bermudez estimated that 20 per cent of the 35,766 jobs created in El Paso, Texas, between 1977 and 1985 could be attributed to the maquiladoras in the state of Chihuahua, especially Ciudad Juarez. Most of the economic growth in El Paso has been in support services, trucking, and warehousing. Maquiladoras spent $70 million in 1987 for transport servces in El Paso and leased or owned most of the city's 27 million square feet of industrial warehouse space. Meanwhile, on the Mexican side of the border, maquiladoras spent $570 million on wages and operating costs in 1987, compared to $488 million in 1986 and $215 million in 1983, according to Cesar Alarcon, director of the Maquiladora Association in Juarez.

According to Michie:

> [The maquiladora program] has been a primary stimulus for economic development in El Paso, but we have not integrated this industry into our community's economic development to the degree that we should have. What we have not seen is the development of more basic infrastructure within our economy, such as tool and die facilities, metal stamping, plastic injection molding, the kinds of basic infrastructure that would support the diversification of our economy. The demand for these services has existed in this area since the mid-1970s but El Paso has not recognized that the demand is there and has not responded to building.
>
> (Downer 1988: 26)

JAPANESE LOCATIONAL PREFERENCES

Japanese firms have also sought non-union environments within the United States to locate new plants. Like US-owned companies, Japanese auto makers have sought locations where local residents – from whom the plant labor force will be drawn – hold unsympathetic attitudes towards unions. However, Japanese firms have not located in the Deep South because they prefer closer proximity to suppliers and customers located further north. Consequently, Japanese auto makers have sought communities within the Upper South and Great Lakes states with relatively low concentrations of unionized workers in general and automotive workers in particular.

Surveying community attitudes

The challenge for Japanese producers has been to identify locations within the interior of the United States where the residents are likely to hold anti-union attitudes. Logically, any rural location would offer better prospects of finding anti-union workers than an inner city, but how can a board of directors sitting 10,000 km away in Japan truly assess the nuances of attitudes displayed in various rural American communities?

Published surveys of consumer attitudes can detect broad differences among people living in different states or regions of the country but not subtle distinctions among neighboring small towns. Given the fundamental importance of identifying a community with a suitable labor climate, Japanese firms considering production in the United States consequently have administered detailed questionnaires.

While Japanese firms desperately seek information about local attitudes, they hold an even greater desire to retain anonymity. After all, with only one site to be selected for the plant, why generate potential resentment among the rejected communities? Surveys are therefore typically distributed on behalf of an unnamed Japanese firm. The identity of the Japanese company is also protected by having a state development agency or utility company administer the questionnaire. Further, to avoid stirring up unwanted publicity and false expectations, Japanese surveys are distributed to key local officials rather than the general population.

The contents of the questionnaires show the distinctive locational concerns which Japanese firms hold. Many of the questions reveal deep concern about the level of quality of the American labor force. On the one hand, Japanese firms want to be assured of the availability of an adequate supply of workers with appropriate skills. Several questions are included in the questionnaire concerning the availability of different categories of workers, such as electrician, clerical, and maintenance, by age and sex, depending on the particular needs of the firm, as well as the prevailing local wages for each type of needed worker. Questions may also be asked concerning the typical recruitment procedures in the community.

On the other hand, they want to avoid competing with other firms for the same workers for fear that shortages will develop and higher wages will have to be paid. Questions are asked concerning the location of the nearest firms and the types of products manufactured. Firms are also interested in the prevailing wage structure, including starting salaries and average annual increases. The surveys reveal particular concern for information about the nearest Japanese-owned manufacturers, because a Japanese firm prefers not to compete for workers with an already established Japanese company. As Japanese firms avoid locating next door to other Japanese firms, once a small town is picked by one Japanese firm, it rarely gets a second one. Thus, labor-related concerns lead to a spatial tension between the goal of accessibility to large labor markets and the goal of isolation to avoid competition and higher wages.

Resolving the tension in the direction of spatial isolation for many Japanese firms is another characteristic of the local labor force considered critical, the extent of unionization in the community. The questionnaires reveal that Japanese firms are highly concerned with avoiding labor unions. Respondents are asked the name of the strongest union in the community and the history of local work stoppages. The surveys may include more detailed questions related to union activity, such as the presence of closed shops, the number of members in different unions, the reasons for recent strikes, and the location of the office of the most important local union.

The surveys reveal concern for other characteristics of the workers, in addition to the extent of union activity. Respondents may be asked to provide distributions by race, religion, and handicapped status. That the presence of these groups in the community is a labor force issue is revealed by questions concerning the applicability of local regulations concerning preferential hiring practices. Thus, on balance, labor concerns point towards selecting an isolated location; the advantages that derive from the absence of competing firms, unions, and minorities outweigh the disadvantage of a limited labor supply in the community.

At the same time, Japanese firms are worried about the quality of the local labor force. This concern is revealed through a series of questions relating to the level of experience of different categories of workers, as well as the numbers of public, private, and vocational schools and colleges in the community. Surveys also ask about the extent of absenteeism at existing firms and attitudes among the local labor force towards overtime.

Japanese firms are also concerned with financial issues in addition to labor costs. These include local tax rates, availability of incentives, house prices, and the cost of living. Other issues are conveniently grouped under the concept of quality of life. Questions are asked concerning the numbers of recreational and medical facilities, hotels, restaurants, churches, banks, cultural events, such as concerts, and service workers, such as fire and police officers. Questions are also asked about weather. Concern for the quality of life on balance has increased the importance of accessibility rather than isolation. Locations near where other Japanese companies have gone are likely to have passed the test of having attractive amenities.

Many of the questions are beyond the ability of local officials to answer for a variety of reasons. In some cases, local officials may not have access to the needed information and must rely on state agencies or power companies for assistance. Questions may relate to the conditions in the region as a whole, rather than the local community. For example, what local government official knows the capacity of usable telephone lines or how long the snow remains on the ground?

The intent of a number of questions is unclear. Part of the problem derives from the format of the survey, which may consist of phrases rather than questions. The blank space next to a phrase such as 'distance to nearest city' can be filled in with some confidence, but under the category of 'Working Condition regulation, law, custom,' what should be placed next to the phrase 'Shift worker'? Terms may be unclear as a result of difficulties in the use of English words by the Japanese authors of the survey. The words may be unfamiliar, such as luminescence under the heading of 'building restrictions' or 'VISA' under the heading of 'Japanese employee.'

Other words in the questionnaire may be found in an English dictionary but the meaning in the context of the phrase may be unclear. For example, under the heading 'Foundation,' what is meant by the phrase 'less expense for foundation, requested min by law'? What is intended by the phrase 'rate value' under the heading 'labour productivity'? Some financial terms are unfamiliar to American local officials, such as 'material payment rule (L/C, D/A, D/P),' 'Lease Fee (triple net) in $\$/m^2/year$,' and 'STD contracts, agreements, and payments between employee and company.'

Other phrases may require local officials to interpret the type of information being requested. Examples of open-ended requested information include 'general idea for spare time,' 'Japanese perception,' and 'common problems for workers.' And how does a local official respond to the phrase 'Local Attitude toward Japanese company'? Is the more suitable answer a lengthy essay extolling the broad-mindedness of local citizens or merely the word 'excellent?'

The underlying significance of these hard-to-interpret questions is unclear. Are Japanese firms utilizing poor translators? Given the deliberateness with which Japanese companies approach the decision to invest in the United States, dependence on inferior translators appears to be uncharacteristically risky. Alternatively, are the Japanese being deliberately ingenuous? By asking questions which appear innocently simple, perhaps they generate unusually candid replies.

Whereas General Motors claimed that freight costs were the most important factor in locating its Saturn plant, Japanese firms may not even ask any questions on the subject. A more relevant transport factor to the Japanese is access to the plant site by airplane. Surveys ask for the total amount of time required to reach the community from New York City, including travel time in the air and on the ground.

Based on the sites actually selected, Japanese firms seem unconcerned with proximity to interstate highways and have no hesitancy in locating

50 km from the nearest interchange. It is not certain whether this isolation is accidental or deliberate. Have Japanese, like many foreigners, misjudged the scale of distances inside the United States? Do they consider uncrowded two-lane roads in rural areas to be adequate for their shipping needs compared to the congestion they normally experience in Japan? Do they assume that because the United States is so large, any location is far from other places, rendering relative access an unimportant concept?

Japanese preference for peripheral sites

In short, Japanese plants desire low-cost locations, which avoid strong union centers, yet have access to an adequate labor pool. The result of the pull between the desires for isolation and proximity has been to create a pattern of clustering at a regional level but dispersal within a substate – or multicounty – scale. Compared to General Motors' priorities, as revealed in its Saturn site-selection process, Japanese firms pay less attention to transport and utility costs and more attention to labor climate.

Honda, as the first Japanese auto maker to build a US plant, set the locational pattern by selecting Marysville, Ohio, 50 km from downtown Columbus and 25 km from the outer ring road. Compared to Ohio's other large metropolitan areas, such as Cleveland, Cincinnati, Dayton, and Toledo, Columbus had the lowest percentage of the labor force engaged in manufacturing, according to the 1978 census. With the state capital and the publicly-owned Ohio State University both located in Columbus, state government is the city's leading employment sector. Columbus also had fewer automotive workers than Ohio's other major metropolitan areas. Finally, Columbus was widely regarded as the least unionized metropolitan area in Ohio.

Given its location, Honda has had little difficulty in keeping its Ohio plants non-union. The United Auto Workers launched an unsuccessful organization drive in the early 1980s, soon after the plant opened. The union tried again in 1986, but shortly before the scheduled date of the election, the union withdrew its application, in the face of certain defeat.

The most openly anti-union Japanese auto maker operating in the United States has been Nissan. Nissan officials had harsh words for the American labor force before the Smyrna plant was built. Takashi Ishihara, president of Nissan, bluntly explained that the company decided to build trucks rather than cars in the United States at first, 'because we are unsure about the American workers and we are concerned about quality – and if the vehicles are not going to be up to our quality, then at least we want them to be our lowest-selling item.' (Halberstam 1986: 621).

To run the Smyrna plant, Nissan chose Marvin T. Runyon, a former vice-president at Ford in charge of assembly plants. Feeling frustrated by union rules, as well as Ford's bureaucracy, Runyon had decided to take early retirement at age 55 and take a new job in the south. Given a free rein to organize the Nissan plant, Runyon took to wearing the same uniform worn by production workers, with the name 'Marvin' stitched on the shirt (Halberstam 1986: 622). Runyon strongly opposed unionization:

> A lot of avenues of communication are just not open when you have a union shop. You can't build a quality product with an adversary relationship between unions and management. You just can't. You won't get the cooperation necessary to build a quality product with the union.
>
> ('Nissan doesn't want union at Smyrna, Runyon says', *Automotive News* 1981: 42)

The choice of rural Tennessee fit Nissan's strong anti-union orientation. The nearby Nashville metropolitan area had little automotive employment. More importantly, Tennessee was a so-called right-to-work state, meaning that the state had passed legislation outlawing labor contracts which required workers to join a union as a condition of employment. Right-to-work legislation is common in southern and western states, but Tennessee was the right-to-work state closest to the traditional core region of automotive production.

Pro-union activists bitterly attacked Nissan even before the Smyrna plant was built. Groundbreaking ceremonies in February 1981 were marred by protests from hundreds of members of building-trade unions against the company's selection of a non-union firm, Daniel Construction Company, from Greenville, South Carolina, to build the plant. Boos and cries of 'Go home, Japs,' greeted Nissan speakers at the groundbreaking, including president Runyon, and executive vice-president Masataka Okuma. Tennessee governor Lamar Alexander's speech was interrupted with shouts of 'Sit down, Rat!' Union members carried signs saying 'Boycott Datsun' (the model name used in the United States by Nissan at the time); an airplane circled overhead with a streamer saying 'Boycott Datsun. Put Tennesseans back to work' (Plegue 1981: 181).

Protesters prevented Nissan officials from carrying out the actual earth-moving portion of the groundbreaking ceremony. Runyon, substituting for Okuma, tried to drive a Datsun 4×4 truck equipped with a snowplow to break the earth, but the vehicle was surrounded by union supporters, ten to twelve deep, according to one witness. 'They couldn't move the truck more than three or four inches in either direction. State troopers moved them back just a bit and the truck moved about six inches, and that was the groundbreaking.' The truck ended up with three flat tires and 'Boycott Datsun' bumper stickers affixed. Rocks were thrown at vehicles, and one policeman was hit in the head, although no one was arrested. Despite the protest, Okuma had 'no second thoughts at all' about building the plant. 'This is the ideal location for Nissan' (Plegue 1981: 181).

Construction workers also staged protests at Toyota's Georgetown, Kentucky, plant site, a few years later, because the general contractor, a Japanese firm named Ohbayashi, employed largely non-union workers. Ohbayashi insisted that subcontract work was given to the lowest qualified bidder regardless of whether the bidder was unionized and refused union requests to refer workers to construction jobs through union halls (Noble 1986: A-18).

The UAW opened a campaign in early 1988 to organize the Nissan plant. Before the National Labor Relations Board will conduct an election to determine whether workers want a union, at least 30 per cent of the potentially eligible members of a union must sign cards requesting the election. The union set a goal of obtaining cards from 60 per cent of the workers, but more than a year had mustered barely half that total. Nonetheless, the union filed the necessary cards on 18 May 1989, and the election was set for 26–27 July.

With wages comparable to other assembly plants, issues revolved around the line speed, safety standards, and other plant working conditions. Nissan management waged an anti-union campaign described as 'aggressive and sophisticated,' including frequent strongly anti-union broadcasts on the plant's closed-circuit television network (Levin 1989d: A-1). The 2,400 eligible voters overwhelmingly rejected UAW representation, by a margin of 1,622 to 718, or 69 to 31 per cent. Jerry Benefield, Runyon's successor as head of Smyrna, and also a former Ford executive, observed that 'there must be a good reason for employees to want to pay $30 a month in dues to the UAW. If the union can't come up with a reason, they're going to have a difficult time organizing this plant and other plants like it around the country' (Chappell 1989e: 2).

Unionized Japanese plants

Despite setbacks at Honda and Nissan, the UAW has gained the right to represent workers at Japanese-managed US assembly plants, including Mazda, Diamond-Star, and Nummi. Not by coincidence, US auto makers have an interest in all three of these plants. In deference to American partners, Japanese managers in joint venture plants have not actively opposed union organizing efforts.

Nummi, the joint venture between General Motors and Toyota, was the first Japanese-managed US plant to sign an agreement recognizing the UAW as the bargaining agent at its Fremont, California, plant. The joint venture was a marriage of convenience for the two giant car makers. GM took the opportunity to observe first-hand Japanese production methods and management procedures, many of which the company hoped to apply to its new Saturn Corporation. The arrangement also gained GM a domestically built subcompact model which could compete in quality and price with Japanese products, because, after all, the car actually was a Toyota in all but name.

Toyota felt compelled to enter the agreement, in order to catch up with its two main Japanese competitors in the United States – Honda and Nissan – both of whom had already opened American plants. As the best-selling Japanese producer in the United States every year since 1976, Toyota saw its position threatened by Honda and Nissan's ability to add several hundred thousand domestically produced vehicles to their import totals (Rehder, Henry, and Smith 1985: 38). Toyota also used the project to assess the level of ability of US workers and suppliers prior to investing in its own plants in Kentucky (Snyder 1989: 57).

Nummi decided to recognize the union in 1983 before the plant opened in order to gain more public support for a highly controversial joint venture between the largest car makers in the United States and Japan. Wages and benefits would be set at prevailing US industry rates, and workers would be selected from those laid off when GM had closed the Fremont plant the previous year. In return, UAW Local 2244, which had represented the workers when GM ran the plant, agreed to accept new job classifications, seniority, and other work rules in accordance with Toyota's production system. The first page of the seventeen-page agreement signed by the UAW and Nummi recognized the importance of the compromise.

> In this regard, both parties are undertaking this new proposed relationship with full intention of fostering an innovative labor relations structure, minimizing the traditional adversarial roles and emphasizing mutual trust and good faith. Indeed, both parties recognize this as essential in order to facilitate the efficient production of a quality automobile at the lowest possible cost to the American consumer while at the same time providing much needed jobs at fair wages and benefits for American workers.
>
> (Rehder, Hendry, and Smith 1985: 36–37)

Peaceful contract negotiations were also held between Nummi and the UAW local in 1985 and 1988. In the 1985 contract, Nummi management and labor both pledged that they 'will exhibit mutual trust, understanding and sincerity and, to the fullest extent possible, will avoid confrontational tactics.'

Mazda – one-fourth owned by Ford and the only transplant in the automotive industry's 'home' state of Michigan – felt compelled to recognize the union at its Flat Rock plant. The UAW, for its part, wanted to show that it could negotiate a flexible contract with a Japanese company. The two sides agreed in 1985 on the terms of the first contract which would go into effect when the plant opened in 1988. The first contract called for a starting wage rate to be no greater than 85 per cent of the prevailing industry level. Wages were to reach prevailing levels by 1991. Mazda estimated that the agreement would save the company $6 an hour through lower wages, plus another $1.50 from more flexible plant operations (Fucini and Fucini 1990; Sorge 1985a: 1).

Mazda officials expressed no difficulties in working with the union. James Gill, Mazda Motor Corporation spokesman, 'The union is a resource for dealing with problems instead of dealing with 3,000 individuals' (Levin 1988a: 27). Union officials sit in on management production planning meetings. Yet, 80 per cent of the workers hired by Mazda had no prior experience in the automotive industry (Mandel 1989: 55).

Diamond-Star, the joint venture between Chrysler and Mitsubishi, did not give the union automatic recognition. However, the union won a representation election in December 1988, receiving votes from 70 per cent of the 872 eligible workers (Gelsanliter 1990).

Job application procedures

Japanese-managed plants in the US – even if unionized or joint ventures - generally screen job applicants very meticulously. The application process typically requires five types of tests. First, workers must take a battery of general aptitude tests to determine reading, writing, and mathematical skills. Then, they are subjected to tests of their mechanical skills and performance capabilities. Japanese managers point out that employers do not need to obtain information on aptitude and performance levels in Japan, because vocational schools maintain accurate records.

Third, applicants are tested for their work-place attitudes in situations of assembly plant conditions. Do they work well with others and adapt well to change? Are they flexible of mind and comfortable with ambiguous work situations? Fourth, applicants must pass medical tests. Finally, applicants are interviewed to determine ability to communicate, as well as verify the results from observations of work-place attitudes.

Some of the tests may be administered by public agencies or private consultants. For example, Diamond-Star only considers applicants who first passed a general aptitude test given by the Illinois Department of Employment Services. Applicants next go to the Diamond-Star Assessment Center, managed by an outside consultant, to determine work-place attitudes. Mazda's testing program was set up by Arthur Young, a Detroit consultant.

The interviewing process is lengthy and time-consuming. Mazda applicants must spend fifteen hours and make five trips to the test site over a two-month period – unpaid, of course. As a result of careful screening, companies only hire a small percentage of applicants. Mazda received 100,000 applications for 3,500 jobs, while Diamond-Star selected 872 workers from a pool of 30,000 applicants (Kertesz 1987: 1).

Once hired, workers at Japanese plants undergo lengthy training sessions. At Mazda, for example, training for the initial workers lasted ten to twelve weeks. The first three weeks comprised an orientation to Japanese management techniques and corporate philosophy. The next seven weeks involved classroom and hands-on training in basic skills, such as welding. Recruits are then assigned to work areas for more specific training, based on assessments of particular capabilities (Kertesz 1987: 1; 1989a: 52).

Are the rigorous Japanese recruitment and training practices necessary? Nummi has been the only Japanese-managed assembly plant brave enough to forgo them. Under GM management, the Fremont workers were known for high rates of absenteeism, substance abuse, and plant sabotage, while producing cars with high rates of defects. Between its opening in 1963 and its closure in 1982, Fremont had been forced to shut four times as a result of strikes and 'sickouts'. Before being rehired, workers had to undergo a three-day pre-employment assessment program, which utilized American-style interviews and role playing techniques with Japanese-style concern for values, philosophies, and attitudes. UAW members played an active role in assessing applicants. Toyota, for its part, selected a car for the Fremont plant which was relatively easy to assemble.

The Fremont workers responded by producing cars which invariably have among the fewest – if not the fewest – number of defects in GM's monthly internal quality audits. The number of hours needed to produce a car at Fremont was twenty under Nummi, compared to thirty-four under GM and twenty-eight for all GM assembly plants. The per cent of the labor force absent without an excuse was 2 per cent under Nummi, compared to 20 per cent under GM and 9 per cent for all GM plants. During the last year under GM, 4,000 Fremont workers produced 200,000 cars in two shifts. Under Nummi, the plant produced about the same number of cars and ran two shifts, but only 2,200 employees were needed.

If the turnaround at Nummi reflected favorably on the quality of the workers it reflects unfavorably on the quality of the managers when the Fremont plant was under GM's control. GM's former production workers were rehired, but the former plant managers were not.

IMPACT OF SITE SELECTIONS ON MINORITY HIRING

The selection of sites on the basis of non-union environments has affected racial patterns in the automotive industry. Japanese car makers have been accused of choosing small towns outside the industry's traditional core region in part to avoid hiring large numbers of blacks. The black workforce as of 1990 was 10.6 per cent at Honda, 14 per cent at Toyota, 17.6 per cent at Nissan, and 18.7 per cent at Mazda (Chappell 1990b: 3). In comparison, in the Detroit-area plants, blacks comprised about 25 per cent of the workforce at GM and Chrysler and 35 per cent at Ford (Widick 1989: 222).

Relatively few blacks have been hired at Japanese-owned plants because the catchment areas around the plants, where most of the workers live, are predominantly white. The proportion of minorities is 1.7 per cent around the Subaru/Isuzu plant in Lafayette, Indiana, 4 per cent around

Diamond-Star's Normal, Illinois, plant, 10 per cent in the three counties surrounding Nissan's Smyrna, Tennessee, plant, and 11 per cent near Honda's Marysville, Ohio, plant. Mazda's Flat Rock plant, between Detroit and Toledo, alone has a relatively high percentage of blacks nearby, 29 per cent.

In contrast, the US motor vehicle industry has long been an important source of employment for blacks. In the 1960s, 'all blacks could get was bus boy, car wash, or a job in the auto factory,' recalled Joe Wilson, president of the UAW local at a GM plant in Detroit. After the plant closed in 1987, Wilson commented, 'you can still get the bus boy, and you can still get the car wash. What's changed is you can't get anything in the auto industry' (Schlesinger 1988: 1).

The number of blacks employed in the US automotive industry declined from 180,000 in 1979 to 140,000 in 1988, although the proportion of jobs held by blacks remained constant at around 17 per cent, because employment of whites also declined during the decade. However, the percentage of black auto workers was expected to decline in the 1990s because of the likely distribution of plant openings and closures. According to a 1988 study by University of Michigan professors Robert E. Cole and Donald Deskins, blacks comprised 20 per cent of the population in communities with older plants and 15 per cent in communities with new or retooled plants (Cohen 1989: AD9). The proportion of blacks exceeded 60 per cent in Detroit and 40 per cent in other traditional automotive centers, like Flint, Pontiac, and St Louis. The Fleetwood and Clark Cadillac plants, closed by General Motors in December 1987, were located in black and Hispanic neighborhoods of Detroit and had mostly black and Hispanic workforces. One-third of the workers at the two plants were women, mostly single parents.

The Japanese have claimed that sites with few minorities living nearby were selected for reasons other than race. Rural locations have been selected, they say, because of availability of abundant, low-cost land and access to good transportation. However, selected communities have also boasted relatively high average educational levels and what is termed a positive 'work ethic,' characteristics which the Japanese have clearly associated with low levels of minorities. According to Michael J. Kane, head of the US–Japan International Management Institute at the University of Kentucky, near Toyota's Georgetown plant, the Japanese 'say a fact is a fact, and traditionally minorities have a low level of education' (Schlesinger 1988: 28).

Mark Davis, as president of Greater Lafayette Progress Inc, a non-profit development corporation in the Subaru/Isuzu plant's hometown, has defended the Japanese site-selection practice. 'What kind of public education system does Detroit have? What kind of technological capacity in terms of college graduates does Detroit have? There's no God-given right big cities should be viewed as the automotive centers' (Cohen 1989:

AD9). However, insensitive statements by Japanese leaders have fueled the charges of discrimination. Most frequently cited is a 1986 comment by then Japanese Prime Minister Yasuhiro Nakasone that Japan's racial homogeneity made it a more 'intelligent society' than the US 'where there are blacks, Mexicans and Puerto Ricans' (Schlesinger 1988: 28).

Regardless of the intentions of the Japanese firms, the effect of their site-selection procedures has been to make it harder for minorities to obtain employment simply because relatively few minorities live near most of the plants. According to Cole and Deskins, the Japanese 'behavior with regard to site location and hiring practices heightens rather than eases the dilemma of blacks who are being rapidly displaced from their traditional American employers' (Cohen 1989: AD9).

Defenders of Japanese firms against charges of racial discrimination point out that affirmative action policies have resulted in a black labor force which exceeds the percentage of blacks in the surrounding communities. However, the US Equal Employment Opportunity Commission (EEOC) alleged that at Honda, hiring practices effectively screened out black applicants, even though the company had an affirmative action policy. The government action against Honda was precipitated by complaints concerning the company's strict rule that individuals seeking employment had to live within twenty miles (later thirty miles) of the plant when filing the application. The restriction excluded residents of the City of Columbus, thirty miles to the southeast, where virtually all of the region's blacks lived. The Urban League's Columbus chapter received complaints from seven black residents of the city, most of whom had been laid off from a Rockwell International Corporation plant near Honda when work on a B-1 bomber contract was finished. According to the President of the Columbus Urban League, the applicants were 'even committed to move. Honda said move first. Who's going to take a chance like that?' (Schlesinger 1988: 1).

Honda responded that its share of black workers – 5.1 per cent – merely reflected the portion of applicants who were black, 2.8 per cent. The residency restriction was designed to strengthen ties to local communities at a time when employment was declining at other nearby firms. Nonetheless, Honda reached a settlement with the EEOC in March 1988. Honda agreed to give $6 million in back pay to 370 black and female employees who should have been hired sooner, according to the government. Honda also agreed to recruit minorities more actively by changing its promotion practices and expanding its hiring area to include predominantly black areas of Columbus.

A spokesperson for Honda pointed out that the company didn't admit any wrongdoing in the settlement, and that it had already decided on its own to implement the actions ordered by the government, raising its minority share from 2.8 per cent in 1987 to 10.6 per cent in 1990. Furthermore, according to the Honda spokesperson, the EEOC had brought charges of gender and racial discrimination against General Motors and Ford, as well, during the 1980s. GM settled for $44.5 million and Ford for $23 million, both far higher figures than Honda.

According to UAW official Wilson, 'Auto shops were really the bridge that brought black people from the lower class to the middle class. What are you going to do for the people left behind, locked up in the ghetto who have no place to go and no car to get out to rural areas?' (Cohen 1989: AD9).

10 Whipsawing existing plants

The 6,500 auto workers of Kenosha, Wisconsin, cheered Chrysler's announcement of 9 March 1987 that it was acquiring 46 per cent of American Motors from Renault for $1.5 billion and was offering to buy up other shares in order to obtain majority control. Renault had bought AMC stock in 1979 for $405 million and had arranged for the Kenosha plant to start building its cars three years later. During the five years of production at Kenosha, more than 600,000 Renaults were assembled, including over 200,000 during the 1984 model year, but the arrangement never proved profitable to the French company. After years of uncertainty while AMC struggled to survive, the future of the company's massive 'home' assembly plant in Kenosha finally appeared safe in Chrysler's hands.

A few months before the takeover, in June 1986, Chrysler had struck a deal with AMC to have its full-sized cars built on one of Kenosha's two assembly lines. Still looking for additional assembly plant capacity at the time it bought AMC, Chrysler was negotiating to have its subcompacts built at Kenosha, as well. The takeover announcement came the morning after an unsuccessful all-night bargaining session among Chrysler, AMC, and the local union over the terms of a package of concessions, commitments, and incentives to facilitate production of subcompacts at Kenosha. The union was resisting Chrysler's request to reduce job classifications at Kenosha from 162 to 35. Even the newly elected governor of Wisconsin had joined the negotiations in an attempt to break the logjam.

CHRYSLER WHIPSAWS KENOSHA AND TOLEDO

When it acquired AMC, Chrysler gained five additional assembly lines at four plants. AMC assembled automobiles at three locations, including two lines at Kenosha, Brampton, and Bramalea, Ontario. The company also had one truck assembly plant, at Toledo, Ohio, where Jeeps were built. Chrysler gained needed plant capacity, plus AMC's profitable Jeep product line. However, at a stroke Chrysler went from having insufficient to excessive plant capacity. 'We bought four plants,' said Iacocca, 'but we really only have business for two' ('Chrysler still may have too many plants', *Automotive News* 1988: 3).

One of the four former AMC plants – Bramalea – was not threatened with closure. In June 1984, American Motors had reached an agreement with the governments of Canada and Ontario to build the Bramalea plant. The national and provincial governments each loaned over $100 million to American Motors, so that the financially-strapped company could afford to build the $764 million plant (both figures in Canadian dollars). American Motors agreed to pay the two governments a royalty amounting to 1 per cent of the sales price of every car built at the plant. Production began at Bramalea in 1987 (Sorge 1984b: 2).

As American Motors' other three plants were old – Toledo and Kenosha dated from the turn of the century, Brampton from the 1930s – Chrysler looked for reasons other than age to target closures. Brampton's attractions included Canada's lower costs of doing business, especially health insurance, and Canadian laws making plant closures difficult and expensive. Further, Brampton scored second, behind only Bramalea, in Chrysler's 1988 quality audit of the cars produced at each assembly plant. That narrowed the choice to Toledo and Kenosha. Toledo built Jeeps, the model which most enticed Chrysler to buy AMC, while Kenosha had experience building Chrysler products, as well as Jeep engines.

The Kenosha plant was Chrysler's oldest and least efficient. Bodies were welded and painted in a five-storey former mattress factory dating back to the 1890s, located near Lake Michigan. Bodies were then carried several kilometers through the streets of Kenosha in open trucks to the main plant, which opened in 1902. The City of Kenosha had offered to assist in construction of a new plant in the mid-1980s, but Renault (AMC's partner at the time) vetoed the plan. However, the Chrysler takeover lulled Kenosha workers into an attitude of business as usual. After all, even before it owned the plant, Chrysler had already invested $100 million in Kenosha to retool for production of full-sized cars. With its sales at the highest level since 1977, Chrysler bought AMC primarily to expand its production capacity.

Meanwhile, workers at AMC's Toledo Jeep plant faced an uncertain future in March 1987. In February 1986, American Motors had announced its intention to replace both the turn-of-the-century Toledo and Kenosha plants with one new facility. The announcement set off a bidding war among states, and the governor of Wisconsin was publicly committed to offering the most lucrative package of incentives to ensure that the plant would be built in his state. Adding to Kenosha's advantage, Jeep engines were already being built in the city before being shipped to Toledo for final assembly. AMC's majority stockholder, Renault, temporarily shelved the new plant idea because of opposition from French unions to additional investment in North America, but the threat of of closure still hung over Toledo (Johnson 1986: 1).

Poor labor–management relations further weakened the Toledo plant's prospects for survival. The Toledo local union had offered AMC wage concessions in the 1982 contract in exchange for a fixed repayment from the company for each vehicle produced during the life of the contract, known as a 'wheel tax' (Connelly 1986a: 2). By 1986, though, AMC, alarmed at the high cost of the Toledo wheel tax, sent a letter to plant employees threatening closure on 30 June unless a less-costly profit-sharing method of repayment were adopted instead (Connelly 1986b: 1). Toledo remained open as the 30 June deadline passed, but the union authorized a strike in early 1987, charging that bargaining had been conducted in bad faith, while the company sued the local union for alleged breach of contract (Clark 1987a: 6).

Acquiring AMC gave Chrysler an antiquated plant mired in labor disputes at Toledo, but local union leaders, recognizing the plant's perilous future, went to Chrysler within days of the takeover with an offer to renegotiate the local contract. A local with a long-established reputation for militancy was volunteering concessions to the new plant owners! The Toledo local voted to reduce job classifications and grant other concessions in return for a commitment from Chrysler to keep the plant open during the contract's five-year life, two years longer than typical automotive industry contracts. Wages were set at 2 per cent below Chrysler's average. The 'wheel tax' was suspended, after Toledo workers had recouped only 55 per cent of their original concessions (Mandel 1988: 18).

Given the choice between two antiquated plants, Chrysler retained Toledo, with its concessionary contract, over Kenosha, where workers reverted to an attitude of 'business as usual' after the takeover. Nonetheless, Chrysler's announcement, ten months after the AMC acquisition, that Kenosha would close, came as a genuine shock to the workers, who thought that Chrysler had made a firm commitment to retain the plant. The bitter president of the Kenosha local called Iacocca a liar. Kenosha closed in December 1988.

Chrysler's second closure after the takeover was one of its own plants rather than a former AMC facility. In early 1990, the company targeted its Fenton No. 1 plant, which was assembling aging, slow-selling sports models. However, Chrysler wasn't finished with assembly plant closures; although Iacocca had stated in 1988 that the company had two plants too many, by 1990 the numbers had changed. According to Richard E. Dauch, executive vice-president of world-wide manufacturing, 'At the time Chrysler bought AMC, we needed one more assembly plant of capacity. We didn't buy one. We bought four' (Connelly 1990c: 3). For closure number three, Chrysler went back to the two AMC acquisitions of Brampton and Toledo. The survivor proved to be Toledo; once again, concessions by employees, this time covering the period after the expiration of the five-year agreement, apparently proved critical in the choice.

Unions use the term 'whipsawing' to describe tactics such as those used by Chrysler in choosing between Kenosha or Toledo. Whipsawing means that an employer plays off the workers at one plant against those at another in order to extract concessions. The whipsawing process allegedly follows this scenario: The company tells the workers at plant A that 'declining sales are forcing the closure of one plant, and yours is the logical one to select, because of higher production costs and lower quality ratings. However, if you offer us enough cost-saving "give backs," then it may be worth our while to retain your plant instead of another.' Then the company allegedly goes to plant B and tells the workers, 'plant A has made these concessions; if you don't match or exceed them, then we'll be forced to close your plant and keep A open'. The company then bounces back and forth between plants A and B until it obtains the best package of concessions.

Automotive producers deny that they practise whipsawing. Instead, they claim, they are simply presenting the workers with the cold, hard facts about the relative performance of individual plants. The selection of a plant to close from among several alternatives is dictated by long-term differences in operating costs, not short-term differences in concessionary contracts. Regardless of whether or not whipsawing is explicitly practised, the automotive industry nonetheless provides several stark examples of decisions to retain or close competing plants on the basis of local labor climate.

REDUCING LABOR COSTS THROUGH OUTSOURCING

To reduce the cost of components, during the late 1970s and early 1980s US producers expanded the practice of 'outsourcing.' General Motors and Ford in particular switched orders for some components from in-house divisions to independent suppliers.

Forcing in-house divisions to compete

In-house divisions no longer are guaranteed that they will continue to produce components for the passenger car divisions. Instead, when a new automobile model is planned, the division responsible for production seeks bids from both in-house divisions and outside companies to design components and then manufacture them for the entire five-to-ten-year period during which the model will be sold. Saturn officials were especially aggressive at 'outsourcing' when the initial models were planned in the late 1980s. GM's components divisions were told that as a matter of policy they would not receive preferential treatment in the bidding process. Unsuccessful bidders were told why they had failed to receive Saturn contracts (Rowand 1989: E24).

In-house divisions have been shocked to lose contracts for production of components for which they had formerly been exclusive suppliers. Delco

Moraine, for example, had long supplied GM's passenger car divisions with brakes, but in the late 1980s contracts to make brakes were awarded to Robert Bosch for the Cadillac Allante, to Brake and Clutch Industries for the Corvette, to the Alfred Teves Division of ITT for some Pontiacs, and to Kelsey-Hayes for some trucks. Independent companies received these contracts because they could supply anti-lock brakes, whereas Delco Moraine could not. Delco Moraine was committed to producing its own anti-lock brakes, but lack of funds for research prevented the division from developing the technological capability as rapidly as the independent companies. Meanwhile, GM divisions which wanted anti-lock brakes immediately had to obtain them elsewhere; and once the contract was let to an outside supplier, the business was lost to Delco Moraine for the life of the model.

In-house components divisions feel handicapped in competing with independent suppliers. On the one hand, in-house divisions are expected to continue to supply unprofitable parts which no outsider is interested in making. At the same time, in-house divisions can lose contracts to produce profitable parts. As a result, the balance sheet becomes even less favorable for the in-house division. Yet, components divisions do not control their budgets for research and development; they are not in a position to decide if short-term profits should be sacrificed in order to invest in research for new products, unless central management gives approval.

The United Auto Workers union made expanded 'outsourcing' a major issue during the 1982 negotiations, especially with General Motors. As part of the contract that year, GM agreed to give the union sixty-days' notice of a decision concerning sourcing of a component which could result in the layoff at a plant of 100 workers or 10 per cent of the workforce, whichever was lower. If during the sixty-day notification period, the union offered work-rule changes to make the plant competitive GM would continue to produce the component in-house. GM also agreed not to close for two years any plants operated by components divisions because of outsourcing, and to review the possibility of resuming in-house production of parts outsourced during the period of the 1979 labor contract.

GM 'triages' its components

General Motors then launched a systematic study to determine which components produced by its in-house divisions could not compete in price and quality with independent suppliers. Each part made by GM was given a green-, yellow-, or red-light classification. A green-light component was being produced by a GM division at a lower price and higher quality than by competitors. A yellow-light part was not being competitively produced by GM currently but could be in the future. A red light meant that GM should stop making the part, because the company had little hope of matching a competitor's price and quality. To be classified as red light, a component had to cost roughly 15 per cent more for an in-house division to produce than for an independent supplier; the cost disadvantage for a yellow-light component was roughly 5–10 per cent.

GM's initial review in 1987 classified approximately 10 per cent of the components each as red light and green light, leaving the bulk as yellow light. Red-light parts – the prime candidates for outsourcing – comprised either technologically mature ones which could be produced by a number of suppliers with little technology or highly specialized ones where one large independent supplier had made a technological breakthrough. Even a red-light part still might not be outsourced if, for example, it were highly integrated into GM's design and engineering processes, or if independent suppliers lacked capacity to expand production.

However, with most parts classified as yellow light, the main purpose of the exercise was to put the company's components plants on notice that costs had to be reduced or they risked losing the work. The cost disadvantage of a yellow-light part could be reduced in many ways, such as improved design and modified manufacturing processes, but GM officials made it clear that the greatest savings were likely to come from higher labor productivity. An essential element in this strategy would be to move work to plants where workers were willing to make reforms to be competitive (Krebs, Bohn, and Versical 1987: 1).

Plants producing red-light parts recognized the need for immediate action. Delco Moraine, for example, produced a red-light part, dual-servo drum brakes, which Bendix was selling for 12 per cent less. GM opened its books to the union to show that the disadvantage came mostly from a 20 per cent higher cost on brake shoes installed in the drums. The union asked members for suggestions. The eight dual-servo assembly lines were broken into ten teams, known as U-cells, because of the shape of the work areas. One of the tasks taken on by the teams was inspection, saving the cost of hiring separate inspectors to check parts coming off the old assembly lines. Total savings from all of the reforms amounted to $3.4 million a year, enough to make the dual-servo drum brakes competitive. Warned during 1987 negotiations that leading-trailing drum brakes were also red-lighted, the union accepted more flexible work rules in exchange for an agreement to delay outsourcing (Roberts 1988b: 7).

MEASURING LOCAL LABOR CLIMATE

Two dimensions of the local labor climate have become critical to the retention decision – comparative quality ratings and attitudes towards adopting more flexible work rules.

Quality ratings

For a number of years, auto makers tinkered with indicators to measure the relative performance of each plant's workforce. Traditionally, these measures tracked employee behavior, such as unexcused absenteeism, number of grievances, and amount of time taken to perform specific tasks. However, the existence of these rating systems – let alone the results – was rarely publicized, even inside the companies.

In recent years, as American producers finally acknowledged that the Japanese built better as well as cheaper cars, ratings have been introduced comparing the quality of the products turned out at each plant. Not only are these comparative quality ratings circulated regularly to the workforce, they are on occasion shared with newspapers and plant visitors as well.

Chrysler's quality ratings are based on random checks of vehicles as they come off the assembly lines. Auditors look for four types of flaws – customer acceptance, water test, specification audit, and driveability audit. The important category of demerits is known as customer acceptance standards, covering flaws which would be readily apparent to customers, such as rattles and squeaks, fit and finish, and operation of door and window controls. The water test sprays a large volume of water at high pressure to expose leaks. The specification audit checks measurements of the body, components, and subassemblies. The driveability audit tests the engine's starting and performance, steering, brakes, lights, and other vehicle operations. A lower score meant fewer demerits.

Chrysler's audits reveal one reason that the company has been shifting a higher percentage of production to Canadian plants. In 1989, the median number of demerits for Chrysler's fourteen asssembly plants in operation at the time was 19.0. However, the three assembly plants with the lowest number of demerits – and consequently, the highest quality – were all Canadian. The median score for the four Canadian plants was 14.8, compared to 20.3 for the ten US facilities (Kertesz 1989b: 57).

Similarly, General Motors' Customer Oriented Vehicle Evaluation audits a random sample of cars and trucks as they leave the final assembly lines. The COVE audit documents the number of discrepancies per vehicle at each assembly plant. Two patterns have emerged from GM's audit. First, GM's joint venture plants, such as Nummi in California, produce relatively high quality products. Reinforcing this perception, when GM has applied its audit procedures to competing cars it has found that Japanese models, especially Toyota's, have fewer discrepancies, while models built by Ford and Chrysler have more.

Second, the results of GM's COVE audits match independent surveys of customer satisfaction. For example, during the late 1980s and early 1990s GM's Buick LeSabre model ranked at or near the top in a series of J.D. Power's surveys of customers' perceptions of quality and satisfaction. GM publicized these results, and sales of the LeSabre increased rapidly, in sharp contrast to the declines suffered by most of the company's other vehicles. However, few of the LeSabre's highly satisfied buyers would have been aware that the assembly plant in Flint where the model was built received some of the best scores on GM's unpublicized internal COVE audit.

The legacy of 'Fordist' work rules

The labor climate in US automotive plants has also been influenced by acceptance of concessions on new work rules inspired by Japanese management practices. These new work rules are replacing so-called 'Fordist' arrangements that date back to the installation of the moving assembly line. Acceptance of more flexible Japanese-inspired work rules has sometimes been made a condition – although not a guarantee – of keeping open the plant.

Early cars were individually crafted by tradespeople who were skilled at making molds, fashioning parts, building engines, constructing wooden bodies, and stitching leather seat covers. Because they possessed essential – but scarce – skills, workers influenced the pace of production and conditions in the factory. Craftspeople determined how many hours they would work and how many cars they would produce in a day, and they prepared lists showing the price that employers were obliged to pay for each task. Even when more tasks became routinized, employers could not exercise complete control over the pace of production, because people could not be forced to push bins, carry heavy blocks, or fetch parts beyond certain speeds (Babson 1986: 18–19, 29).

The balance of power between automobile workers and management was overturned by Ford's moving assembly line. Prior to the installation of the moving assembly line, one skilled mechanic, plus a few helpers, would construct an entire engine by hand. Instead, engines were built by hundreds of unskilled workers, each performing specialized tasks at fixed work stations along the line. 'One would ream bearings, one every seven seconds, all day long; the next would file bearings, one every 14 seconds, all day long; and the next would put bearings on camshafts, one every 10 seconds, all day long' (Babson 1986: 30). Frederick Taylor, the nation's leading theorist on scientific management, inspired Ford to chop factory operations into hundreds of discrete activities, so that unskilled workers could be hired to perform one simple task on a repetitive basis. Henry Ford boasted that by the early 1920s, 85 per cent of his employees could learn their jobs in two weeks, 43 per cent in one day (Babson 1986: 49).

As Ford's need for labor expanded rapidly – from 450 workers in 1908 to 18,000 in 1915 and 62,000 in 1920 – the principal source of recruitment was the large pool of immigrants recently arrived in Detroit from the US south and Eastern Europe. Detroit's population nearly quadrupled from 466,000 in 1910 to 1,720,000 in 1930; foreign-born or children of foreign-born accounted for two-thirds of the increase, blacks another 10 per cent. However, while Ford easily attracted unskilled workers, the company had difficulty retaining them. Overwhelmed by the pressures of the moving assembly line, Ford's average worker in 1913 quit within fourteen weeks. Constant retraining of new employees was costly and inefficient (Babson 1986: 27; Nevins and Hill 1957: 687).

Ford's solution in early 1914 to attracting and retaining workers was to more than double wages, from $2.25 to $5 a day. The additional expenditures were more than offset by savings generated by lower levels of absenteeism, turnover, and training. To attract workers, other automobile producers were forced to match Ford's wages. Because workers were not paid when a plant shut – typically a several-month period between the end of production of one model and retooling for a new one – $5 per day translated into annual wages of less than $1,000.

The Depression of the 1930s destroyed the uneasy peace which had reigned in the automotive industry following Ford's adoption of the $5-a-day policy. Average daily wages had crept up to $7 by the late 1920s, but fell during the 1930s to $4 or even $3 for the minority who managed to retain their jobs. Ford's hourly workforce was slashed from 171,000 in 1929 to 46,000 four years later (Nevins and Hill 1957: 687). Largely because of declining automobile production, unemployment in Michigan reached 46 per cent in 1933 and over 80 per cent in some communities.

To restore profits, the companies increased line speeds to a killing pace (Cray 1980: 292). Workers were afflicted with nervous disorders called 'Forditis' or 'the shakes.' A study by Dr. I.W. Ruskin described auto workers as 'men with aching bodies and shivering nerves, with their mental faculties warped and deteriorated . . . cast out of industry without a qualm' (Nevins and Hill 1962: 153). When they could no longer keep up with the pace, older men were summarily fired and easily replaced from the legions of unemployed younger men (Cray 1980: 292).

Conditions if anything were worse in the supplier firms, where most workers were still paid on the basis of piece-work rather than hourly rates. A holdover from when most jobs were skilled, the piece-work system required workers to stay at the factory until their daily quotas were filled. A 1925 memorandum from Dodge to its sales force explained that:

> Whenever possible all work is done on piece work basis. A day's work of a large part of the employees consists in turning out a day's production, whatever that may be. Today it is 1,100 cars, and the workman's hours are not from 8 to 5, but from 8 until 1,100 cars run off the assembly line under their own power.
>
> (Wilmer 1925)

One of the largest suppliers, Briggs Manufacturing, which made bodies for Ford and Chrysler, increased the daily quota of sedan backs from eight to twenty after the onset of the Depression, while reducing pay from $1 to 35 cents an hour (Babson 1986: 61). In reality, earnings fell to as low as 10 cents an hour by early 1933, because under the piece-work system

Briggs did not pay for 'dead' time, when materials were unavailable, even though workers had to remain at their work stations (Nevins and Hill 1962: 36).

The unionization movement restored relatively high wages to the US automotive industry. Echoing Henry Ford's $5-a-day strategy, automobile producers after World War II bought labor peace by offering wage rates higher than those prevailing in other industries. However, contracts uniformly applicable to plants throughout the country severely restricted the ability of auto makers to adopt locational strategies that could minimize labor costs. Saddled with uniform national contracts, US auto makers have searched for other measures to determine the relative efficiency of the labor force at different plants. The search for comparative indicators has become more urgent in recent years as firms decide which plants to close.

FLEXIBLE WORK RULES

In recent years, while wage and benefit levels remain identical among plants in accordance with national contracts, local unions have had more latitude in negotiating innovative work rules with plant management. Japanese-style rules encourage 'jidoka,' which is the construction of a superior quality product through giving their all-out best. Workers are encouraged to pursue the concept of 'kaizen,' the process of constantly improving in order to come closer to the impossible goal of perfection.

Characteristics of flexible work rules

Five types of work rules distinguish the Japanese approach. First, workers are organized into teams, as few as four or as many as eighteen per team. Each team member is trained to perform a variety of production operations, as well as support tasks, such as inspecting products, housekeeping, keeping records, changing tools and dies, and maintaining and repairing equipment. Workers rotate through all of the jobs in the team, changing perhaps every two weeks, in order to balance the workload over the long term and minimize boredom from performing repetitive tasks. The team concept requires other team members to perform the tasks of an absent worker. In principle, the work of a lazy co-worker must be covered as well, although in reality strong peer pressure reduces the likelihood of such behavior.

One worker in each team is appointed leader and paid an additional fifty cents per hour. The leader is expected to complete the team's administrative work, fill in for an absent member, and help those having trouble finishing their jobs. Team leaders play a difficult role, because they must be one part supervisor and part colleague to the other team members. The closest equivalent to the traditional foreman – known as a group leader – supervises several teams.

The second distinctive Japanese work rule is greater control over the immediate work environment. The Japanese system encourages the pursuit of 'wa,' which is belief in the existence of harmony between workers and machines. Since the era when Henry Ford was inspired by Taylor's time-study theories, tasks in US-managed automotive plants were designed by company-hired industrial engineers. In contrast, under the concept of 'wa,' teams take responsibility for designing their own work stations, such as the arrangement of machines, under the assumption that people actually performing tasks know the best way to organize them. Tasks are designed to be performed at comfortable heights, and machines are modified to eliminate bending, stretching, and heavy lifting.

When the arrangement of machines on the factory floor needs to be changed, management first asks line workers how they think things should flow. When tooling changes take too long, management asks hourly workers rather than expensive outside consultants how to fix the problem. Workers are encouraged to fashion short-cuts and home-made parts to make otherwise obsolete machines useful. However, workers' control is limited to the immediate environment around the work station: the company still decides how much work must be done by each team.

Complaints mounted at GM's Delco Moraine plant in Dayton concerning injured wrists – diagnosed as carpal tunnel syndrome – as a result of wrestling thousands of stiff rubber diaphragms onto power-brake boosters every day. The complaints were buried in GM's bureaucracy for years, until responsibility for day-to-day relations with vendors was delegated to the workers. After receiving a telephone call from union representative Kay Hoover, the vendor visited the plant the next day and found a solution: lubricate the diaphragms.

When it retooled the Wayne, Michigan, plant in 1990 for assembly of the 1991 Escort, Ford emulated as closely as possible the organization of Mazda's plant in Japan which produced the 323, a model then essentially identical mechanically to the redesigned Escort. Work stations where the team concept had been accepted, such as body and stamping, were redesigned by the workers.

Third, Japanese-style rules encourage problem-solving at the lowest possible level of decision-making. In a traditionally organized plant, an individual who has a problem first meets with the union representative. If the union official considers the complaint justified, a formal grievance is filed with the company. Thousands of grievances can pile up in a plant and take months to resolve. In contrast, under Japanese-style rules, a worker is expected to look for solutions through direct consultations with team and group leaders. Even in the unionized Japanese-managed plants in the United States, a union representative will arrive at the scene of a problem together with someone from the company's labor relations staff, and the two will try to solve the problem on the spot.

Workers in Japanese plants have the right to stop the assembly line by pulling a cord. A chime sounds, lights flash on a panel to indicate where the cord was pulled, and team and group leaders converge at the site of the problem. If the cord is not pulled again within one minute, the line shuts down until the problem is fixed. Stopping the assembly line is encouraged in order to identify and solve a problem which could adversely affect the quality of the final product. In fact, workers who fall behind or can't finish the assigned operation are not merely encouraged – they have the 'right and obligation' to pull the cord to prevent defective cars from continuing down the line. If one team pulls the cord frequently, then work may be shifted to another team or more people added to the team. However, workers may feel pressure not to pull the cord too often for fear of being perceived as trouble-makers or slackards.

The fourth characteristic of Japanese-inspired management is sharing with workers information traditionally considered sensitive or secret. Japanese management practice encourages solving problems through development of a consensus among all affected parties. Collaborative problem-solving is only possible in an atmosphere of mutual trust, created through open communication of corporate information. Management and workers meet together to discuss problems and solutions. Newsletters once reserved for descriptions of retirees' dinners and United Way drives are filled with detailed data concerning production costs, product planning, and sales projections.

At Nissan's Smyrna plant, groups of workers meet for ten minutes at the start of the day to present ideas and complaints, and fifty-five 'involvement circles' discuss quality, safety, and productivity. At the unionized plants, such as Mazda, union officials sit in on planning meetings and make suggestions on line speed and other production issues. At Ford, groups of workers meet regularly on company time to think up ways of doing things better and more efficiently. For example, blue-collar workers are pulled off the line to advise designers on problems of translating ideas for future products into reality.

Fifth, Japanese management practices call for leveling actions to minimize differences in status among workers. Japanese-managed plants typically have only a handful of job classifications for hourly workers, compared to as many as several hundred at US plants. Workers are called associates, even by union officials. Managers at Japanese plants eat in the same cafeterias and participate in the same athletic and social activities. Offices are open, separated perhaps by partitions, but not by doors. Nobody is entitled to a reserved parking space closer to the building, and everyone wears the same uniform, with the first name sewn above the shirt pocket. In plants where uniforms have not been adopted, 'white collar' workers shun coats and ties in favor of sport shirts.

The leveling practices require managers to adopt substantially different attitudes, because they are forgoing many of the perks provided in American plants.

> Significantly, although the Nummi executives selected hourly workers almost exclusively from the displaced UAW workforce as agreed in the letter of intent, the overwhelming majority of the salaried staff was chosen from other sources. It was critical that all participants in building the new organization, whether workers, managers, or union representatives, be tuned in to the collaborative problem-solving atmosphere necessary to Nummi's development.
>
> (Rehder, Hendry, and Smith 1985: 40)

Reinforcing the Nummi decision, Roger Smith, GM chairman during the 1980s, believed that the company's main obstacle was its 'frozen middle.' 'Tens of thousands of managers hampered both by hubris in past glories and an unshakable belief that techology can solve all the problems.' However, disagreeing with Smith's analysis was H. Ross Perot, who became GM's largest stockholder when his company Electronic Data Services was acquired. According to Perot, GM's 'middle management and down understands the problems and would dearly love to change. At the top, where change can be made, they don't want to do it. And that was the core of our problem' (Rehder 1988: 54–55; Levin 1989a).

Adopting flexible work rules at US plants

The United Auto Workers first asked for a formal program of involving employees in defining the work-place environment during its 1970 contract negotiations with General Motors, but the concept was put on the back burner that year. Three years later, such programs were inserted in the Big Three contracts, known as 'quality of work life' (QWL) at GM, 'employment involvement' at Ford, and 'quality work circles' at Chrysler.

At first, American producers did not regard worker involvement as a significant method of improving quality. However, GM and the union agreed to establish a pilot project to improve what was regarded as the worst operation – glass installation – at one of the company's worst assembly plants – Tarrytown. Virtually all of the thirty workers in the glass installation area were recruited to attend a paid thirty-hour training program devised by an outside consultant. Within eight months, the percentage of scrap – once as high as 80 per cent – was minimal, relations between supervisors and workers had improved, and grievances, absenteeism and tardiness were lower. Workers in adjoining areas at Tarrytown clamored for similar programs, while others were brought into Tarrytown from other plants for two-day programs to observe how QWL operated. Tarrytown, which had previously ranked lowest in quality among GM's assembly plants, rose to the highest position. A few years later, GM awarded assembly of its first plastic APV minivans to Tarrytown.

In addition to Tarrytown, reforms of work rules were also introduced under 1982 contracts at Chrysler's Huntsville electronics plant and Ford's Dearborn glass plant. Two newly opened GM plants – Shreveport truck assembly and Buick-Oldsmobile-Cadillac Group's Livonia engine – were also selected in the early 1980s for introduction of innovative work rules at a plant-wide scale, because workers had not become used to operating under old rules. Part of the 1984 agreement between the UAW and GM, known as Document 85, encouraged management to approach local unions to reopen contracts early 'to explore innovative wage structures.' National agreements signed by GM and Ford in 1987 further encouraged local union leaders and plant managers to find ways to create more productive work environments. By the late 1980s, 45,000 GM workers were estimated to be covered by modern operating contracts, including assembly plants in Tarrytown, Lordstown, Janesville, Leeds, Van Nuys, and Arlington, as well as a number of components plants.

Six of Chrysler's assembly plants as of 1989 had so-called modern or progressive operating agreements, which institute teams and pay for knowledge programs and reduce the number of job classifications from over 100 to approximately a dozen. Workers at two Chrysler assembly plants – Sterling Heights and Belvidere – agreed only to an intermedidate stage, in which job classifications are reduced but the other reforms not introduced.

US producers, rebuffed in their attempts to open nonunionized plants in the south, embraced Japanese-style work rules as a way to lower production costs, primarily labor. Unions for their part were ready to adopt Japanese-style work rules in order to save jobs. Thus, Big Three management and UAW leaders reached a compromise: job security for work rule concessions. The trade-off was accepted explicity at some plants: under 1982 contracts, 80 per cent of the workers at a particular plant could be offered lifetime job security. Ford and the UAW chose the Livonia Transmission plant as a pilot for the new project in May 1982. Six months later four GM plants signed on, including Buick assembly in Flint, New Departure Hyatt in Bristol, Fisher Body soft trim in Grand Rapids, and Delco Electronics in Kokomo.

A critical element in the changed attitudes among Fremont workers was Nummi's no-layoff policy. Workers were willing to accept the stresses of the Japanese system in exchange for guaranteed jobs. The no-layoff policy was put to the test only a few years after Nummi began operations. Annual sales of GM's Nummi car, originally marketed as the Chevrolet Nova, were supposed to reach 200,000 a year, but by 1988 had fallen to 100,000, and daily production fell from 910 to 650. Heavy competition, a weak advertising campaign, and an uninspiring design took its toll on the Nova's sales. Instead of layoffs, Nummi took 100 workers at a time and sent them – with full pay – for training in problem-solving and interpersonal relationships. Nummi lost an estimated $100 million in 1988, largely as a result of the no-layoff policy. Nonetheless, workers had more faith in Nummi's ability to protect their jobs than GM's. When John Krafcik, former Nummi quality-control engineer, later at MIT, asked Nummi workers whether they would transfer to a GM plant if one were built across the street, the response was uniformly no (Chethik 1988: E-6).

Some Japanese supplier plants in the United States have emulated the no-layoff policy, as well. Yamakawa Manufacturing in Portland, Tennessee, found that job security was the biggest factor in attracting workers from the north. Brenda Williams, hired by Yamakawa at half her former pay after a six-month layoff from a GM plant in Ohio, said: 'I like the feeling that this place will be here a while' ('How Japan is winning Dixie', *US News & World Report* 1988: 54).

Opposition to flexible work rules

Japanese-style management is viewed by detractors in the automotive industry as 'management by stress.' Traditionally, a US worker had a fixed amount of time to perform a specific task at an assigned work station along a constantly moving line. While the line never stopped moving, a worker given, say, sixty seconds to perform a task could enjoy twenty seconds of relaxation every minute if the task actually took only forty seconds to complete. More importantly, through the practice of 'getting ahead' – that is, performing an assigned task several times in rapid succession by walking down the line rather than remaining at one station – the worker could grab a longer break, perhaps several minutes, enough time to use the toilet, smoke a cigarette, or read the newspaper.

Deteriorated working conditions in Japanese plants allegedly stem primarily from the concept of 'kaizen,' the never-ending pursuit of improvement. Under 'kaizen,' jobs are constantly redefined to ensure that a worker actually spends the full sixty seconds working. A worker with twenty seconds of relaxation every minute has a poorly defined job. The Japanese system discourages 'getting ahead' in order to get a breather: by constantly redefining tasks, all sixty seconds of every minute will be filled. Increasing the pressure on workers is the virtual elimination of repair bays along the line in Japanese-managed plants. Once a product starts down the line, it stays in sequence until the end. If a part is delivered in defective condition or is installed incorrectly, the entire line shuts until the problem is fixed.

At first glance, 'kaizen' carries a serious flaw: because a worker who figures out a way to reduce the time needed for one task is 'rewarded' with the burden of additional tasks to perform as well, what is the inducement for suggesting improvements? To create the necessary incentive, Japanese plants award financial bonuses to workers with good ideas. Further stimulating the process, an individual can also receive cash for identifying ways to make the jobs of other workers more 'efficient'.

'Kaizen' inevitably results in stress, critics charge, because breakdowns in the production process are deliberately encouraged in order to discover the weaknesses in the system. If workers never need to pull the cord stopping the line, their jobs have been designed to be insufficiently challenging. The ideal condition in a Japanese-managed plant is for workers to alternate frequently between periods of falling behind and periods of implementing redesigned assignments. Consequently, 'kaizen' is a thinly-disguised method of increasing line speed, that is the number of products completed in a unit of time, long a cherished goal of automotive producers and bitterly opposed by unions (Parker and Slaughter 1988b: 2).

As a result of the brisk pace, workers in Japanese-managed plants may be suffering from relatively high injury rates. Safety was the principal issue in the UAW's unsuccessful campaign to organize Nissan's Smyrna plant in 1989. Nissan released figures showing that its injury rate was lower than the average in the US automotive industry, but refused to turn over its complete injury list to either the union or Tennessee officials. The union claimed that Nissan was counting only workers whose injuries had kept them from work for more than eight days (Chappell 1989c: 57; 1989d: 6).

Mazda's Flat Rock plant reported higher injury rates than American plants in the Detroit area, primarily from performing repetitive work tasks. Michigan Injured Workers, a nonprofit advocacy and support group, received a number of the complaints from Mazda workers about injuries on the job. In response, the Michigan Labor Department carried out an inspection and cited the Mazda plant for numerous safety violations. Two explanations for the high incidence have been offered: inadequate training and an insufficient number of workers, reducing rotation of jobs (Kertesz 1989a: 1).

In American plants, a worker returning after an injury can be given a less physically demanding job, but the Japanese team concept does not provide 'easy' jobs. A worker recovering from an injury could debilitate a team by being unable to rotate among all of the required jobs. On the other hand, workers in Japanese plants may feel compelled to continue working even when injured to avoid being fired. Further, Japanese work rules may include tough policies concerning absences. At the Nummi plant, a worker who is absent on three occasions within a ninety-day period is charged with an offense. Anyone committing four offenses in a year is fired. No distinction is made as to the reason for the absences, even unexpected illness (Holusha 1989: 1).

The elimination of 'easy' jobs in Japanese plants creates stress primarily for older workers. In unionized American plants, older workers take advantage of the seniority system to claim jobs which are less physically demanding, such as inspection, repair, and material handling. Stories abound of workers in their fifties spending their days reading newspapers while on call in case a machine breaks or a new tool is needed from the storage room. Job rotation – an integral part of the team concept – may require older workers to return to the more physically demanding jobs they had supposedly passed on to younger colleagues. Some older workers may agree to sacrifice to save the plant, but others would rather see the plant closed than relinquish their seniority privileges.

Workers may be sharply divided on the merits of Japanese-inspired work rules, but they agree on the principal motivation for adopting them: fear. Both defenders and critics are afraid of losing jobs, which pay five times more than flipping hamburgers in a fast-food restaurant. One former GM worker rehired by Nummi effectively summarized prevailing attitudes towards Japanese-style rules. 'I can't honestly say I like it better; but I'm working, and that's better.' Workers in Fremont were lucky: they got a second chance under Nummi soon after GM had closed the plant. However, auto workers throughout the country see the spectre of nearby plant closures lurking in the shadows. Uniformly, they are thinking, 'my plant could be next' (Holusha 1989: 10).

Union attitudes towards flexible work rules

The United Auto Workers Union encourages locals to negotiate changes in work rules, but to minimize whipsawing prohibits locals from agreeing to increase line speeds or to reduce wages, cost of living adjustments, or other benefits. Innovative ideas originating from workers may not result in layoffs. New work rules have been embraced at some plants and resisted at others.

Trading security for work rules has proved controversial within the UAW. The 1982 GM contract was ratified by only 52 per cent of the voters, with less than half of the eligible members voting. Workers threatened with closure, especially in components plants, overwhelmingly supported the contract, but most assembly plant workers opposed it, claiming it could lead to a bidding war between locals faced with either providing cost-saving proposals or plant closure. Locals in assembly plants called instead for combatting 'systematic management attacks on local agreements and practices' (Sorge 1982a: 2; 1982b: 4).

Given widespread opposition among its members towards Japanese-style management practices, UAW leaders have been ambivalent. The union supports new work rules to make US producers more competitive, but only if they are adopted in an atmosphere of cooperation. Problems have arisen in plants where managers try to impose on unwilling workers some of the cost-saving Japanese-style rules, such as the team concept and fewer job classifications, without the other elements, notably turning over to the workers more responsibility and information. At Buick City, for example, within a year of the implementation of flexible work rules, workers had voted to replace every team leader.

Dissidents within the United Auto Workers opposed to management cooperation then formed a group called New Directions, which scored several victories in the late 1980s against entrenched mainstream union leaders by arguing that too many concessions had been made. The strongest opposition to new work rules was clustered in the Detroit suburbs. New work rules instituted at GM's Orion assembly plant when it opened in 1983 were suspended after eighteen months, while workers at Chrysler's new Sterling Heights plant turned down new work rules.

An early dissident leader in the Detroit suburbs was Don Douglas, president of Local 594, which represented GM's Truck and Bus Plant at Pontiac. Douglas ran for director of the UAW's combined regions 1 and 1B, which covered southern Michigan, challenging incumbent Bob Lent. Douglas lost the election in 1989 by 719 votes to 450, but finished first in some of the plants, including Pontiac and Sterling Heights. Local 594 claimed a victory in 1987 against GM's attempt to impose new work rules at the Pontiac Truck and Bus plant. The union had resisted a request from GM in the fall of 1986 to reopen the Pontiac contract so that new work rules could be adopted. When GM unilaterally introduced the team concept at Pontiac Plant 6 in early 1987, the local called a strike. The strike was settled quickly, once GM agreed to restore job classifications eliminated when teams were introduced.

However, the 'victory' at Pontiac was shortlived: within a year of the strike, GM had dismantled most of the plant's functions. The company sold its bus operations to Greyhound, eliminating 550 jobs at Pontiac; moved medium-duty truck production to Janesville, eliminating 2,000 more jobs at Pontiac; and launched a joint-venture with Volvo to produce heavy trucks, eliminating 1,600 other jobs at Pontiac (Krebs, Bohn, and Versical 1987: 1). Then, the Pontiac West truck plant was closed, effective in 1993. Given GM's expanding truck production capacity elsewhere in the country, Pontiac's remaining workers faced an uncertain future.

New Directions also showed strength in the St Louis area. Workers at GM's new assembly plant at Wentzville, outside of St Louis, supported mainstream UAW leaders, while leaders at the old truck and bus plant in the central city preferred New Directions. When GM closed the in-town plant in 1987, many workers exercised their seniority rights and transferred to Wentzville, where disputes between the union factions exacerbated the normal tensions of integrating workers from a closed plant.

A New Directions leader Jerry Tucker was elected in 1988 as director of the UAW's Region 5, which covered Missouri, as well as Arkansas, Colorado, Kansas, Louisiana, New Mexico, Oklahoma, and Texas. Tucker had lost to the incumbent Ken Worley in 1987 by a fraction of a percentage, but the US Department of Labor secured a court ruling invalidating the result and ordered a new election, won by Tucker. In 1989, Tucker lost his reelection bid to a mainstream representative Roy Wyse by 438 votes to 234. According to Wyse, workers 'saw New Directions had nothing to offer' (Kertesz 1989c: 18).

The New Directions 'victory' at Pontiac – even if short-lived – was made possible because the strike disrupted production of strong-selling products. Workers at plants building strong-selling products have felt emboldened to express their opposition to new work rules. On the other hand, workers in plants threatened with outsourcing or building slow-selling models have felt more vulnerable. Workers at the Janesville assembly plant agreed to new work rules, because of fear that otherwise the plant would not get a new product to replace the aging compact J-body cars. Shortly after the agreement was concluded, Janesville was rewarded with a new product, a medium-sized pickup truck. The retooled plant was transferred to GM's Truck and Bus Group and appeared to have a secure future.

The compromisers, who dominate the leadership of the United Auto Workers argue that adopting flexible work rules increases the probability that a plant will remain open. Dissidents argue that militant confrontation is the best way to retain a plant. The debate grows more bitter, because both sides have powerful ammunition gathered from the growing collection of plant closures.

GM WHIPSAWS NORWOOD AND VAN NUYS

In 1986, General Motors was building the sporty Chevrolet Camaro and its twin Pontiac Firebird at two plants. One was located in Norwood, Ohio, an industrial suburb of Cincinnati. The Norwood plant was opened in 1923 to assemble Chevrolets for the Ohio River Valley and the Upper South. The other plant was in Van Nuys, California, in the San Fernando Valley, north of Los Angeles. Van Nuys was built immediately after World War II to build Chevrolets for the rapidly expanding southern California market.

Norwood had been building exclusively Camaros and Firebirds since 1967, shortly after the models were introduced. Sales fluctuated wildly, from a peak of over 300,000 in 1969 to less than 100,000 in 1972. However, sales increased steadily through the rest of the 1970s, reaching an all-time high of half a million during the 1979 model year. To meet the growing demand for Camaros and Firebirds, General Motors added Van Nuys as a second assembly site, beginning with 1976 models.

Camaro and Firebird sales reverted to an unstable pattern during the early 1980s. From 500,000 in 1979, sales declined to 200,000 in 1981, recovered to 400,000 in 1984, and slumped to 300,000 in 1986. With sales forecasts for the aging models showing further declines during the late 1980s and early 1990s, GM officials concluded that production should terminate at one of the two plants.

The two plants held somewhat different assets and liabilities. Norwood was ideally located in the heart of the nation's emerging automotive producing region, which – with considerable Japanese influence – was moving south from Michigan. Barely 1 km from the plant was an interchange with I-75, the region's most important highway corridor. Van Nuys was the last surviving American-owned west-coast assembly plant. GM claimed that building Camaros in California added $400 per car in freight charges.

On the other hand, the Van Nuys plant was more modern than Norwood. Van Nuys was a two-storeys. 250,000-m^2 building, while Norwood comprised 300,000 m^2 stacked on three storeys. Further, the 25-hectare site of the Norwood plant was landlocked, that is surrounded by other buildings or major highways on all sides, preventing expansion. Although more accessible to suppliers than Van Nuys, the congested Norwood site had difficulty accommodating the large number of truck arrivals and departures generated by just-in-time delivery.

Given the tradeoff between Norwood's age and Van Nuys' remote location, the decision concerning which plant to retain for Camaro and Firebird production would be made primarily by comparing the labor climate in the two plants. As neither plant was in a position to compete effectively within General Motors for replacement products, the plant that failed to retain production of the Camaro and Firebird would be forced to close altogether.

GM closes Norwood

General Motors announced on 6 November 1986 that production would end at Norwood in mid-1988. The closures of ten other plants were announced as well, including a stamping plant in nearby Fairfield, Ohio, as well as seven in Michigan and one each in Illinois and Missouri. The eleven plants employed 26,000 hourly workers, of which 4,300 were at Norwood and 2,600 at Fairfield. The announcement represented the largest single loss of jobs ever recorded in the Cincinnati area, about 3 per cent of the region's manufacturing employment at the time.

Although never explicitly acknowledged by GM officials, the most critical factor in selecting Norwood for closure was the plant's relatively poor reputation for labor–management relations. UAW's Local 674, which represented the Norwood workers, was one of the last bargaining units to approve a contract in 1970, the last national strike against GM. Two years later, the Norwood plant was closed by a 174-day strike associated with the takeover and consolidation of a nearby Fisher Body plant by GMAD. Local UAW officials revealed that absenteeism was running at 12 per cent at the time of the closure announcement, roughly one-third higher than the company-wide average. In contrast, word filtered east from California that Van Nuys workers had voted to introduce the team system and eliminate most job classifications.

Workers tended to dismiss the announcement as a negotiating ploy on the part of the company. 'I will believe it when I see it,' claimed Charles Basdon, an electrician at the plant. 'I think they are using this as a reason to reopen the contract. Unless they stop making Camaro and Firebirds, this plant will never close.' Bernice Jones, an assembly line worker, was defiant. 'They want concessions. I am not for giving up anything in the contract. If they close I'll go home and be a housewife and a mother' (Howard 1986: A-16). But local government officials who were briefed by General Motors believed the plant would be closed. Said one official, GM 'didn't even leave a crack in the door' (McCarty 1986: A-1).

Other workers took the closure threat seriously but did not display concern, either because of generous unemployment benefits or the expected long lead time before the plant would actually close. According to Bill Mobley, who had worked for GM for twenty years, 'I would take a transfer to finish out my 10 years. It hurts, but it is not the end of the world.' Tony Hankins, a carpenter in the plant, wasn't worried yet. 'We got two years and we still got a chance to show 'em we can still run this plant. It might change their minds. That's what we are going to try to do anyway' (Howard 1986: A-16).

Local union officials believed they could save the plant by offering concessions. On 9 November, the Sunday following the closure announcement, 674 union members gathered at the nearby Cincinnati Gardens, in a meeting closed to the press and public, and authorized the union's bargaining committee to propose new work rules. One month later, on 8 December 1986, the plant management received the suggested changes. The principal concession in the package was to permit a 10 per cent increase in the speed of the assembly line, although with the understanding that no production job would require more than 51.8 minutes of work an hour. The union also offered to discourage absenteeism, an effort which proved so successful that the company laid off sixty workers in February 1987 because of higher attendance rates.

Even at this late date, though, union officials were reluctant to make more radical changes. Ron Rankin, President of Local 674, rejected the team concept. And the union felt compelled to couple the contract concessions with other cost-saving measures, including the closure of an inefficient production booth and paint-baking oven, conservation and recycling of steam energy, and transfer of some operations to periods of lower off-peak electricity rates. According to the union, the package would save the company $112 million in 1987 and $200 million in 1988. GM officials offered no public comments on the package.

Hope that Norwood could be saved continued into early 1987. On 26 January, GM President F. James McDonald told a Senate labor subcommittee hearing called by Ohio Senator Howard Metzenbaum that the company would listen to the union's cost-cutting plans to make the plant competitive. Citing McDonald's commitment, union officials pressured company officials for an audience, requests which grew more insistent after GM announced in mid-February that it was starting a second shift at Van Nuys to produce Camaros and Firebirds. On 27 February, eight representatives of Local 674 travelled to C-P-C headquarters in Warren, Michigan, accompanied by three UAW international officers, to meet Robert Schultz, GM Vice President and head of C-P-C, as well as other division officers. The union proposals were flatly rejected. Union leaders tried unsuccessfully to plead their case directly with McDonald, and a few days later, on 3 March, General Motors announced that Norwood's closing date was moved up from mid-1988 to 26 August 1987.

Plant manager Herb Stone played a leading role in trying to save the plant. Stone, 47, was sent to Norwood in November, 1985, at a time when management and union leaders had reached such a deadlock that both sides were asked to leave the plant. Stone instituted training programs, improved quality, and met regularly with employees, but didn't have enough time to turn the plant around before the axe fell. When union leaders made their last-ditch appeal in Warren, Stone went with them. Once the closing date was set, Stone spent half of his time on the factory floor talking with workers about their future plans. Union leaders regarded Stone as a hard worker who gave his best shot at saving the plant.

The plant's final months were quiet. The Christmas shutdown was extended two extra weeks into January, and the plant closed for one week in February and two weeks in March because of sluggish sales and a shortage of parts. According to plant manager Herb Stone, workers 'are concerned about losing their jobs. But they've got to look to the future.' Stone found that workers were working hard in the final few weeks because they realized 'the jobs are going to go to those that do a good job.' Stone declined to release GM's final audit of quality at Norwood in early August, but he said the report showed 'continuing improvement.' He reported no incidents of sabotage at the plant (Boyer 1987a: B-1).

One week before the 26 August closure, employees held farewell picnics during the lunch hours of each of the two shifts. By then the mood was somber. According to one worker, Connie Waits, 'People are quiet and kind of sad.' Montgomery, who had become co-director of UAW-GM Human Resources Center found the 'mood real quiet and subdued . . . like they're waiting for an execution' (Boyer 1987a: B-1). The last Firebird was bought by the nearby Norwood Quality Inn motel and contributed to a raffle to raise funds for a Norwood student scholarship. Each employee received mementos selected by a twenty-person union-management transition team, including a color 11×14 aerial photograph of the plant, a smaller photograph of the main employee entrance, and a commemorative plate lettered in gold with the plant's opening and closing dates (Boyer 1987c: A-8).

The last car of all, a bright red Camaro IROC, rolled off the assembly line about 8.30 p.m. on 26 August. The keys to the car were presented by plant manager Stone and UAW Local President Rankin to Wendell Spurlock, a nineteen-year Norwood employee, who won it in a raffle. Given the final day off to accompany his new Camaro on its route down the assembly line, Spurlock shook the hand of everyone who had worked on the car. Several hundred workers and their families watched as Spurlock's red Camaro rolled off the line. They cheered and applauded and shook each others' hands (Boyer 1987b: A-1).

Reasons for selecting Van Nuys

Workers at the Ohio plants thought that they had learned a lesson in whipsawing. GM went to both Norwood and Van Nuys and threatened closure unless more flexible work rules were adopted. The local at Van Nuys capitulated, while Norwood resisted; consequently, Norwood was closed. Maryann Keller's highly regarded analysis of GM's problems during the 1980s was similar: 'Two plants would be pitted against one another – the one with the most efficiency would remain open and the losing plant would be closed' (Keller 1989: 136). However, 4,000 km west in California, militant Van Nuys workers held a sharply different perspective of why their plant survived the whipsawing.

In 1982, Van Nuys workers were told by the management that the plant was on GM's 'danger list' of targeted closures. The fate could be averted only by implementing the team concept, reducing absenteeism, and building higher quality cars. Soon after the warning, the second shift was laid off. In reaction, local union leaders chose not to accede to company demands. Instead, the local threatened General Motors that if Van Nuys were closed, a boycott would be implemented in Los Angeles County, the nation's largest automobile market. Thousands of sympathizers sent letters to GM Chairman Roger Smith in support of the proposed boycott. The letter typically included a list of all organizations to which the writer belonged and a warning that the organizations would be urged to join the boycott. With over half of the Van Nuys workers Hispanic, the grape boycott organized by Cesar Chavez served as a model, and civil rights groups joined the effort. The president of the Van Nuys local, Pete Beltran, was the first Chicano to hold the office.

The second shift was recalled at Van Nuys in May 1983, as Camaro sales increased, but union militants kept up the pressure. In January 1984, they met with GM President McDonald and extracted a promise to keep open the plant for two years, although he refused to extend the commitment to five years. But tensions escalated again in 1986. Newly arrived plant manager Ernest Schaefer initiated a strong push for flexible work rules, amid a new round of warnings about plant closures. Schaefer reported that GM refused to consider bids for new products at Van Nuys until after the team concept was adopted.

Pressure for acceptance of the team concept split the local union. The Fighting Back faction, led by local president Beltran, opposed concessions, while Responsible Representation, led by bargaining committee chairman Ray Ruiz supported them. In April 1986, Responsible Representation supporters swept the elections to choose delegates to the UAW national election. Shortly thereafter, local members voted to accept flexible work rules, but only by a 53 per-cent to 47 per-cent plurality. And the vote placed a critical contingency on the agreement: new rules would be implemented only after a promise from the company to retain the plant for ten years.

But in July, the second shift was laid off again, and Ruiz unilaterally agreed to implement the team concept without a firm commitment to keep open the plant. That and other actions by Ruiz viewed as high-handed by the rank-and-file resulted in his being voted out of office in October 1986 within a week of GM's Norwood closure announcement. Beltran chose to run for Ruiz' former job as chairman of the bargaining committee and narrowly won in June 1987, although supporters of new work rules retained some of the offices. Ruiz left town for a job with the union in Detroit, while Beltran was fired for absenteeism in April 1988 (Mann 1987). In 1989, the Van Nuys workers again turned towards the dissidents; five of the seven delegates elected from the plant to attend the UAW's triennial convention were New Directions supporters.

From the perspective of Van Nuys, why did GM choose them over Norwood? Was it because supporters of flexible work rules had won two elections at the plant during the summer of 1986, at the time the closure decisions were being made? Or was it because of fear of a boycott led by a predominantly Chicano work force in the nation's largest car market?

Despite the conflicting evidence from Los Angeles, mainstream union leaders remained committed to a strategy of cooperation so that domestic producers could compete more effectively with the Japanese. New Directions was criticized as failing to offer a constructive alternative, merely opposition to the necessary changes. As evidence, workers at Mazda's Flat Rock plant elected a New Directions candidate, Philip Keeling, as president in 1989. Yet critics of New Directions noted that after his election Keeling stopped attacking Mazda's Japanese-style management practices. Acording to Harry C. Katz, professor of labor relations at Cornell, the main concern of New Directions leaders is opposition to the high-handed political style of mainstream UAW leaders rather than deep-seated philosophical differences (Levin 1989c: 5).

GM's 'Fiero fiasco' energizes the dissidents

Essential to the union support for flexible work rules is faith in the principle that concessions are being traded for jobs security. As long as the weight of evidence clearly pointed to the fact that plants which had made concessions were being retained while militant ones were being closed, the wisdom of the approach was clear to the vast majority of auto workers. But then the 'Fiero' incident undermined for many workers the credibility of adopting new rules.

'Fiero' became the rallying cry of union dissidents. The Pontiac Fiero was a moderately priced sports car sold by General Motors first as a 1984 model. Its most distinctive feature was a mid-body engine. A former Fisher Body plant in Pontiac was retooled to build the Fiero. The Fiero had been first planned during the 1970s as a lightweight, energy-efficient subcompact, but the gap between original design and final execution was too long. The energy crises of the 1970s had receded from public memory, and Americans were ready to buy sportier models instead. GM restyled the Fiero for contemporary tastes but ended up with a poorly designed compromise, neither an econobox nor a true sports car.

GM sold 137,000 Fieros in 1984 but couldn't make a profit. After only five years of production, GM decided in March 1988 to close the plant. The closure provided dissidents with powerful ammunition, because the plant was a model of the new cooperative spirit between labor and management. Workers participated in nontraditional tasks, such as choosing colors, advising on production schedules, and conducting customer evaluations. UAW vice president for GM Donald Ephlin called the Fiero fiasco 'penny wise and pound foolish' (Kertesz 1989d: 18). 'The closing of Fiero had a significant impact because it had been so touted as a model of the way to do things. It had serious implications for morale. I was shocked' (Levin 1989b: D-1). Less than a year later, Ephlin resigned his union position, citing political turmoil and poor planning in both GM and the UAW.

Other factors enter into the choice between two plants. Age is clearly significant; older plants like Norwood may be expensive to renovate, although GM's oldest plant, at Tarrytown, was successfully modernized in the late 1980s to build plastic minivans. When a model's life is ended, the producer may find it more efficient to close the plant where the model was assembled than retool it and transfer a product from elsewhere. But the willingness of the local labor force to negotiate more flexible work rules has proved critical in close calls.

Epilogue: GM announced in 1991 that Van Nuys would also close.

11 Conclusion

This book has applied geographic perspectives to understanding recent changes in the distribution of motor vehicle production within the United States. For geographers, this case study of an important industrial sector may shed light on the on-going debate concerning appropriate locational theory. For people working in the US auto industry, this book illuminates the relevance of geographic concepts in past and current production decisions.

Within geography, the structuralist approach has dominated industrial location theory in recent years. Structuralists understand decisions concerning the location of production within the context of broader changes in the organization of capitalism. That the automotive industry is at the forefront of these organizational changes is evident from the terms 'Fordist' and 'post-Fordist' that structuralists employ to designate the dominant mode of capitalist production in the past and the recent changes. Like other industrial sectors, auto makers have replaced 'Fordist' methods of mass-producing large batches of standardized vehicles through routinized work tasks inspired by Frederick Taylor, with 'post-Fordist' flexible production techniques. According to structuralists, changes in the location of production derive from these organizational changes.

The structuralist explanation has some validity. Auto makers have restructured the workplace so that some unskilled jobs can be relocated to lower-wage sites, especially in Mexico. On the other hand, Japanese-inspired flexible work rules place a higher demand on employing a better-educated, skilled labor force, characteristically found in the US industrial heartland.

In reality, labor issues help to explain the micro-scale question of why auto makers have selected specific communities within the interior of the United States rather than the macro-scale issue of why production has reconcentrated in the interior. Auto makers have located new plants in communities within the I-65 and I-75 corridors where they find the local labor climate suitable for the introduction of a flexible work environment. Normally, a suitable community is one that does not have a tradition of unionization.

However, local labor conditions do not adequately explain why US auto production has clustered along the I-65 and I-75 corridor. Structural changes in automotive products, especially the need to manufacture smaller batches of a wider variety of platforms, have triggered a spatial reordering. But Alfred Weber, whose locational theory has been rejected by many contemporary geographers, came closer to the mark a century ago, when he emphasized the importance of minimizing aggregate transport costs.

Assembly plants that once produced identical models for a local area now have responsibility for distributing products throughout North America. If the critical factor of production is minimizing the cost of distributing the product to a national market, then the optimal location for that plant is near the I-65 and I-75 corridors. For producers of components, the critical locational factor is accessibility to the principal customers, the assembly plants that have clustered in the I-65 and I-75 corridors.

The restructuring of the North American automotive industry will continue to evolve. New plants may be constructed, especially to replace older ones; with certainty, some plants still functioning in the early 1990s will not survive the decade. This book can only offer a snapshot from the early 1990s of an industry in the midst of transformation.

Mergers, acquisitions, and bankruptcies will change the number of producers. Will Chrysler remain an independent US-owned company, or will it be taken over by a foreign firm? Will it continue to produce passenger cars, or will it specialize in light trucks? Can General Motors survive as a vertically integrated corporation, or will it be dismantled in order to become more efficient and competitive? Will Ford merge with a foreign firm in order to assure its place as a major world auto maker? The answers to these questions – which the reader may already possess – will alter the geography of North American motor vehicle production.

Further changes are likely because the most fundamental problem dominating North American automotive production had not been solved in the early 1990s, namely a substantial surplus in the capacity to produce vehicles in North America compared to demand. US and Canadian assembly plants were capable of turning out two million more automobiles and light trucks per year than consumers in the two countries were willing to buy. To bring productive capacity in line with projections of consumer demand, auto makers needed to close at least 10 per cent of the North American assembly lines in operation during the early 1990s.

The burden of closing the additional plants will fall on US-owned companies, while Asian-owned plants in North America operate at or near capacity. By retaining the capacity to build more vehicles than they can sell, US-owned companies – in particular the ones holding most of the excess plants – incur additional costs which raise the price of their products compared to Japanese competitors. But closing a plant may be nearly as expensive, once costs are totalled for compensating laid-off workers, dismantling the equipment, and reshuffling production among surviving plants, not to mention the hostility engendered in communities which would have to bear substantial economic and social costs.

Will US-owned auto makers find new markets for their products, especially among consumers under age 45, half of whom now buy foreign cars, or will they lose an even greater share of the market to Asian competitors? Will venerable American models survive only as names attached to Asian-built products, as occurred in the electronics industry?

US firms recognized that Japanese automobiles would capture an even higher market share during the 1990s and were planning production capacity accordingly. That the Japanese success was based principally on building higher quality vehicles had not yet been universally accepted among American officials. For two decades, American firms have believed that the Japanese were increasing market share because of unfair advantages. 'Put us on a level playing field,' claim US auto makers, 'and we'll be competitive with the Japanese.'

According to the 'level playing field' theory, US firms will recover their lost market share once an unfair advantage held by the Japanese is eliminated. During the 1970s, American firms believed that the Japanese held an unfair advantage because they concentrated on fuel-efficient cars. 'Wait until energy prices decline,' the Americans claimed; 'then US companies will recover their lost market share.'

During the 1980s, the following reasons were offered in turn to account for the 'unfair' advantage held by the Japanese:

1 The Japanese dump cars in the United States at a loss in order to increase their market share. Wait until Japanese exports to the United States are restricted by quotas. In reality, Japanese producers increased their market share while the quotas were in effect, because sales of American-made cars declined.
2 The Japanese specialize in subcompact models. Wait until Americans prefer to buy compacts and intermediates again. In reality, Japanese producers built larger models which proved just as attractive to American consumers.
3 The Japanese build cars for $2,000 less than Americans, because of lower labor costs. Wait until they are forced to build cars in the United States with American workers. In reality, the Japanese transplants in North America pay wages comparable with American-owned plants.
4 The Japanese transplants in North America have lower production costs because they utilize foreign-made components. Wait until they buy expensive US-made parts. In reality, importing components from Asia became more expensive, but the transplants brought along their suppliers to North America. The Hondas built in Ohio had a higher percentage of North American components then several Ford and General Motors models.
5 The Japanese operate nonunion plants in the United States. Wait until

> they are forced to accept union rules. In reality, the unionized transplants, such as Mazda and Mitsubishi, did not suffer from a competitive disadvantage compared to the nonunionized Japanese firms.

By the early 1990s, American producers complained that the Japanese transplants had an unfair advantage because they were newer facilities and employed younger workers. 'Wait until the Japanese are forced to make the same level of contributions to pension funds, and their plants need modernization.' Meanwhile, the Japanese market share continued to increase.

Even if a balance were achieved between capacity and demand, North American auto makers face a constant round of replacement and modernization of older plants in order to remain competitive. Improving the paint shop, meeting pollution standards, reorganizing work stations, delivering components directly to the assembly line – these can be expensive undertakings, especially in older plants. Ford, although the only US-owned auto maker without excess capacity in the early 1990s, possessed the oldest plants; with the exception of the Avon Lake minivan plant, Ford had not built any new assembly plants in the United States since the 1950s. Will replacements for aging plants be located in the same communities or elsewhere?

Not only are the number of actors and plants likely to change in the near future, the spatial distribution documented in this book may change, as well. The interior of the United States attracted an increasing percentage of motor vehicle production during the 1980s and early 1990s, but that pattern could change again. The interior of the United States is the optimal location for a firm wishing to supply a product to consumers throughout North America from one plant, given that the most critical economic factor is minimizing the cost of shipping the product from the factory to the consumers.

What happens if one of the two elements in the formula changes? What if minimizing freight costs from the factory to consumers is no longer the most critical locational factor? With deregulation of the trucking and rail industries in the United States, freight haulers have more flexibility in structuring rates.

More realistically, though, changes may occur in the other assumption in the formula, that a firm wishes to ship one product from one plant to consumers throughout the continent. Increasing diversity of products led auto makers to abandon the branch assembly plant concept during the 1970s. Instead of building identical models for regional distribution at branch assembly plants, producers allocated one or at most two plants to assemble each product, in response to consumer demand for a wider variety of models.

Assembly lines generally operate most efficiently when they build between 200,000 and 250,000 vehicles per year on one platform. As the market further fragments into a series of specialized 'niches,' production strategies are being formulated to utilize fully assembly plant capacity. Platforms are being designed to carry greater varieties of bodies so that customers perceive more choices. Increasingly, platforms built at particular assembly plants are being shared by more than one firm, each of which holds independent responsibility for designing the body. As long as the most efficient production configuration remains one-quarter million units of a single platform, the I-65/75 corridor is likely to continue to attract assembly plants. However, if one platform became so successful that it could be built at three or four plants rather than one or two, then auto makers may consider locating branches in coastal communities again.

While the optimal annual output may remain one-quarter million, an assembly line may be modified to build several platforms rather than just one. With materials increasingly handled by robotics and governed by computerized delivery schedules, assembly plants can become more flexible. Locating assembly plants near consumers may again prove efficient, because a mix of products could be built in response to local demand, perhaps even based on precise orders from customers. An assembly plant in California could build different body-styles than one in Michigan, because consumers don't buy the same models in the two states.

Alternatively, what if a smaller assembly plant – with an annual output of less than 100,000 – could profitably build moderately priced vehicles, not just expensive sports cars like Corvettes? Plants producing low-volume 'niche' products would remain in the interior to facilitate national delivery, but more popular models could be built at smaller-scale coastal branch plants. Local differences in preferences for particular models could also be accommodated at smaller-scale branch assembly plants.

Within the I-65/75 corridor, communities like Marysville, Smyrna, and Spring Hill, not traditionally associated with motor vehicle production, attracted new assembly plants during the 1980s, while plants were closed in Detroit, Kenosha, Norwood, Pontiac, and St Louis. Local labor climate clearly played a role in some cases, but other factors enter into locational decisions. Workers at rural plants in the future may embrace unionization to solve problems at Asian-managed plants. In any event, as plants increasingly cluster in rural interior communities, labor conditions may become more uniform and less important in locational decisions. Instead, decisions to open, retain, or close plants may be based on age, extent of government subsidies, and product planning, not just freight and labor factors.

This study concentrated on the North American automotive industry, with only limited reference to global changes. North America, Europe, and East Asia have evolved as the three principal markets; while producers based in one market have been able to export some vehicles to the other two markets, successful long-term penetration into new markets has occurred primarily through construction of plants. Given overcapacity at a global, not just North American, scale, will the strategy continue? Will European companies build plants in North America? Will Japanese companies export to Europe from North America? Alternatively, will global automotive markets become increasingly integrated? Will Japanese firms take responsibility for producing most of the world's subcompact cars, while Europeans concentrate on compacts and North Americans on intermediates and light trucks?

The arrival of hundreds of Japanese automotive firms has changed the landscapes of small Midwestern communities which long viewed themselves as the 'real' America, free from the alien cultural influences dominating the large east- and west-coast metropolitan areas. Americans have made a number of adjustments to accommodate the Japanese. Supermarkets stock squid, octopus, and seaweed, and local video stores rent samurai films. Sashimi has joined steak and potatoes on restaurant menus, while sake can be ordered at bars instead of beer.

Builders have made modifications in constructing homes to meet Japanese preferences. The Japanese want homes that are made bright by overhead lights and many large open windows. The hoods of stoves must be ventilated directly to the outside to carry away smoke from stir-frying. Big bathtubs are preferred. At the ground-breaking ceremony for Mazda's Flat Rock plant, the governor of Michigan bowed before a Japanese altar, and the local high school band played the Japanese national anthem; then the site was blessed by a Shinto priest.

The thousands of Japanese working at automotive plants in North America have adapted to cultural differences as well. They have learned how to shop in large American-style supermarkets, make long distance telephone calls, obtain driving licenses and automobile registrations, pump gas at a self-service station, and choose salad dressings. Engaging in small talk at cocktail parties has proved difficult; the Japanese can respond to direct questions but don't know how to initiate discussions. The families of Japanese workers must make even greater adjustments to North America. Wives may have limited – if any – ability to speak English and are not allowed to hold jobs because of visa restrictions. They must figure out how to find a doctor and where to leave their children while they shop. (Japanese department stores often have nurseries to mind the children.)

Japanese families are advised by their employers not to cluster in one neighborhood in order to maintain a lower visibility and blend in better. However, housing alternatives may be limited within a thirty-minute driving range of a rural factory – the maximum distance from work preferred by most Japanese as well as American commuters. Most of the suitable housing within a half-hour of Honda's Ohio plants, for example, are located in a handful of suburbs on the northwestern side of the Columbus metropolitan area. Fear of crime keeps Japanese families from even walking around – let alone living – in downtown areas.

Special schools have opened in North America to educate the children of Japanese workers. Dozens of 'Saturday' schools have been established to provide the children of Japanese workers with supplemental education so that they do not fall behind their cohorts back home. The Japanese Ministry of Education, Science, and Culture has accredited a number of elementary and secondary schools in North America, so that students can get the education they need to pass rigorous entrance examinations for Japanese universities.

Communities have recovered from the loss of an automotive plant. GM's assembly plant in St Louis has been converted into a mix of office, warehouse, and industrial space, called the Union-Seventy Center because of its proximity to Interstate 70. Clark Properties acquired the vacant plant in November 1988 for $500,000 and spent $21 million to renovate the property; the city of St Louis provided a 25-year tax abatement on the value of the improvements. The first two tenants were Norcliff Thayer and Mercantile Bank. Norcliff Thayer, a division of Beecham, a British pharmaceutical company which makes Tums antacids, took 10,000 m^2 for a warehouse; the bank occupied 3,200 m^2 to house records. The principal attraction of the site was the low rent – $20–40 per m^2 m, compared to $40–50 per m^2 in suburban complexes (McGuire 1989: 25).

At Norwood, the former GM assembly plant was demolished in 1989 to make way for a complex of buildings named Central Parke, developed by Belvedere Corporation. As in St Louis, Central Parke includes a mix of office, warehouse, and industrial uses. The first tenant was Star Banc Corporation, which moved its retail and credit card operations into a 3,700-m^2 two-storey building. The next two structures erected on the site included a 9,000-m^2 five-storey office building and a 7,000-m^2 warehouse.

Two years after the closing of Norwood and Fairfield, only 600 of the 6,300 workers had found jobs at other GM plants, including approximately 50 at Saturn. Another 300 had gone directly from the closed plants to other jobs. Approximately 2,600 went into a training program, including 600 at local vocational institutions and 1,700 in a week-long, job-search skills-training program at a job retraining and placement center opened in the Cincinnati area by the UAW-GM Human Resources Center; many of those individuals would eventually find jobs outside the automotive industry. Fourteen hundred retired or took early retirement through a cash buyout of their GM benefits, amounting to over $50,000 for some workers. After two years, the remaining 1,400 still hoped to find jobs with General Motors, because they had high seniority levels (Boyer 1989: D-1).

During the first six months after Norwood closed, no evidence was detected of an unusually high incidence of social disorders. Ten alcohol-related problems were referred to the United Way and a similar number to the Cincinnati Care Unit. Five former employees committed suicide during the first six months, including three retirees and one who had left the plant two years before it closed. Five other potential suicides were referred to social service agencies (Elkins 1988: B-4).

Gus Howard, a spray painter at Norwood, summed up the future after the closure:

> This is my livelihood. I don't know how to do anything else. I have done this for 16 years. At age 50, where can I find another job like this? I have children, house and car notes. I guess I will have to go back to frying hamburgers. That's what I did before I came here.
>
> (Howard 1986: A-16)

Local business leaders in Kenosha also envisioned a healthy economic life after Chrysler closed its assembly plant. The main reason for local optimism was the prospect of lower wage rates. When the assembly plant was in operation, employers were forced to pay between $12 and $14 an hour to attract workers; with the plant closed, starting wages had declined to $8 to $12 (Levin 1988: D1).

This study carries several lessons for geographers. First, geographers must build a collection of case studies concerning locational patterns in a variety of industrial sectors. Industrial location concepts have been evolving with only a limited empirical base. Typically, the starting point is an elucidation of structuralist location theory, followed by specific examples which either sustain or refute the theory. Rather than starting from a theoretical base, geographers should become immersed in a series of in-depth studies of key industrial sectors. Then, let geographical theories emerge from the evidence.

Second, geographers should not automatically reject the locational theories developed nearly a century ago, even though the assumptions are overly simplistic. The automotive industry demonstrates that freight charges are as critical as any other factor in explaining the changing distribution of plants. Automotive firms do calculate the point which minimizes aggregate transport costs and do consider this factor in locating plants. Alfred Weber may not have had all of the answers, but he was not completely irrelevant to analyzing locational decisions in the North American automotive industry.

Above all, the automotive industry demonstrates that location matters. Geographers have long contributed to locational decisions in retailing; major chains employ geographers to determine the optimal location for new stores, because a few hundred meters can be critical to attracting enough customers. Selecting the optimal location for a factory has not traditionally required such pinpoint precision, but as competition among automotive firms grows more intense, location will become a more critical element in the competitiveness of firms.

Early in the twentieth century, Henry Ford emerged from among hundreds of automotive firms to dominate North American production by adopting a bold new locational strategy. A century later, will the handful of surviving firms be those who figure out the best locations for their plants?

Bibliography

Alonso, W. (1967) 'A reformulation of classical location theory and its relationship to rent theory', *Papers of the Regional Science Association* 19: 23–44.

'Anderson tales: From Perry Remy to Delco-Remy' (1983) *Automotive News*, 58, 16 September: 203–207.

Association of American Railroads (annually 1966–1986) *Railroad Facts*, Washington: Association of American Railroads.

'Autumn decision due on US Toyota plant', (1985) *Automotive News* 60, 8 July: 3.

Babson, S. (1986) *Working Detroit: The Making of a Union Town*, Detroit: Wayne State University Press.

Bachelor, L.W (1990) 'Flat Rock, Michigan, trades a Ford for a Mazda: state policy and the evaluation of plant location incentives', in E.J. Yanarella and W.C. Green (eds), *The Politics of Industrial Recruitment: Japanese Automobile Investment and Economic Development in the American States*, New York: Greenwood Press.

Bailey, L.S. (1971) 'The other revolution: the birth and development of the American automobile', in Editors of *Automobile Quarterly* (eds), *The American Car Since 1775*, New York: Dutton.

Beesley, N. (1947) *Knudsen: A Biography*, New York: McGraw-Hill.

Bernstein, J. (1983) 'Chrysler to build G-24 in St Louis', *Automotive News* 58, 3 January: 8.

—— (1985) 'Kansas City makes pitch to Toyota', *Automotive News* 60, 22 July: 22.

Biederman, R.A. (1989) 'Mexico's maquiladora industry: production sharing that works!' advertisement in *The New York Times Magazine* 11 June: 53–54.

'Black unionists slap Nader effort to halt GM plant' (1981) *Automotive News* 56, 2 March: 30.

Bladen, V. (1961) *Report, Royal Commission on the Automotive Industry*, Ottawa: Queen's Printer.

Bloomfield, G.T. (1978) *The World Automotive Industry*, Newton Abbot, London, and North Pomfret, VT: David & Charles.

—— (1981) 'The Changing spatial organization of multinational corporations in the world automotive industry', in F.E.I. Hamilton and G.J.R. Linge (eds), *Spatial Analysis, Industry and the Industrial Environment; Volume 2: International Industrial Systems*, New York: John Wiley.

Boas, C.W. (1961) 'Locational patterns of American automobile assembly plants', *Economic Geography* 37: 218–230.

Boyer, M. (1987a) 'GM plant won't go with a bang', *The Cincinnati Enquirer* 23 August: B-1.

—— (1987b) 'Handshakes and wishes, and it's over', *The Cincinnati Enquirer* 27 August: A-1.
—— (1987c) 'Few autos remain at plant after closing', *The Cincinnati Enquirer* 27 August: A-8.
—— (1989) 'Center to help closure victims is set to close', *The Cincinnati Enquirer* 28 August: D-1.
Bradley, J.J., and Langworth, R.M. (1971) 'Calendar year production: 1896 to date', in Editors of *Automobile Quarterly* (eds) *The American Car Since 1775*, New York: Dutton.
Braverman, H. (1974) *Labor and Monopoly Capital*, New York: Monthly Review Press.
Buckley, J. (1988) 'How Japan is winning Dixie', *US News & World Report*, 9 May: 43–59.
Burns, J.F. (1987) 'Canada loan spurs G.M. to save plant', *The New York Times* 1 April: D-13.
Chappell, L. (1989a) 'US suppliers need joint ventures to expand, says study', *Automotive News* 64: 29 May: 30.
—— (1989b) 'Transplants are target of move to curtail trade zones', *Automotive News* 64, 12 June: 25–53.
—— (1989c) 'Nissan won't turn over injury list', *Automotive News* 64, 26 June: 57.
—— (1989d) 'Tennessee fines Nissan in plant-injury rhubarb', *Automotive News* 64, 10 July: 6.
—— (1989e) 'U.A.W. battered in Nissan vote, finds old pitches don't work', *Automotive News* 64, 31 July: 2.
—— (1990a) 'Subaru-Isuzu, Indiana at odds over plant training funds', *Automotive News* 65, 15 January: 11.
—— (1990b) 'Toyota praised for minority hiring at Kentucky plant', *Automotive News* 65, 2 April: 3.
Chethik, N. (1988) 'NUMMI in no-layoff bind', *The Cincinnati Enquirer* 10 January: E-6.
Chrysler, W.P. (1937) *Life of an American Workman*, New York: Dodd, Mead & Company.
'Chrysler cleanup costs climbing' (1989) *The Detroit News* 12 July: 1A.
'Chrysler plans to close old Hamtramck plant' (1979) *Automotive News* 54, 4 June: 3.
'Chrysler still may have too many plants' (1988) *Automotive News* 63, 7 March: 3.
Clark, G.L. (1986) 'The crisis of the Midwest auto industry', in A.J. Scott and M. Storper (eds) *Production, Work, Territory: The Geographical Anatomy of Industrial Capitalism*, Boston: Allen & Unwin.
Clark, L. (1987a) 'AMC offers Jeep for Toledo concessions', *Automotive News* 61, 2 February: 6.
—— (1987b) 'Bramalea plant carries AMC-Renault hopes', *Automotive News* 61, 9 February: 49.
Cohen, S. (1989) 'Japanese auto boom moves into heartland', *The Hamilton Journal-News* 19 February: AD9.
Connelly, M. (1986a) '"Wheel-tax" choice perils Toledo Jeep plant', *Automotive News* 60, 21 April: 2.
—— (1986b) 'AMC warns of a June 30 Toledo closing', *Automotive News* 60, 19 May: 1.
—— (1990a) 'GM-Chrysler joint venture looks for outside business', *Automotive News* 65, 12 February: 6.
—— (1990b) 'Monaco a sales dud; Dodge looks for spark', *Automotive News* 65, 12 April: 2.

—— (1990c) 'Ax could fall on another Chrysler assembly plant', *Automotive News* 65, 23 April: 3.

Cooperman, A. (1989) 'Plant closing takes its toll', *The Cincinnati Enquirer* 12 February: D-3.

Crabb, R. (1969) *Birth of a Giant: The Men and Incidents That Gave America the Motorcar*, Philadelphia, New York, and London: Chilton Book Company.

Cray, E. (1980) *Chrome Colossus: General Motors and Its Times*, New York: McGraw-Hill Book Company.

Curtice, H.H. (1955) 'Statement before Subcommittee on Antitrust and Monopoly of the US Senate Committee on the Judiciary', 2 December, General Motors Institute Alumni Foundation Collection of Industrial History, 77–13.17.

Cyert, R.M., and March, J.G. (1963) *A Behavioral Theory of the Firm*, Englewood Cliffs, NJ: Prentice-Hall.

Dammann, G.H. (1986) *75 Years of Chevrolet*, Sarasota: Crestline Publishing.

DeLorenzo, M. (1984) 'Toyo Kogyo studies 3 US sites', *Automotive News* 59, 19 March: 2.

—— (1985) 'Toyota strategies for US plant', *Automotive News* 60, 15 July: 1.

Dicken, P. (1971) 'Some aspects of the decision making behavior of business organizations', *Economic Geography* 47: 426–437.

'Did Ford Rouge set pattern for Toyota?' (1983) *Automotive News* 58, 4 April: 42.

Dodge et al. v. Commissioner of Internal Revenue (1927a) 'Petitioners' statement of facts', 18 April, Accession 96, Box 3, Archives and Library, Henry Ford Museum and Greenfield Village, Dearborn, MI.

Dodge et al. v. Commissioner of Internal Revenue (1927b) 'Reply brief of Respondent', Accession 96, Box 17, Archives and Library, Henry Ford Museum and Greenfield Village, Dearborn, MI.

Doss, H.C. (1926) 'Reminiscences', Accession 94, Box 12, Archives and Library, Henry Ford Museum and Greenfield Village, Dearborn, MI.

Downer, S. (1988) 'Boom town on the border: Mexico's low labor costs attract huge business', *Automotive News* 63, 2 May: 1, 20–26.

Dunbar, W.F. (1980) *Michigan: A History of the Wolverine State* (revised edition by G.S. May), Grand Rapids, MI: William B. Eerdmans.

Durant, W.C. (1915) 'Letter from W.C. Durant to L.G. Kaufman, Chathan and Phenix National Bank', 27 August, General Motors Institute Alumni Foundation Collection of Industrial History, D74–2.8b.

—— (1940) 'Letter from W.C. Durant to A.P. Sloan', 13 September, General Motors Institute Alumni Foundation Collection of Industrial History, D74–2.1c

—— (1944) 'Letter to Clarence B. Hayes, President Hayes Industries, Jackson', General Motors Institute Alumni Foundation Collection of Industrial History, D74–2.1c.

—— (undated) 'The true story of General Motors', autobiographical notes, General Motors Institute Alumni Foundation Collection of Industrial History, D74–2.1a and D74–2.1b.

Elkins, R.M. (1988) 'Impact of GM plant closing mild, so far', *The Cincinnati Enquirer* 20 March: B-3.

Feron, J. (1987) 'G.M. to build mini-vans in New York', *The New York Times* 4 February: B-2.

Fiordalisi, G. (1989) 'Did Kentucky overpay for Toyota?', *Automotive News* 63, 7 July: 18.

Fleming, A. (1982) 'Helping Detroit-area auto suppliers', *Automotive News* 57, 22 February: 174.

—— (1983) 'Silver streaks and indian chiefs', *Automotive News* 59, 16 September: 115–120.

Flink, J.J. (1970) *America Adopts the Automobile, 1895–1910*, Cambridge, MA, and London: The MIT Press.

—— (1988) *The Automobile Age*, Cambridge, MA, and London: The MIT Press.

Ford, H. (1926) 'What I have learned about management in the last 25 years', in collaboration with Samuel Crowther, *System, The Magazine of Business*, January: 37–40, 103–106, Accession 96, Box 18, Archives and Library, Henry Ford Museum and Greenfield Village, Dearborn, MI.

'Ford blames imports for plant shutdown' (1982) *Automotive News* 58: 29 November: 9.

Ford Motor Company (1903–1919) *Secretary's Record*, Accession 85, Box 1, Archives and Library, Henry Ford Museum and Greenfield Village, Dearborn, MI.

Forney, Paul (1987) 'Why General Motors selected Spring Hill, Tennessee, for its Saturn plant', paper presented at the Applied Geography Conference, Knoxville, TN.

Fucini, J.J., and Fucini, S. (1990) *Working for the Japanese: Inside Mazda's American Auto Plant*, New York: The Free Press.

Gabe, C. (1985) 'GM pushes tax cut for expansion', *The Cleveland Plain Dealer* 24 April: 12–A.

Galbraith, J.K. (1967) *The New Industrial State*, Boston: Houghton-Mifflin.

Gawronski, F.J. (1985a) 'Toyota plants due in US, Canada', *Automotive News* 60, 29 July: 1.

—— (1985b) 'Toyota plant sites to be selected by "open bidding"', *Automotive News* 60, 19 August: 2.

Gelsanliter, D. (1990) *Jump Start: Japan Comes to the Heartland*, New York: Farrar Straus Giroux.

General Motors Corporation (1921) 'Extract from minutes of meeting of executive committee', 13 January, General Motors Institute of Alumni Foundation Collection of Industrial History, P76–5.74.

—— (1935) *Annual Report*, Detroit: General Motors.

Glasmeier, A.K., and McCluskey, R.E. (1988) 'US auto parts production: an analysis of the organization and location of a changing industry', *Economic Geography* 64: 142–159.

Glasscock, C.B. (1937) *The Gasoline Age*, Indianapolis and New York: The Bobbs-Merrill Company.

Glickman, N.J., and Woodward, D.P. (1988) *The New Competitors*, New York: Basic Books.

'GM freight charges to be same for all US' (1981) *Automotive News* 57, 17 August: 57.

'GM plant funds OK'd for Detroit' (1981) *Automotive News* 56, 23 February: 40.

'GM-Toyota decision due in fall, Smith says', (1982) *Automotive News* 57, 28 June: 42.

'GM, railroads use just-in-time at Lansing plants' (1985) *Automotive News* 60, 21 January: 24.

'GM's chief: strike may shut all plants'(1986) *Dayton Daily News* 20 November: 1.

Goodenough, L.W. (1925) 'Statement on behalf of Messrs. David and Paul Gray, and Philip Gray, Deceased before the solicitory of internal revenue in re: valuation of Ford Motor Company stock as of March 1st, 1913', Accession 84, Box 2, Archives and Library, Henry Ford Museum and Greenfield Village, Dearborn, MI: 165–184.

Greenhut, M.L. (1956) *Plant Location in Theory and in Practice*, Chapel Hill, NC: University of North Carolina Press.

Gustin, L.R. (1973) *Billy Durant: Creator of General Motors*, Grand Rapids, MI: William B. Eerdmans.

Halberstam, D. (1986) *The Reckoning*, New York: William Morrow.

Hamilton, F.E.I., ed. (1974) *Spatial Perspectives on Industrial Organization and Decision-Making*. London: Wiley.

—— (1978) 'The changing milieu of spatial industrial research', in F.E.I. Hamilton (ed.) *Contemporary Industrialization*, London: Longman.

Hamilton, K. (1982) 'GM's Fremont plant most likely for Toyota assembly, Smith says', *Automotive News* 57, 9 August: 2.

—— (1985) 'McCurry details Toyota site hunt', *Automotive News* 56, 2 September: 8.

Harder, D.S. (1959) 'Interview', 12 November, Accession 975, Box 1, Archives and Library, Henry Ford Museum and Greenfield Village, Dearborn, MI.

Harney, K. (1989) 'Supplier shakeout: Global competition spells significant changes for OE suppliers', *Automotive News* 64: 12 June: E30–E32.

Harper, D.V., and Johnson, J.C. (1987) 'Potential consequences of deregulation of transportation revisited', *Land Economics* 63, 2: 137–146.

Hartley, J. (1980) 'Toyota's Kato rips protectionism', *Automotive News* 56, 8 September: 16.

—— (1982) 'Toyota–GM pow-wow surprises the Japanese', *Automotive News* 57, 5 April: 2.

—— (1985) 'Chrysler, Mitsubishi make it official', *Automotive News* 60, 22 April: 1.

Hastings, C.D. (1926) 'Memorandum of interview with Mr. Charles D. Hastings, president of the Hupp Motor Company with Mr. Sidney T. Miller', 15 November, Accession 96, Box 11, Archives and Library, Henry Ford Museum and Greenfield Village, Dearborn, MI.

Hawkins, N.A. (1925) 'Affidavit before the solicitor of internal revenue in re: valuation of Ford Motor Company stock as of March 1st, 1913', Accession 84, Box 2, Archives and Library, Henry Ford Museum and Greenfield Village, Dearborn, MI: 189–214.

Hayes, T.C. (1987) 'Mexico's border plants thrive on weak peso', *The New York Times* 23 November: D-1.

Henrickson, G.R. (1951) *Trends in the Geographic Distribution of Suppliers of Some Basically Important Materials Used at the Buick Motor Division, Flint, Michigan*, Ann Arbor: University of Michigan Institute for Human Adjustment.

Henry, J. (1989) 'New York raises the roof and other things for GM', *Automotive News* 64, 5 June: 41.

Hoffman, K., and Kaplinsky, R. (1988) *Driving Force: The Global Restructuring of Technology, Labor, and Investment in the Automobile and Components Industries*, Boulder, San Francisco, and London: Westview Press.

Holmes, J. (1983) 'Industrial reorganization, capital restructuring and locational change: an analysis of the Canadian automobile industry in the 1960s', *Economic Geography* 59: 251–271.

—— (1986) 'The organization and locational structure of production subcontracting' in A.J. Scott and M. Storper (eds) *Production, Work, Territory: The Geographical Anatomy of Industrial Capitalism*, Boston: Allen & Unwin.

Holusha, J. (1986) 'Chrysler, Mitsubishi pick Illinois site', *The New York Times*, 8 October: A-1.

—— (1989) 'No Utopia, but to workers it's a job', *The New York Times*, 29 January, 3: 1–10.

'Honda moves towards US car assembly' (1980) *Automotive News* 55, 14 January: 1.

'Honda paves way for US auto plant' (1980) *Automotive News* 55, 21 January: 6.
'Honda's Ohio car plant confirmed' (1978) *Automotive News* 54, 9 October: 1.
Hoover, E.M. (1937) *Location Theory and the Shoe and Leather Industries*, Cambridge, MA: Harvard University Press.
—— (1948) *The Location of Economic Activities*, New York: McGraw-Hill.
Howard, A. (1986) 'Disappointed workers plan to work to keep plants from being closed', *The Cincinnati Enquirer* 7 November: A-16.
Hurley, N.P. (1959) 'The automobile industry: a study in industrial location', *Land Economics* 35: 1–14.
Interstate Commerce Commission (1913–1980) *Annual Reports*, Washington: ICC.
Irvin, R.W. (1979a) 'Chrysler to build new compacts at Jefferson Ave.', *Automotive News* 54, 11 June: 1.
Isard, W. (1956) *Location and Space Economy*, Cambridge, MA: The MIT Press.
—— (1979b) 'UAW talks get off on wrong foot', *Automotive News* 54, 23 July: 3.
—— (1980) 'States woo Nissan as firm hunts US plant site', *Automotive News* 55, 9 June: 1–46.
Jensen, C. (1985) 'Mitsubishi-Chrysler ventures eye S. Ohio', *The Cleveland Plain Dealer* 16 April: 1–E.
Johnson, R. (1985a) 'Ford may buy 50 pct. of Mazda US output', *Automotive News* 60, 18 February: 2.
—— (1985b) 'Chrysler drops its lawsuit to halt GM–Toyota venture', *Automotive News* 60, 15 April: 2.
—— (1985c) 'Illinois expected to be Mitsubishi plant site', *Automotive News* 60, 7 October: 2.
—— (1985d) 'Toyota said to favor site east of Toronto', *Automotive News* 60, 4 November: 3.
—— (1986) 'AMC plant is opposed by Renault', *Automotive News* 60, 14 April: 1.
Kahn, H. (1985) 'Nader slaps GM bid to cut property taxes', *Automotive News* 60, 6 May: 39.
Kamath, R., and Wilson, R.C. (1983) *Characteristics of the United States Automotive Supplier Industry*, Ann Arbor: University of Michigan Joint US-Japan Automotive Study, Working Paper Series No. 10.
Kelderman, J. (1980a) 'Nissan confirms its 1983 plan to build light pickups in US', *Automotive News* 55, 21 April: 1.
—— (1980b) 'Proxmire rips Cadillac plant plans', *Automotive News* 56, 17 November: 3.
—— (1985) 'Toyota joins hands with Kentucky', *Automotive News* 60, 16 December: 5.
Keller, M. (1989) *Rude Awakening: The Rise, Fall, and Struggle for Recovery of General Motors*, New York: William Morrow.
Kertesz, L. (1987) 'Mazda plant near Job 1', *Automotive News* 61, 29 June: 1.
—— (1989a) 'Injury, training woes hit new Mazda plant', *Automotive News* 64, 13 February: 1–51.
—— (1989b) 'Chrysler Canada tops quality audit', *Automotive News* 64, 22 May: 57.
—— (1989c) 'UAW-maker cooperation faces convention fight', *Automotive News* 64, 19 June: 2.
—— (1989d) 'Donald Ephlin: GM market-share cutback like "winning half a game"', *Automotive News* 64, 19 June: 18.
Klingensmith, F.L. (1926) 'Memorandum of conference with Frank L. Klingensmith by Arthur J. Lacy on May 21st, 1926', Accession 96, Box 11, Archives and Library, Henry Ford Museum and Greenfield Village, Dearborn, MI.

Knudsen, W.S. (1926) 'Memorandum of interview with Wm.S. Knudsen, President of Chevrolet Motor Company by Sidney T. Miller and Mr. F.D. Jones', 25 June, Accession 96, Box 11, Archives and Library, Henry Ford Museum and Greenfield Village, Dearborn, MI.

Krebs, M. (1985a) 'GM seeks tax cuts in three states', *Automotive News* 60, 8 July: 26.

—— (1985b) 'Toyota looks at Tennessee as location of US plant', *Automotive News* 60, 5 August: 64.

—— (1989) 'Buick City: From bottom to No. 1', *Automotive News* 64, 18 December: 1.

—— Bohn, J., and Versical, D. (1987) 'New model launches marred at GM, Ford', *Automotive News* 61, 30 March: 1.

Kuhn, A.J. (1986) *GM Passes Ford, 1918–1938: Designing the General Motors Performance-Control System*, University Park: The Pennsylvania State University Press.

Lacy, A.J (1925) 'Statement of Arthur J. Lacy, Esq., on behalf of Messrs. James Couzens, Horace H. Rackham, and John W. Anderson before the solicitor of internal revenue in re: valuation of Ford Motor Company stock as of March 1st, 1913', Accession 84, Box 2, Archives and Library, Henry Ford Museum and Greenfield Village, Dearborn, MI: 80–163.

——Lacy, A.J., and Anderson, J.W. (1926) 'Memorandum of conference in connection with Dodge tax matter arising out of sale of Ford Motor Company Stock', 26 January, Accession 96, Box 12, Archives and Library, Henry Ford Museum and Greenfield Village, Dearborn, MI.

'Last Ford plant in Calif. sold for high-tech park' (1983) *Automotive News* 59, 7 November: 6.

Launhardt, W. (1882) 'Die Bestimmung des Zwockmassigsten Standorts einer Gewerblichen Anlage', *Zeitschrift des Vereins Deutscher Ingenieure* 26: 106–115.

Lever, W.F. (1978) 'Company dominated labour markets: the British case', *Tijdschrift voor Economische en Sociale Geografie* 69: 306–312.

—— (1985) 'Theory and methodology in industrial geography', in M. Pacione (ed.) *Progress in Industrial Geography*, London and Dover, NH: Croom Helm.

Levin, D.P. (1988a) 'U.A.W. making peace with Japanese in US', *The New York Times* 19 December: D-1.

—— (1988b) 'Kenosha looks beyond Chrysler', *The New York Times* 22 December: D-1.

—— (1989a) *Irreconcilable Differences: Ross Perot versus General Motors*, Boston, Toronto, and London: Little, Brown & Company.

—— (1989b) 'Union vote is crucial test for G.M.', *The New York Times* 16 May: D-1.

—— (1989c) 'U.A.W.'s challenge from within', *The New York Times*, 18 June, 3: 5.

—— (1989d) 'U.A.W. bid to organize Nissan plant is rejected', *The New York Times*, 28 July: A-1.

Lienert, P. (1978) 'Honda plans US car plant by '81', *Automotive News* 53, 24 April: 1.

Linge, G.J.R., and Hamilton, F.E.I. (1981) 'International industrial systems', in F.E.I. Hamilton and G.J.R. Linge (eds), *Spatial Analysis, Industry and the Industrial Environment; Volume 2: International Industrial Systems*, New York: John Wiley.

Lively, R.A. (1949) *The South in Action: A Sectional Crusade Against Freight Rate Discrimination*, Chapel Hill: University of North Carolina Press.

Losch, A. (1954) *The Economics of Location*, translated by W.H. Woglom, New Haven: Yale University Press.

Maher, W.B. (1989) 'Reform Medicare: the rest will follow', *The New York Times*, 9 July: C-3.

Mahoney, J.H. (1985) *Intermodal Freight Transportation*, Westport, CT: Eno Foundation for Transportation Inc.

Malecki, E.J. (1986) 'Technological imperatives and modern corporate strategy', in A.J. Scott and M. Storper (eds) *Production, Work, Territory: The Geographical Anatomy of Industrial Capitalism*, Boston: Allen & Unwin.

Mandel, D. (1988) 'One plant dies; one blossoms', *Automotive News* 63, 28 November: 1.

—— (1989) 'Mazda: two-year assessment finds workers learned auto jobs, car quality matches Japan', *Automotive News* 64, 7 August: 55.

Mann, E. (1987) *Taking on General Motors: A Case Study of the UAW Campaign to Keep GM Van Nuys Open*, Los Angeles: Center for Labor Research and Education, Institute of Industrial Relations, UCLA.

Massey, D. (1979) 'A critical evaluation of industrial-location theory', in F.E.I. Hamilton and G.J.R. Linge (eds) *Spatial Analysis and the Industrial Environment* 1, New York: Wiley.

Massey, D. and Meegan, R. (1982) *The Anatomy of Job Loss: the How, Where, and Why*, London and New York: Methuen.

—— (1985) 'Profits and job loss', in D. Massey and R. Meegan, *Politics and Method: Contrasting Studies in Industrial Geography*, London and New York: Methuen.

May, G.S. (1975) *A Most Unique Machine: The Michigan Origins of the American Automobile Industry*, Grand Rapids, MI: William B. Eerdmans.

—— (1977) *R.E. Olds: Auto Industry Pioneer*, Grand Rapids, MI: William B. Eerdmans Publishing Co.

—— (1990) *Encyclopedia of American Business History and Biography: The Automobile Industry*, volume 1: 1896–1920; volume 2: 1920–1980, New York: Bruccoli Clark Layman.

'Mazda: $1 billion to suppliers' (1989) *Automotive News* 64, 23 October: E29.

McCarty, J.F. (1986) 'Closings leave cities with financial crisis', *The Cincinnati Enquirer* 7 November: A-1.

McCormick, J. (1983) 'Suppliers adjusting to just-in-time', *Automotive News* 59, 22 November: 40.

McCosh, D. (1985) 'Iacocca plans to use Mitsubishi as a spur', *Automotive News* 60, 14 October: 2.

McGuire, J. (1989) 'Developers plan bold rebirths for industrial dinosaurs', *The New York Times* 7 May: 25.

'Midwest for site' (1985) *Automotive News* 60, 11 March: 1.

Miller, R. (1952) 'Ford Division assembly plant histories', Accession 429, Box 1, Archives and Library, Henry Ford Museum and Greenfield Village, Dearborn, MI.

'Mitsubishi plant choice narrowed to 4 states' (1985) *Automotive News* 60, 28 January: 263.

Moritz, M., and Seaman, B. (1981) *Going for Broke: The Chrysler Story*, Garden City, NY: Doubleday & Company.

Motor Vehicle Manufacturers Association (1957) *Motor Vehicle Facts and Figures*. Detroit: Motor Vehicle Manufacturers Association.

'Murphy: Cadillac will renew Detroit' (1980) *Automotive News* 56, 10 November: 3.

National Register of Historic Places (1982) 'Nomination form for Ford Motor Company Edgewater assembly plant, 309 River Road, July 15, 1982'.

Nevins, A. (1954) *Ford: The Times, the Man, the Company*, New York: Charles Scribner's Sons.

Nevins, A., and Hill, F.E. (1957) *Ford: Expansion and Challenge 1915–1933*, New York: Charles Scribner's Sons.

—— (1962) *Ford: Decline and Rebirth, 1933–1962*, New York: Charles Scribner's Sons.

'New hiring procedure for GM southern plants' (1978) *Automotive News* 54, 18 September: 3.

Newmark, J.H. (1936a) 'My 25 years with W.C. Durant', part 2, *Commerce and Finance*, 30 May.

—— (1936b) 'My 25 years with W.C. Durant', part 3, *Commerce and Finance*, 13 June.

—— (1936c) 'My 25 years with W.C. Durant', part 6, *Commerce and Finance*, 25 July.

—— (1936d) 'My 25 years with W.C. Durant', part 12, *Commerce and Finance*, 17 October.

Niemeyer, G.A. (1963) *The Automotive Career of Ransom E. Olds*, East Lansing: Michigan State University Business Studies.

'Nissan doesn't want union at Smyrna, Runyon says' (1981) *Automotive News* 57, 21 September: 42.

'Nissan Motors US production material suppliers' (1984) *Automotive News* 60, 27 October: E-26.

'Nissan production-approved US suppliers' (1984) *Automotive News* 59, 12 March: E-46.

'Nissan to build in US?' (1980) *Automotive News* 55, 14 April: 3.

Noble, K.B. (1986) 'Town's industrial rebirth mired in labor dispute', *The New York Times* 8 September: A-18.

Norbye, J.P. (1971) 'The Race to produce: Automobile manufacturing in the United States', in Editors of *Automobile Quarterly* (eds) *The American Car Since 1775*, New York: Dutton.

Ohio Historic Preservation Office (1988) 'Ford Motor Company Cincinnati plant', National Register of Historic Places Registration Form.

Palander, T. (1935) *Beit rage Zur Standortstheirie*, Uppsala: Almqvist & Wilksells Boktrycken.

Parker, M., and Slaughter, J. (1988a) *Choosing Sides: Unions and the Team Concept*, Boston: South End Press.

—— (1988b) 'Management by stress: behind the scenes at Nummi Motors', *The New York Times* 4 December, 3: 2.

Parlin, C.C., and Youker, H.S. (1914) 'Report of investigation by Charles Coolidge Parlin, manager, and Henry Sherwood Youker, assistant manager, division of commercial research of advertising department, the Curtis Publishing Company', Accession 96, Box 3, Archives and Library, Henry Ford Museum and Greenfield Village, Dearborn, MI.

Pastor, S. (1986) '2 Japanese auto makers to build a $500 million plant in Indiana', *The New York Times* 3 December: A-1.

Peet, R. (1984) 'Class struggle, the relocation of employment, and economic crisis' *Science and Society* 48(1): 38–51.

—— ed (1987) *International Capitalism and Industrial Restructuring: A Critical Analysis*, Boston: Allen & Unwin.

Perez, J.A. (1989) 'The Maquiladora option', advertisement in *The New York Times Magazine* 11 June: 60–62.

Pfau, H. (1971) 'The master craftsmen: the golden age of the coachbuilder in America', in Editors of *Automobile Quarterly* (eds) *The American Car Since 1775*, New York: Dutton.

Plegue, J. (1981) 'Tennessee buzz-saw rips Datsun', *Automotive News* 56, 9 February: 3.

—— (1983) 'GM-Toyota auto to bow as a 1985 model', *Automotive News* 58, 21 February: 1.

Plotkin, A.S. (1984) 'Turning it around at Framingham', *Automotive News* 59, 20 February: 41.

—— (1986a) '"Radical" GM plan stunned supplier', *Automotive News* 60, 26 May: 55.

—— (1986b) 'GM plant is axed in zoning tiff', *Automotive News* 61, 14 July: 1.

Pound, A. (1934) *The Turning Wheel*, Garden City, NY: Doubleday, Doran & Company.

Pred, A. (1967) 'Behavior and location: foundations for a geographic and dynamic location theory, Part 1', *Lund Studies in Geography*, Series B, 27.

Rackham, H.H. (1926) 'Memorandum of conference had by Judge Lacy and Joseph E. Davies with Horace H. Rackham', 26 February, Accession 96, Box 19, Archives and Library, Henry Ford Museum and Greenfield Village, Dearborn, MI.

Rae, J.B. (1965) *The American Automobile: A Brief History*, Chicago: University of Chicago Press.

'Railroads hike share of auto hauling' (1985) *Automotive News* 60, 28 January: 36.

'Reagan sees recovery sign at St Louis Chrysler plant' (1983) *Automotive News* 58, 7 February: 6.

Rehder, R.R. (1988) 'Japanese transplants: a new model for Detroit', *Business Horizons* 31, 1: 52–61.

——Hendry, R.W., and Smith, M.M. (1985) 'Nummi: the best of both worlds? *Management Review* 74: 36–41.

Ribe, A.J. (1944) 'Kangaroos in freight rates', *Nation's Business* 32, February: 1.

Risen, J. (1987) 'Lordstown's blues mellow with age', *The Cincinnati Enquirer* D-1.

Roberts, A. (1988a) 'Sweatshop image fails to fit "maquiladora"', *Dayton Daily News* 27 January: 1, 4.

—— (1988b) 'Delco Moraine counting on anti-lock brake', *Dayton Daily News* 28 January: 7.

Rowand, R. (1981) 'Ford-Toyota decision due in July', *Automotive News* 56, 9 March: 2.

—— (1982a) 'GM and Toyota talk, but Detroit is skeptical', *Automotive News* 57, 15 March: 2.

—— (1982c) 'Did rail tieup reveal flaw in "just-in-time"?', *Automotive News* 58, 27 September: 1.

—— (1982d) 'New "just-in-time" systems called boon for Midwest', *Automotive News* 58, 11 October: 1.

—— (1983) 'The dynamic duo: Delco and Kettering', *Automotive News*, 16 September: 183–192.

—— (1989) 'Alan Perriton: Saturn's materials manager reveals how the auto maker makes its purchasing decisions', *Automotive News* 63, 12 June: E20.

Rubenstein, J.M. (1986) 'Changing distribution of the American automobile industry', *Geographical Review* 76: 288–300.

—— (1987) 'Further changes in the American automobile industry', *Geographical Review* 77: 359–362.

—— (1988a) 'Changing distribution of American motor-vehicle-parts suppliers', *Geographical Review* 78: 288–298.

—— (1988b) 'The changing distribution of US automobile assembly plants', *Focus* 38, 3: 12–17.

—— (1988c) 'The changing distribution of US motor vehicle parts suppliers', *Focus* 38, 4: 10–14.

—— and Reid, N. (1987) *Ohio's Motor Vehicle Industry*, Oxford, OH: Miami University Department of Geography, Geographical Research Paper Number 1.
Rutti, R. (1985) 'GM seeks big tax cut for its Parma plant', *The Cleveland Plain Dealer* 28 March: 1–A.
Sayer, A. (1983) 'Theoretical problems in the analysis of technological change and regional development', in F.E.I. Hamilton and G.J.R. Linge (eds), *Spatial Analysis, Industry and the Industrial Environment; Volume 3: Regional Economics and International Systems*, New York: John Wiley.
Scheinman, M.N. (1990) 'Maquiladoras in the automobile industry', in K. Fatemi (ed.) *The Maquiladoras Industry: Economic Solution or Problem?* New York: Praeger.
Schlesinger, J.M. (1988) 'Fleeing factories: shift of auto plants to rural areas cuts hiring of minorities', *The Wall Street Journal* 12 April: 1.
Scott, A.J. (1988) *New Industrial Spaces*, London: Pion.
Scott, A.J. and Storper, M. (1986) 'Industrial change and territorial organization' in A.J. Scott and M. Storper (eds), *Production, Work, Territory: The Geographical Anatomy of Industrial Capitalism*, Boston: Allen & Unwin.
Segal, H.P. (1988) '"Little plants in the country": Henry Ford's village industries and the beginning of decentralized technology in modern America', *Prospects* 13: 181–223.
Serrin, W. (1973) *The Company and the Union*, New York: Alfred A. Knopf.
Shannon, H.H. (1931) 'History of freight classification', *Traffic World* 31 January: 283.
Sinsabaugh, C. (1940) *Who, Me? Forty Years of Automobile History*, Detroit: Arnold-Powers, Inc.
Sloan, A.P. (1941) *Adventures of a White-Collar Man*, New York: Doubleday.
—— (1953) 'Testimony at trial *US v. E.I. duPont de Nemours and Company et al.* (126 FSupp 235)', 9–10 March.
—— (1964) *My Years with General Motors*, Garden City, NY: Doubleday.
Smith, D.M. (1966) 'A theoretical framework for geographical studies of industrial location', *Economic Geography* 42: 95–113
—— (1981a) *Industrial Location*, 2nd ed., New York: Wiley.
—— (1981b) 'Modelling industrial location: towards a broader view of the space economy', in F.E.I. Hamilton and G.J.R. Linge (eds), *Spatial Analysis, Industry and the Industrial Environment; Volume 1: Industrial Systems*, New York: John Wiley.
Smith, H. (1971) 'The neighboring industry: growth of the motorcar in Canada', in Editors of *Automotive Quarterly* (eds), *The American Car Since 1775*, New York: Dutton.
Smith, P.H. (1970) *Wheels Within Wheels: A Short History of American Motor Car Manufacturing* (2nd ed.), New York: Funk & Wagnalls.
'Smith 90 pct. sure of GM–Toyota tie' (1982) *Automotive News* 58, 6 September: 2.
Snyder, J. (1989) 'Nummi: Higashi says talks with GM delayed Toyota's decision to build trucks at Fremont', *Automotive News* 64, 7 August: 57.
Sorenson, C.E (1956) *My Forty Years with Ford*, New York: W.W. Norton.
Sorge, M (1979a) 'UAW election set in new GM plant', *Automotive News* 54, 2 July: 6.
—— (1979b) 'UAW authorizes mini-strikes at 7 key GM plants', *Automotive News* 55, 27 August: 41.
—— (1980a) 'Dodge Main site eyed for Cadillac plant', *Automotive News* 55, 30 June: 2.
—— (1980b) 'Detroit, Hamtramck join in bid for Cadillac plant', *Automotive News* 55, 7 July: 42.

—— (1981a) 'Nader joins battle against new GM plant in Detroit', *Automotive News* 56, 12 January: 6.
—— (1981b) 'Chrysler plant to close in April, UAW confirms', *Automotive News* 56, 26 January: 3.
—— (1981c) 'GM vs. Poletown: a way of life is vanishing', *Automotive News* 56, 9 February: 26.
—— (1981d) 'Vega, Pinto called 'damn jokes', *Automotive News* 56, 2 March: 8.
—— (1982a) 'New GM contract squeaks through', *Automotive News* 57, 19 April: 2.
—— (1982b) 'No concessions, say GMAD locals', *Automotive News* 57, 28 June: 4.
—— (1983a) 'GM centralizing, decentralizing', *Automotive News* 58, 14 February: 2.
—— (1983b) 'Chrysler buys VW plant', *Automotive News* 58, 11 April: 1.
—— (1983c) 'Poletown: a community and a way of life change as work proceeds on new GM plant', *Automotive News* 59, 3 October: 15.
—— (1983d) 'GM-Toyota moving ahead while awaiting FTC approval, building stamping plant next door', *Automotive News* 59, 3 October: 62.
—— (1984a) 'Any way you slice it – no simple cure', *Automotive News* 59, 26 March: D-1.
—— (1984b) 'AMC to build plant in Canada for mid-size car to bow in '88', *Automotive News* 59, 18 June: 2.
—— (1985a) 'Mazda to have $6 wage edge at Michigan plant', *Automotive News* 60, 25 February: 1.
—— (1985b) 'Chrysler, Mitsubishi near accord on US ventures', *Automotive News* 60, 8 April: 1.
—— and Bernstein, J. (1983) 'Chrysler rear-drives to be built in St Louis', *Automotive News* 58, 31 January: 2.
—— and DeLorenzo, M. (1983) 'GM–Toyota pact only a beginning', *Automotive News* 58, 28 February: 1.
Stevens, W.K. (1987) 'Impasse at truck plant in Pennsylvania drags on', *The New York Times* 7 February: A-24.
Storper, M., and Walker R. (1989) *The Capitalist Imperative: Territory, Technology, and Industrial Growth*, London: Basil Blackwell.
Sundstrom, G. (1984) 'Studies can't quantify Foreign Trade Zones', *Automotive News* 59, 19 March: 51.
'The Freight car situation' (1912) *Automobile Trade Journal* 16, 10, 1 December: 79.
Thomas, C.M. (1987) 'Japanese "export" supplier pacts', *Automotive News* 62, 4 March: 20.
'Toyo Kogyo said to eye US assembly' (1983) *Automotive News* 59, 21 November: 1.
'Toyota near decision on US plant site' (1985) *Automotive News* 60, 28 October: 1.
'Toyota near decision to build US plant' (1982) *Automotive News* 57, 15 February: 2.
'Toyota officials in Knoxville on plant search' (1985) *Automotive News* 60, 18 November: 5.
'Toyota orders study on assembly in US' (1980) *Automotive News* 55, 14 April: 2.
'Toyota still studying reports on possibility of US plant' (1982) *Automotive News* 57, 1 February: 6.
'Toyota to Kentucky; new subsidiary likely' (1985) *Automotive News* (1985) 60, 9 December: 1.
'UAW to demand St Louis reopening' (1982) *Automotive News* 57, 9 August: 1.

'UAW win at OKC' (1979) *Automotive News* 54, 23 July: 46.

US Department of Transportation (1981) *The US Automobile Industry, 1980*, Washington: US Department of Transportation.

—— (1982) *Profiles of Major Suppliers to the Automotive Industry*, Volumes 1–7, Washington: National Highway Traffic Safety Administration Office of Research and Development.

US Senate (1956) *Automobile Marketing Practices*, Washington: Congressional documents.

US-Canadian Automotive Agreement Policy Research Project, *The US-Canadian Automotive Products Agreement of 1965: An Evaluation for the Twentieth Year*, Austin: University of Texas LBJ School of Public Affairs, Policy Research Project Report 68.

Vartan, V.G. (1986) 'Car part shift aids suppliers', *The New York Times*, 28 July: D-6.

Versical, D. (1989) 'Clinton Lauer: Ford's vp-purchasing & supply looks at his supply base from a global standpoint these days', *Automotive News* 64, 12 June: E12.

—— (1990) 'Small-car content shift leads GM's CAFE plan', *Automotive News* 64, 26 June: 61.

Wager, R. (1975) *Golden Wheels: The Story of the Automobiles Made in Cleveland and Northeastern Ohio, 1892–1932*, Cleveland: Western Reserve Historical Society Publication, no 15.

Walsh, J. (1983) 'New "just-in-time" plan seen boon to suppliers', *Automotive News* 59, 10 October: 16.

Ward's Communications (annually) *Ward's Automotive Yearbook*, Detroit: Ward's Communications.

Weber, A. (1929) *Theory of the Location of Industries* (translated by C.J. Fredrich), Chicago and London: The University of Chicago Press.

Weisberger, B. (1979) *The Dream Maker: William C. Durant, Founder of General Motors*, Boston: Little, Brown.

Widell, C.E. (1940) 'Freight rates with a southern accent', *Nation's Business* 28 January: 1.

Widick, B.J. (1989) *Detroit: City of Race and Class Violence*, Detroit: Wayne State University Press.

Wilmer, E.F. (1925) 'Report of lecture by Mr. Edward F. Wilmer, Chairman Board Goodyear Tire & Rubber Co. delivered at the Bankers Club 4/18/25 meeting of members of New Yorks sales force of Dillon Read & Co.', Accession 96, Box 19, Archives and Library, Henry Ford Museum and Greenfield Village, Dearborn, MI.

Witcover, J. (1977) *Marathon: The Pursuit of the Presidency 1972–1976*, New York: Viking.

Womock, J.P., Jones, D.T., and Roos, D. (1990) *The Machine that Changed the World*, New York: Rawson.

Wylie, J. (1989) *Poletown: Community Betrayed*, Urbana and Chicago: University of Illinois Press.

Index